MOLECULAR DYNAMICS SIMULATION

MOLECULAR DYNAMICS SIMULATION

FUNDAMENTALS AND APPLICATIONS

KUN ZHOU

School of Mechanical and Aerospace Engineering, Nanyang Technological University, Singapore

BO LIU

School of Mechanical and Vehicle Engineering, Hunan University, China

ELSEVIER

Elsevier
Radarweg 29, PO Box 211, 1000 AE Amsterdam, Netherlands
The Boulevard, Langford Lane, Kidlington, Oxford OX5 1GB, United Kingdom
50 Hampshire Street, 5th Floor, Cambridge, MA 02139, United States

British Library Cataloguing-in-Publication Data
A catalogue record for this book is available from the British Library

Library of Congress Cataloging-in-Publication Data
A catalog record for this book is available from the Library of Congress

ISBN: 978-0-12-816419-8

For Information on all Elsevier publications
visit our website at https://www.elsevier.com/books-and-journals

Publisher: Matthew Deans
Acquisitions Editor: Brian Guerin
Editorial Project Manager: Emily Thomson
Production Project Manager: Nirmala Arumugam
Cover Designer: Christian J. Bilbow

Typeset by MPS Limited, Chennai, India

Contents

List of symbols

a	Side length of a cubic simulation box
	Mean nearest-neighbor separation
$\boldsymbol{a}$	Acceleration vector (the second-order time derivative of a position vector)
	Edge vector of a unit cell
$\boldsymbol{a}_{\mathrm{pre}}$	Predicted acceleration vector
$\boldsymbol{a}_{\mathrm{cor}}$	Corrected acceleration vector
a_{GE}	Lattice constant of graphene (2.46 Å)
a_{SE}	Lattice constant of silicene (3.98 Å)
A	A physical quantity
	A material-dependent constant
	Amplitude of surface roughness
	Contact area
	Cross-section area
	Surface area
A_{In}	Area of the interface between a graphene monolayer and a silicene monolayer
$A(\boldsymbol{R}^N)$	Configurational component of the Helmholtz free energy
$\langle A \rangle$	True mean value of a physical quantity A
$\langle A \rangle_{\mathrm{blk}}$	Time-averaged value of a physical quantity A obtained in a block period of an MD simulation
$\langle A \rangle_{\mathrm{sim}}$	Time-averaged value of a physical quantity A obtained from an MD simulation
$A(\boldsymbol{R}^{N_n}, t)$	Amplitude factor of a nuclear wave function
b_{ij}	Bond order in the REBO potential
$\boldsymbol{b}$	Third-order time derivative of a position vector
	Edge vector of a unit cell
B	A physical quantity
B_{ijk}	A constant that denotes interaction strength

BO_{ij}	Bond order between the *i*th and the *j*th atom
c	A fitting parameter
$\mathbf{c}$	Edge vector of a unit cell
c_{Ii}	Linear coefficient for the mapping function of the coordinates of the *I*th coarse-grained sites
$c(t)$	Autocorrelation function of hydrogen bonding
C	A material-dependent constant
C_{GE}	Constant volume heat capacity of a graphene nanosheet
C_{SE}	Constant volume heat capacity of a silicene nanosheet
C_V	Isochoric heat capacity
d	Mean grain size
	Thickness of a nanochannel
D	Diffusion constant of a gaseous or liquid system
	Diameter of the nanowire
	Interlayer distance
	A material-dependent constant
D_c	Diffusion coefficient
D_0	Dimer binding energy in the Tersoff potential energy function
D_e	Potential energy well depth of the Morse potential energy function
$D(\omega)$	Phonon density of states
e	Elementary electron charge ($1.60217662 \times 10^{-19}$ C)
e_i^{Pot}	Potential energy of the atom *i*
E, E_a	Electric field intensity
E_0	Original energy of a wave packet
$\mathbf{E}$	Electric field strength vector
$\mathbf{E}_i$	Electric field vector at the position $\mathbf{r}_i$
ΔE	Interfacial bonding strength of a nanocomposite
E_a	Adsorption energy
	Activation energy
E_{angle}	Potential energy related to the bending of a bond angle in the ReaxFF force field
E_{avg}	Average kinetic energy of a group of atoms
E_{bond}	Potential energy associated with the bonding between two atoms in the ReaxFF force field

E_{coh}	Cohesive energy of the interface
E_{cut}	Highest kinetic energy pertaining to an expansion of a wave function in terms of an electronic orbital set
E_{Coul}	Potential energy corresponding to electrostatic interactions
$\hat{E}_e(t')$	Phase factor of the complete wave function of an atomic system
E_{gap}	Energy difference between the lowest unoccupied orbital and the highest occupied orbital
$E_{\mathrm{H\text{-}bond}}$	Cohesive energy due to hydrogen bonding
E_{matrix}	Energy of the matrix of a nanocomposite
$E_{\mathrm{nanofillers}}$	Energy of the nanofillers of a nanocomposite
E_{over}	Energy penalty term that prevents the overcoordination of atoms in the ReaxFF force field
E_{rep}	Potential energy of an atom interacting with a repulsive wall
E_{st}	Spring potential energy function
E_{specific}	System-specific energy terms in the ReaxFF force field
E_{system}	Total energy of a molecular system in the ReaxFF force field
E_{tot}	Total energy of the nanocomposite
E_{tr}	Transmitted energy of a wave packet
E_{tors}	Potential energy related to the twist of a dihedral bond angle in the ReaxFF force field
E_{vdW}	Potential energy corresponding to nonbonding van der Waals interactions in the ReaxFF force field
f	Number of degrees of freedom of an MD system
$\boldsymbol{f}$	Resultant atomistic force acting on an atom
$f_{\mathrm{A}}(r_{ij})$	Attractive atomic interaction function of the Tersoff potential
$f_{\mathrm{c}}(r_{ij})$	Cutoff function for nearest-neighbor interactions
f_{d}	Driving force
f_{FVF}	Free volume fraction
f_{ij}	Interatomic force imposed by the *j*th atom in the wall on the *i*th atom in the fluid
$f_i(\boldsymbol{r}-\boldsymbol{r}_i)$	Radial distribution of the remaining charge outside the core of the *i*th atom
f_{iw}	Force exerted on the *i*th fluid particle by the channel wall
f_{p}	Force arising from the pressure difference at the boundary of the "pump" region

$f_R(r_{ij})$	Repulsive atomic interactions of the Tersoff potential
f_w	Force exerted on a channel wall by the fluid particles
$\langle f_z \rangle$	Net electrostatic force averaged over all the water molecules
$\boldsymbol{F}_i(\boldsymbol{R}^N)$	Force acting on the ith particle
$\boldsymbol{F}_I$	Force acting on the Ith coarse-grained site
$\boldsymbol{F}_{ij}$	Force imposed on the atom i due to its interaction with atom j
$\boldsymbol{F}_{ijk}$	Force on the atom i due to the three-body interactions among atoms i, j, and k
$\boldsymbol{F}_j(ijk)$	Three-body force term for the calculation of heat flux
$F[n_i(\boldsymbol{r}_i)]$	Function of the local electron density $n_i(\boldsymbol{r}_i)$ at the atomic site $\boldsymbol{r}_i$
g	Gradient of the potential energy
$g(r)$	Radial distribution function
$g(\theta_{ijk})$	Bond angle distribution function of the Tersoff potential energy function
$g_{MN}^{\text{ref}}(R)$	Target radial distribution function
$g_{MN}^{k}(R)$	Paired radial distribution function for the effective potential obtained from the kth IBI iteration step
$g_{MN}^{\mu VT}(R_{IJ})$	Radial distribution function in the μVT ensemble
G	Interfacial thermal conductance
G^+	Interfacial thermal conductance of the system corresponding to the forward heat flux direction
G^-	Interfacial thermal conductance of the system corresponding to the opposite heat flux direction
$\boldsymbol{G}$	Metric tensor given by $\boldsymbol{G}=\boldsymbol{h}\boldsymbol{h}^T$, with $\boldsymbol{h}=(\boldsymbol{a}, \boldsymbol{b}, \boldsymbol{c})^T$
G_{MN}^{∞}	Kirkwood-Buff integral in the thermodynamic limit
$G_{MN}^{k}(R)$	Plateau value of the running Kirkwood–Buff integral after the kth IBI iteration step
$G_{MN}^{\text{ref}}(R)$	Plateau value of the running Kirkwood–Buff integral calculated from the reference all-atom system
G_{MN}^{V}	Kirkwood–Buff integral of the radial distribution function between species M and N
$\hbar$	Reduced Planck constant (1.054572×10^{-34} J s^{-1})
$\boldsymbol{h}$	A matrix defined by the edge vectors $\boldsymbol{a}$, $\boldsymbol{b}$, and $\boldsymbol{c}$ of a unit cell
H	Separation distance
	Hamiltonian of a simple one-dimensional harmonic oscillator

$\boldsymbol{H}$	Hessian matrix of potential energy
H_{ij}	Strength of the steric repulsion for the two-body term of the Vashishta potential energy function
H_e	Partial Hamiltonian
H_2	Interlayer distance between the two graphene layers in each electrode
$H(\boldsymbol{R}^N, \boldsymbol{P}^N)$	Hamiltonian of an atomic system
$H_{2,adi}(p_2, x_2)$	Hamiltonian of particle 2 in an adiabatic two-particle system
$\boldsymbol{J}$	Heat flux vector
J^+	Heat flux in the forward direction (from silicene to graphene)
J^-	Heat flux in the reserve direction (from graphene to silicene)
$J_x(y,t)$	Momentum of the flux density
J_μ	Component of the heat flux vector $\boldsymbol{J}$ in the μ direction
k	Force constant in the harmonic function
k_s	Spring constant of an oscillator
	Spring constant of a simple one-dimensional harmonic oscillator
k_0	Wavenumber
$\boldsymbol{k}_0$	Wave vector
k_B	Boltzmann constant (1.380649×10^{-23} J K^{-1})
k_i^b	Material-dependent constants that determine the interaction strength of bond stretching
k_i^θ	Material-dependent constants that determine the interaction strength of bond angle bending
k_χ	Force constant
K	Evaporation constant
K_2	Kinetic energy of particle 2
$K(t)$	Instantaneous kinetic energy
K_{fict}	Fictitious electronic "kinetic energy"
Kn	Knudsen number
K_{st}	Stochastic kinetic energy defined in the Bussi–Donadio–Parrinello thermostat
l	Length of the MFP of the phonons
$\boldsymbol{l}$	Set of variationally determined parameters
l_b	Intrinsic length of the phonon MFP of the bulk material
$l_{i,0}$	Reference value of bond length

l_i	Bond length
l_y	Length of the simulation box in the y-direction
L	Length of an MD simulation system
$\boldsymbol{L}$	Second-order tensor or matrix of the deformation rate
L_P	Period length
m	Mass of a simple one-dimensional harmonic oscillator
m_e	Mass of an electron ($9.109\ 383 \times 10^{-31}$ kg)
m_i	Mass of the ith atom or particle
	Mass of the ith nucleus
m_w	Fictitious mass of the wall
$\min_{x_1} V(x_1, x_2)$	The lowest energy state of the system
M	Mass of a nucleus
	Fictitious mass having the extra degree of freedom Q
M_I	Mass of the Ith nucleus
$\boldsymbol{M}_{\boldsymbol{R}}^N(\boldsymbol{r}^n)$	Mapping function of the coordinates between the sites of a coarse-grained system and the corresponding atoms in a reference all-atom system
$\boldsymbol{M}_{\boldsymbol{P}}^N(\boldsymbol{p}^n)$	Mapping function of the momenta between the sites of a coarse-grained system and the corresponding atoms in a reference all-atom system
$\boldsymbol{M}_{\boldsymbol{R}_I}(\boldsymbol{r}^n)$	Mapping function of the position of the Ith site in a coarse-grained system
$\boldsymbol{M}_{\boldsymbol{P}_I}(\boldsymbol{p}^n)$	Mapping function of the momenta of the Ith site in a coarse-grained system
n	Empirical fitting parameter
	Local electron density
	Adjusting parameter in the Tersoff potential energy function
n_e	Phonon occupation number
$n(\boldsymbol{r}')$	Electron density distribution
n_b	Number of blocks for the block average method
n_k	Number of atoms in the kth slab
n_t	Total number of configurations sampled during the entire course of an all-atom MD simulation
N	Number of atoms
	Average number of water molecules in a nanochannel

N_a	Number of atoms
N_{A}	Avogadro constant ($6.022140 \times 10^{23}\ \mathrm{mol}^{-1}$)
N_{c}	Number of atoms in a cell
N_{e}	Number of electrons
N_{f}	Number of atoms in the "pump" region
$N_{i,\ \mathrm{nei}}$	Number of neighbors of the atom i
N_{m}	Number of molecules
N_{n}	Number of nuclei
N_{s}	Number of atoms on the box surface
N_{t}	Total atom number
N_{transfer}	Total number of velocity exchange events that occur over time t
N_{w}	Number of atoms in the constraint wall
N_0	Half of the total number of C atoms in a graphene nanosheet
N_1	Number of H atoms on one side of a graphene nanosheet
$\min_{x_1} V(x_1, x_2)$	Minimized potential energy $V(x_1, x_2)$ with respect to x_1 for a fixed value of x_2
$\boldsymbol{p}$	Momentum vector of an atom
	Dipole moment of a water molecule
$\boldsymbol{p}_i$	Momentum vector of the ith atom
p_{bo}	Empirical parameter
$\boldsymbol{p}^{\mathrm{elc}}$	Momentum vectors of the electrons
$\boldsymbol{p}^n$	Momentum vectors of the atoms
$\boldsymbol{p}^{N_e}$	Collection of the momentum vectors $\left\{\boldsymbol{p}_1, \ldots, \boldsymbol{p}_{N_e}\right\}$ of an MD simulation system containing N_{e} electrons
$\boldsymbol{p}^{\mathrm{nul}}$	Momentum vectors of the nuclei
$p_p(\boldsymbol{p}^n)$	Probability density of the atomic momenta of an all-atom system described by momentum vectors $\boldsymbol{p}^n = \left\{\boldsymbol{p}_1, \ldots, \boldsymbol{p}_n\right\}$
$p_{rp}(\boldsymbol{r}^n, \boldsymbol{p}^n)$	Probability density of the dynamic states of an all-atom system described by position vectors $\boldsymbol{r}^n = \{\boldsymbol{r}_1, \ldots, \boldsymbol{r}_n\}$ and momentum vectors $\boldsymbol{p}^n = \left\{\boldsymbol{p}_1, \ldots, \boldsymbol{p}_n\right\}$
$p_r(\boldsymbol{r}^n)$	Probability density of the atomic positions of an all-atom system described by position vectors $\boldsymbol{r}^n = \{\boldsymbol{r}_1, \ldots, \boldsymbol{r}_n\}$
$p(t)$	Instantaneous momentum of a 1D oscillator
p^0	Initial momentum of a 1D oscillator

$p(\theta_{jik}, \theta^0_{jik})$	Function for the bending of the bond angle of the Vashishta potential
P	Pressure
P_α	Total momentum of an MD system in the α-direction, with α being x, y, or z
$P_{\text{adi}}(x_2)$	Probability distribution function of x_2 in the adiabatic limit
$P_{\text{adi}}(x)$	Probability distribution function of x in the adiabatic limit
P_{ext}	External pressure
P_f	Pressure exerted on the fluctuating wall
P_{in}	Inlet pressure
$\boldsymbol{P}^{N_n}$	Collection of the momentum vectors $\{\boldsymbol{P}_1, \ldots, \boldsymbol{P}_{N_n}\}$of an MD simulation system containing N_{n} atoms
$\boldsymbol{P}^N$	Generalized spatial momenta of all the particles in an MD simulation system
$P_P(\boldsymbol{P}^N)$	Probability density of the site momenta of a coarse-grained system described by momentum vectors $\boldsymbol{P}^N = \{\boldsymbol{P}_1, \ldots, \boldsymbol{P}_N\}$
$P(q)$	Normalized probability distribution of a specific degree of freedom q
$p_R(\boldsymbol{R}^N)$	Reference probability density of a coarse-grained system determined by its corresponding all-atom trajectories
$P_{RP}(\boldsymbol{R}^N, \boldsymbol{P}^N)$	Probability density of the dynamic states of a coarse-grained system described by position vectors $\boldsymbol{R}^N = \{\boldsymbol{R}_1, \ldots, \boldsymbol{R}_N\}$ and momentum vectors $\boldsymbol{P}^N = \{\boldsymbol{P}_1, \ldots, \boldsymbol{P}_N\}$
$P_R(\boldsymbol{R}^N)$	Probability density of the site positions of a coarse-grained system described by position vectors $\boldsymbol{R}^N = \{\boldsymbol{R}_1, \ldots, \boldsymbol{R}_N\}$
$P(t)$	Instantaneous pressure
$P(x)$	Gaussian distribution of a random variable x
P_{out}	Outlet pressure
q	Charge of a particle
q_i	Partial charge of the ith atom
Q	A physical quantity
	Charge possessed by a core in the shell model
	Fictitious mass for the additional degree of freedom introduced in the Nosé–Hoover thermostat
	Fictitious degree of freedom introduced in the Andersen barostat
	Water flux

Q_{HP}	Estimated flow rate based on the HP equation
$Q(\boldsymbol{R}^N,\boldsymbol{P}^N)$	A physical quantity of an MD system described by the generalized spatial coordinates $\boldsymbol{R}^N$ and spatial momenta $\boldsymbol{P}^N$
Q_0	Estimation of a quantity Q for a bulk material
r	Distance between two atoms
$\dot{\boldsymbol{r}}$	The first-order derivative of the position vector with respect to time
$\ddot{\boldsymbol{r}}$	The second-order derivative of the position vector with respect to time
r_0	Distance corresponding to the minimum potential energy
	Equilibrium distance
	Equilibrium bond lengths
r_{cm}	Position of the center of mass
r_{cut}	Cutoff distance for a potential energy function
$\boldsymbol{r}^{elc}$	Position vectors of the electrons
$\boldsymbol{r}_i$	Position vector of the ith atom
$\breve{\boldsymbol{r}}_i(\Delta t)$	Position of the ith atom without considering the bond constraint between the atoms i and j
$\boldsymbol{r}_{i,c}$	Position of the ith core for the shell model
$\boldsymbol{r}_{ij}$	Displacement vector from the atom i to atom j
r_{ij}	Distance between the atoms i and j
$\boldsymbol{r}_{i,s}$	Position of the ith shell for the shell model
$r_{i,\alpha}$	Component of $\boldsymbol{r}_i$ in the α-direction
$\boldsymbol{r}_i(t)$	Time-dependent position of the ith atom
r_m	Empirical parameter for the LJ potential function
r_{nei}	Neighbor list distance defined for the Verlet neighbor list method
$\boldsymbol{r}^{N_e}$	Collection of the position vectors $\{\boldsymbol{r}_1,\ldots,\boldsymbol{r}_{N_e}\}$of an MD simulation system containing N_e electrons
$\boldsymbol{r}^n$	Spatial coordinates of the n atoms in the detailed atomistic system
$r^{n,i}$	Configuration of the ith all-atom MD simulation system
$\boldsymbol{r}^{nul}$	Position vectors of the nuclei
r_{4s}	Decay length of the two-body term of the Vashishta potential
r_∞	Radius of the inner boundary of the ambient region

R_c	Critical distance of an atom within which its closest neighbors can interact with it
R_I	Ideal gas constant (8.314462 J K^{-1} mol^{-1})
$\boldsymbol{R}_{IJ}$	Relative position vector of the Ith and Jth coarse-grained site
$\boldsymbol{R}_I$	Position of the Ith coarse-grained site
R_{MSD}	Mean square displacement
$\boldsymbol{R}^N$	Generalized spatial coordinates of all the particles in an MD simulation system
$\Delta\boldsymbol{R}^N$	Atomic displacement
$\boldsymbol{R}^{N_n}$	Collection of the position vectors $\{\boldsymbol{R}_1,\ldots,\boldsymbol{R}_{N_n}\}$of an MD simulation system containing N_n atoms
Ry	Rydberg constant (1 Ry=13.605693 eV)
s	An enhancement factor
	A fictitious degree of freedom
$\boldsymbol{s}_i$	A row vector with the components (l_i,m_i, and n_i)
$\boldsymbol{s}_{(k)}$	Direction vector
$\langle S_{map}\rangle_{\boldsymbol{r}^n}$	Average entropy resulting from the degeneracies caused by configuration mapping
S_{rel}	Relative entropy
$S(\boldsymbol{R}^{N_n},t)$	Phase factor for the nuclear wave function
t	Time
Δt	Time step
Δt^{max}	The maximum value of the AIMD timestep
t_{signal}	Time at which the HFACF possesses a magnitude equivalent to that of W
T	Temperature
ΔT	Temperature difference
∇T	Temperature gradient
T_c	Critical temperature
T_d	High threshold displacement energy
T_{elec}	"Fictitious" temperature of the electrons
T_f	Tilt factor that denotes the angle between the xz-plane and the yz-plane
T_f^0	Initial tilt factor at $t=t_0$ that denotes the angle between the xz-plane and the yz-plane

T_{G}	Temperature of the graphene monolayer
T_{G}^{0}	Initial temperature of the graphene monolayer
T_i	Total nuclear charge of the ith atom
ΔT_{In}	Temperature drop at an interface
T_{MD}	Calculated temperature based on MD simulation
T_{nucl}	Temperature of the nuclei
$T_{N-m}(z_k)$	Instantaneous temperature of the kth slab at the $(N-m)$th time step
Tr(.)	Trace of a matrix
T_{S}	Temperature of the silicene monolayer
T_{S}^{0}	Initial temperature of the silicene monolayer
$T(t)$	Instantaneous temperature
$\langle T(z_k)\rangle_M$	Temperature of the kth slab averaged over the final M time steps of the simulation
$\langle \partial T/\partial z\rangle$	Averaged value of the temperature gradient in the z-direction over time
$\boldsymbol{u}_i$	Displacement vector of the ith atom
$\dot{\boldsymbol{u}}_i$	Time derivative of the displacement vector of the atom i
$u(\boldsymbol{r}^n)$	Atomistic interaction potential energy
$U_{\text{bond-bending}}$	Bond bending potential component
U^{Buck}	Buckingham potential
U_{cross}	Cross potential function terms for the couplings among bond stretching, bond angle bending, and bond torsion
U_{CG}	Effective potentials of the CGMD model
U^{Coul}, U_{el}	Potential energy due to Coulombic interactions
U^{ES}	Potential energy due to electrostatic interactions
$\Delta U^{\mathrm{KBI}}(R)$	KBI–IBI ramp correction term
U^{Morse}	Morse potential function
$U_{MN}^{0}(R)$	Initial guess of the nonbonded two-body potential of a CGMD model
$U_{MN}(R)$	Nonbonded two-body potential of a CGMD model
$U_{MN}^{k}(R)$	Effective potential obtained from the kth IBI iteration step
U_{nb}	Potential energy due to nonbonded van der Waal interaction in the MM force field
$\Delta U_P(R)$	Linear pressure correction term

$U(\boldsymbol{R}^N)$	Potential energy function
$U'(\boldsymbol{R}^N)$	Gradient of $U(\boldsymbol{R}^N)$
$U_{\mathrm{Rigid}}(R^N)$	Potential energy of an MD simulation system with a rigid bond constraint imposed
$U^{\mathrm{SR}}(r_{ij})$	Potential energy due to the repulsion of the electron shells
U^{tot}	Potential energy function
U_χ	Out-of-plane bending interactions that are closely related to the bond torsion term
$U_2(r_{ij})$	Two-body term of the Vashishta potential
$U_3(r_{ij}, r_{ik}, r_{jk})$	Three-body term of the Vashishta potential
v	Velocity of an atom
	Rotational frequency
$\overline{v}$	Average velocity of the water molecules in the nanochannel
v_{C}	Velocity of the center of mass of all the particles in the heat source or the heat sink
v_{com}	Velocity of the center of mass of the atoms in a strip
$v_{i,\alpha}$	Component of the velocity of the ith atom in the α-direction
$\boldsymbol{v}_i$	Velocity vector of the ith atom
$\boldsymbol{v}_{ij}$	Relative velocity of the two points (or atoms) i and j
$v_j(t)$	Velocity of the atom j at time t
v_{max}	Maximum value of the advection velocity
$\langle v_{\mathrm{max}} \rangle$	Time average of the highest atomic velocity
v_{p}	Group velocity of the phonons
	Velocity of the incident wave
v_{s}	Velocity of the surface wave on the solid
v_x	x-component of the local average stream velocity
V	Volume of a system
$V(x_1, x_2)$	Potential of a simple one-dimensional chain model consisting the two particles 1 and 2
$V(\boldsymbol{r})$	Electron–nuclei interaction potential energy
$V(r_{ij})$	Two-body potential that accounts for the interactions among the nuclei
$V_2(r_{ij})$	Two-body potential energy function
$V_3(\boldsymbol{r}_i, \boldsymbol{r}_i, \boldsymbol{r}_k)$	Three-body potential energy function

$V_e^E(\boldsymbol{R}^{N_n})$	Effective potential energy contributed by the electrons
V_{free}	Total free volume
V_{occupy}	Total volume occupied by the nanofillers and the matrix of the composite
V_{vacuum}	Sum of the free volumes that exceed a critical size
$V_H(\boldsymbol{r})$	Hartree potential energy
V_m	Volume of the composite
$V_{ne}(\boldsymbol{r}^{N_e}, \boldsymbol{R}^{N_n})$	Operator for the total potential energy of the nuclei and electrons
V_n	Energy barriers for bond torsion
V^{REBO}	REBO term of the AIREBO potential energy function
V^{LJ}	Lennard–Jones potential energy function
$V^{Torsion}$	Potential energy contributed by bond rotations for the AIREBO potential energy function
V_w	Potential energy contributed by the interaction between the ith particle and the wall
$W(\tau)$	Amplitude of the autocorrelation function
w_A	Wave amplitude of a wave packet
x^0	Initial position of a 1D oscillator
$x_{1,adi}(x_1(0), v_1(0), x_2(0), t)$	Value of x_1 at time t when x_1 evolves adiabatically under the initial conditions of $x_1(0)$ and $v_1(0)$ with x_2 being fixed at its initial value
$x(t)$	Instantaneous position of a 1D oscillator at time t
$x_i(t_0)$	Coordinate of atom i at t=t_0 in the x-direction
y_i	y-component of the position of the center of mass of water molecule
$y_i(t_0)$	Coordinate of the atom i at t=t_0 in the y-direction
Z	Effective configuration partition function
Z_I	Atomic number of the Ith nucleus
$Z_1(x_2, \beta_1)$	Effective configuration partition function of x_2
$Z_{i,eff}$	Effective nuclear charge of the ith ion
z_i	Coordinate of the ith atom in the direction of heat transport
z_k	Position of the center of the kth slab of atoms
z_0	Center of the wave packet
$z(\boldsymbol{r}^{n,i})$	Partition functions of the ith all-atom configuration $\boldsymbol{r}^{n,i}$

$z(\boldsymbol{R}^{N,i})$	Partition functions of the ith coarse-grained configuration $\boldsymbol{R}^{N,i}$
α_i	Electronic polarizability of the ith ion
	Angles at the core of the nanowire for the ith wedge
ρ	A material-dependent constant
$\rho(y,t)$	Mass density at the position y and time t
$\rho(\boldsymbol{R}^N, \boldsymbol{P}^N)$	Probability density distribution of the phases of an MD system described by the generalized spatial coordinates R^N and spatial momenta P^N
ρ_i	Electron density at the site of the ith atom
$\rho_{ij}(r_{ij})$	Electron density of the jth atom at the site of the ith atom
ρ_{N}	Number density of the atoms in an MD system
ρ_N	Number density of coarse-grained particles of the N type
θ	General angle
	Misorientation angle
	Rayleigh angle
θ_{D}	Debye temperature of the system
θ_i	Bond angle
$\theta_{i,0}$	Reference value of θ_i
θ_{ijk}	Angle between two successive bond vectors
θ_{jik}^0	Equilibrium bond angle
$\langle\theta\rangle$	Time-averaged dipole orientation angle of the water molecules
ε	Depth of the potential energy well
	Scaling factor in the SW potential energy function
	Effective spring constant of a simple harmonic spring potential energy function
	Rectification ratio
$\varepsilon_{\mathrm{AB}}$	Potential energy well depth of the Lennard–Jones potential for the two atom types A and B
$\varepsilon_{\mathrm{coh}}$	Cohesive energy density
ε_i	Eigenvalues of the unoccupied orbitals
$\varepsilon_i^{\mathrm{Pot}}$	Potential energy term of the site energy
$\varepsilon_i(k_0)$	Polarization parameter
$\varepsilon_i(t)$	Site energy of the ith atom at time t
$\varepsilon_{\mathrm{fw}}$	Potential energy well (or binding energy) of the LJ potential

ε_j	Eigenvalues of the occupied orbitals
$\dot{\varepsilon}_i$	Atomic strain rate tensor
$\dot{\varepsilon}_{xy}$	Shear strain rate
ε_0	Vacuum dielectric constant (8.854×10^{-12} F m^{-1})
$\Delta\varepsilon$	Amount of heat input into the heat source
ξ	Width of the wave packet
ξ_i	Material-dependent constant
	Noise term in the Langevin equation that imposes a stochastic force on the *i*th atom
ζ	Effective coordination number of the atoms for the Tersoff potential energy function
	Thermal relaxation time
	Velocity rescaling factor
ζ_{ij}	Effective coordination number of the atoms
ς_i	Scaled coordinate vector for the Andersen barostat
π_i	Momentum that conjugates the scaled coordinate vector for the Andersen barostat
Π	Momentum that conjugates the degree of freedom *Q* in the Andersen barostat
η_i	Coefficient that determines the magnitude of the stochastic force imposed on the *i*th atom for the Langevin thermostat
η_{ij}	Exponent of the steric repulsion for the Vashishta potential energy function
μ	Mean of a Gaussian distribution
	Reduced mass of the atom for the Morse potential
	Fictitious mass parameter (in the unit of eV s^2)
υ	Constant that denotes the nonzero collision frequency for the Andersen thermostat
$\delta_{\alpha\beta}$	Kronecker delta
σ	Critical separation distance for a negative potential energy
	Standard deviation of a Gaussian distribution
	Zero-crossing distances of the LJ potential
σ_{AB}	Zero-crossing distance of the Lennard–Jones potential for the two atom types A and B
σ^2	Variance of the Gaussian distribution
$\sigma^{virial}(r)$	Virial stress of a volume element

λ	Positive coefficient for the steepest descent method
	Thermal conductivity
	Time-dependent Lagrange multiplier
λ^+	Thermal conductivity in the forward direction
λ^-	Thermal conductivity in the opposite direction
$\lambda_{\mu v}$	Component of the thermal conductivity tensor $\boldsymbol{\lambda}$
$\boldsymbol{\lambda}$	Thermal conductivity tensor
λ_{GE}	Thermal conductivity of the graphene sheet
λ_{H}	Thermal conductivity of the hybrid graphene/silicene monolayer
λ_k	Lagrange multipliers for bond constraints
$\lambda_{(k)}$	A coefficient
λ_{SE}	Thermal conductivity of the silicene sheet
λ_0	Predefined value of $\lambda_{(k)}$
λ_3	Cutoff parameter in the Tersoff potential energy function
$\lambda_\infty^{\max}$	Upper limit of bulk thermal conductivity
Λ	de Broglie thermal wave length
Λ_{ij}	Lagrange multipliers introduced to dynamically satisfy the condition $\left\langle \psi_i \middle| \psi_j \right\rangle = \delta_{ij}$
γ	Fitting constant for the SW potential energy function
γ_i	Positive atomic friction coefficient
ϕ	A general angle
ϕ_0	Torsion angles at points of minimum potential energy
ω	Vibrational frequency of a bond
	Frequency of motion of a particle
	Strength of a wedge disclination
	Angular frequency
	Phonon frequency
ω_{c}	Critical disclination strength
ω_{D}	Debye frequency that corresponds to the highest phonon frequency allowed in the Debye model
ω_{e}	Vibrational frequency of the electrons
$\omega_e^{\max}$	Highest possible electronic frequency
$\omega_e^{\min}$	Lowest possible electronic frequency

ω_{n}	Vibration frequency of the nuclei
$\omega_n^{\max}$	Highest phonon mode frequency
$\Xi(\boldsymbol{R}^{N_n})$	Wave function of the nuclei in a system
$\Phi(\boldsymbol{r}^{N_e},\boldsymbol{R}^{N_n})$	Wave function of the electrons in a system
$\Phi(\boldsymbol{r}^{N_e},\boldsymbol{R}^{N_n},t)$	Time-dependent wave function that depends on both the nuclear and electronic degrees of freedom
$\Phi(\boldsymbol{r}_i-\boldsymbol{r}_j)$	Pure two-body potential for modeling nuclei
$\Phi^i_{\mathrm{R_c}}$	Set of atoms within a sphere of radius R_c with the atom i at its center
$\Psi(\boldsymbol{R}^{N_n},\boldsymbol{r}^{N_e})$	Total wave function of an atomic system
$\Psi(\boldsymbol{r}^{N_e},t)$	Time-dependent wave function of the electrons in a system
$\chi(\boldsymbol{R}^{N_n},t)$	Time-dependent wave function of the nuclei in a system
Ψ	Wave function
Ψ_i	One-particle orbital subjected to the orthonormal constraint
Ψ_0	Potential energy of electrons at the ground state
Ψ_1	Potential energy of electrons at the first excited state
α	Electronic polarizability of an ion
β	Adjusting parameter in the Tersoff potential energy function
	Isothermal compressibility of a material
χ	Dihedral angle
	Nuclear wave function
χ^2	Objective residual function
τ_{b}	Time steps for the block average method
	Intrinsic relaxation time of the corresponding single-crystal bulk material
τ_{g}	Relaxation time under the effect of grain boundary scattering
τ_{s}	Relaxation time under the effect of surface scattering
τ_{im}	Relaxation time in the presence of impurities
τ_{m}	Upper integral limit in the Green–Kubo expression
$\tau_{m,max}$	Largest numerically meaningful upper limit for the integral in the Green–Kubo expression
τ_P	Coupling parameter that determines how closely a system is coupled with a barostat
τ_T	Coupling parameter of the Berendsen thermostat

$\Delta\tau$	Time interval between two successive collisions for a given atom
Σ	Reciprocal coincidence site density
γ	Atomic friction coefficient of the Langevin thermostat
	Interfacial adhesion energy
$\langle\cdots\rangle$	Average of a physical quantity over all the possible phases of an MD system
$\langle\,\vert\,\rangle$	Inner product of two functions
Ω	A volume element
	System volume
Ω_i	Volume surrounding the atom i
$\Omega_{\mathrm{map}}(\boldsymbol{M}_{\boldsymbol{R}}^{N}(\boldsymbol{r}^{n,i}))$	Degree of degeneracy of $\boldsymbol{R}^{N,i}$
Γ	Transmission coefficient
$\otimes$	Dyadic product of two vectors

Preface

Molecular dynamics (MD) simulation is a powerful computational method for modeling matter on the atomistic scale. It provides detailed information regarding the interactions and movements of the constituent atoms or molecules of a system and allows for the prediction and understanding of its macroscopic properties, thereby complementing physical experiments. Since this method was first applied to model the phase transition of a simple hard-sphere system in 1957, it has undergone extensive development to become a robust, established, and widely adopted tool for the scientific community. Moreover, with the advent of increasingly powerful computers and advanced numerical algorithms, MD simulation is frequently employed nowadays to solve highly complex problems in areas of cutting-edge research, which include nanomaterial development, polymer science, metallurgy, seawater desalination, pharmacy, organic chemistry, catalytic development, and tribology.

A deep understanding of the fundamental principles of MD simulation is indispensable for its utilization. Its theoretical foundation is multidisciplinary in nature and entails key concepts from several subjects, including classical mechanics, molecular mechanics, statistical mechanics, quantum mechanics, and condensed matter physics. Therefore, understanding the rudiments of MD simulation constitutes a daunting task for novice researchers. This book introduces and encapsulates the core principles of MD simulation while emphasizing their significance in the general workflow of the simulation processes. For more experienced practitioners, advanced MD simulation techniques developed in recent years are reviewed along with substantial case studies. The case studies selected feature the wide applicability of MD simulation and include diverse research topics spanning mechanical, thermal, mass transport, and chemical reaction problems.

Chapters 1–4 are devoted to the fundamentals of MD simulation as well as recently developed simulation algorithms. Researchers unfamiliar with MD simulation are encouraged to peruse these four chapters sequentially to attain a solid understanding of these principles. For more experienced practitioners, these chapters serve as a concise handbook in which the theoretical foundations of key concepts and the physical meanings of important simulation parameters are revealed.

Chapter 1 provides a succinct introduction to MD simulation and its roles in contemporary scientific research in terms of potential applications and limitations. The theoretical foundations of MD simulation are illustrated, and its typical workflow is presented with the technical aspects of each step discussed. In addition, a comparison with two other commonly adopted atomistic simulation methods, ab initio calculation and Monte Carlo simulation, is conducted. In Chapter 2 the principles of potential energy functions, which are crucial for describing the atomic and molecular interactions in a system, are explained. The general procedure of constructing a potential function is described, and typical two-body and many-body potentials are explored. Specific potential energy functions for different types of materials are highlighted. Chapter 3 covers the main control methods involved in MD simulation, including the selection of thermodynamic ensembles, efficient temperature and pressure control, and the imposition of common boundary conditions and special system constraints. In Chapter 4 two classes of advanced MD methods, ab initio MD methods and coarse-grained MD methods, are introduced along with their respective theoretical foundations. Understanding these two classes of methods, which extend the range of application of MD simulation to new potential research fields, will keep the reader abreast of its recent developments.

Chapters 5–8 expound the application of MD simulation in various types of scientific problems. In each chapter, the details of model construction are illustrated, and the simulation results are interpreted. Thus, these chapters can be employed as a technical guide for researchers who wish to extend their knowledge of simulations in the corresponding fields.

Chapter 5 delves into the application of MD simulation in a collection of mechanical problems. In particular, simulation models for common mechanical tests, such as tensile, compressive, shear, nanoindentation, and nanoscratching tests, are explored. Popular simulation frameworks and analysis methods for investigating tribological behaviors, interfacial effects in nanocomposites, and the effects of defects in nanocrystalline materials are evaluated. In Chapter 6 the two most commonly utilized MD simulation models for calculating the thermal conductivity of a solid, the direct method and the Green–Kubo method, are studied. Additionally, the application of MD simulation in modeling interfacial thermal transport is demonstrated. Chapter 7 illustrates the application of MD simulation in solving various mass transport problems, including fluid flow in nanoconfinement, nanofiltration with porous thin films, and liquid–vapor transitions. The unique mass transport behaviors that occur on the nanoscale are highlighted with their underlying mechanisms. Chapter 8 introduces the application of MD simulation in chemical reactions by involving reactive force fields capable of modeling chemical bond formation and breakage. The phenomena of irradiation processes and material crystallization are also examined.

The completion of this book could not have been possible without the excellent teamwork shown by the members of our research group, with whom our discussions have always been fruitful. We greatly appreciate the efforts of Liming You, Lichun Bai, Yujia Tian, Terence Zhi Xiang Hong, Devesh Kripalani, Jiazhao Huang, Kiran Raj, Weiming Ji, Asker Jarlöv, Rita Babicheva, Julia Baimova, Hui Zhang, and Kieu Trung Hieu, who contributed valuable feedback during the preparation of this book. We are also deeply grateful to our collaborators who have rendered us tremendous support in conducting research pertaining to MD simulation. Finally, we acknowledge the financial support from the Nanyang Environment and Water Research Institute (Core Fund) at Nanyang Technological University, Singapore and the National Natural Science Foundation of China (No. 11802087).

Kun Zhou and Bo Liu

CHAPTER

1

Fundamentals of classical molecular dynamics simulation

This chapter presents the fundamentals of molecular dynamics (MD) simulation. It begins with highlighting the necessity of computer simulation in scientific research and focuses on the significance of atomistic simulation. The characteristics of MD simulation are then summarized and compared with those of two other atomistic simulation methods, namely, ab initio calculation and Monte Carlo simulation. The roles of MD simulation in contemporary scientific research are presented in terms of its potential applications and limitations. The theoretical foundation of a classic MD simulation is illustrated, and its workflow is presented. The technical aspects of each step of an MD simulation are demonstrated, such as the periodic boundary conditions, neighbor list, time integration algorithms, and statistical ensembles. Finally, popular MD software packages and their hardware requirements are discussed comprehensively.

1.1 Introduction

Computer simulation has been indubitably established as a potent tool for modern scientific research activities. In the past, scientific research was carried out solely based on the interplay between theory and experiment. For a theoretical framework, a set of mathematical equations is established to model an actual system with simplifications and assumptions made. The validation of a theory is usually assessed by its capability to describe the physical behaviors of certain specially selected cases in which all the complexities that could prevent solving the equations are eliminated. In experiments, a system is subjected to direct measurements and observations. Unfortunately, high experimental costs, inaccessibility of experimental conditions, and limitations of experimental equipment restrict most experiments to a few simple and

Molecular Dynamics Simulation
DOI: https://doi.org/10.1016/B978-0-12-816419-8.00006-4

special circumstances. In contrast, physical phenomena in the real world are usually much more complicated.

Computer simulation bridges the gap between theory and experiment by providing a solid routine to investigate the static and/or dynamic response of a system under given initial conditions. In computer simulations, a system model is constructed based on relevant theories, and the solution of the equations is calculated by a computer following certain algorithms. Thanks to the rapid evolution of computer science and technology, highly efficient personal computers (PCs) and other powerful computing resources are presently available to complete such calculations at increasingly higher rates. Hence, greater complexity can be introduced to system models to depict real processes more accurately. In addition, increasingly advanced graphical user interfaces (GUIs) have been facilitating the visualization of the simulated processes and results. Consequently, computer simulation has become analogous to actual experiments because direct observation of the behavior of a system is currently possible with the aid of a GUI.

1.1.1 Atomistic simulation

Compared with experiments, computer simulations can tackle problems of great complexity, save time and expense, exercise better control over important intermediate processes, and track system responses on a temporal and spatial scale inaccessible by conventional experiments. In the family of computer simulation methods, atomistic simulation occupies an important position and has garnered tremendous interest among members of the scientific community, in particular, physicists, chemists, and materials scientists. Atomistic simulation provides a methodology for the detailed modeling of systems on the atomic scale. The nature of matter is determined by the structural and dynamic behaviors of its constituent atoms/molecules. As a result, a large portion of the key questions that physicists and chemists are devoted to solving can be traced back to the relationship between the bulk properties of matter and the underlying interactions among its constituent atoms/molecules.

Ab initio calculation is a quantum mechanics method that relies on quantized descriptions of interatomic interactions. Following the advent and rapid development of the quantum mechanics framework in the 20th century, almost all the molecular properties of a system, such as its structure, conformation, dipole moment, ionization potential, electron affinity, and electron density, can be determined by quantum mechanics calculations. The calculation results usually exhibit good agreement with actual experiments, and on some occasions, an analysis of these results may even provide insightful information that experiments are unable to furnish.

Using ab initio calculations, the characteristics of a system of molecules can be understood before its synthesis, thereby providing chemists and pharmacists a powerful tool in the design of advanced materials or medicine.

The principles of quantum mechanics are based on the delocalization of electrons, and their behaviors are described by their wave functions. According to the Born rule, the probability distribution of a system of electrons is proportional to the modulus square of its wave function. To obtain this wave function, one needs to solve the famous Schrödinger equation. However, this process is extremely complicated, and exact solutions only exist in a small number of cases, typically involving one or a few electrons. In the case where the number of atoms (and electrons) is large, traditional methods for solving the Schrödinger equation quickly fall apart. A myriad of approximations and assumptions are made in the development of quantum mechanics methods to produce solutions that accurately depict the system. Ab initio calculations employ the variational principle and express the entire electron wave function as the superposition of different molecular orbitals, which are approximated by a combination of certain mathematical functions, for example, Gaussian functions. Ab initio calculation provides a high degree of accuracy but is computationally inefficient, and it is usually adopted to treat systems with fewer than 100 atoms. To increase the efficiency of ab initio calculations, a class of semiempirical molecular orbital methods has been developed by introducing some experimental measurement results as input parameters during the calculation. Although these semiempirical methods can tremendously increase computational efficiency, the system size is still limited to hundreds of atoms.

Hence, quantum mechanics methods are only applicable to simple molecules and systems with a small number of electrons. Unfortunately, most atomistic systems encountered in nature and industries, such as biomolecules (e.g., protein, nucleic acid, and polysaccharose) and polymers (e.g., rubber, fat, and oil), contain a large number of atoms and electrons. In addition, in some condensed systems, such as metals, polymers, solid composites, and nanomaterials, the interactions among the individual molecules/atoms and the collective properties of the entire system are of the same importance as the characteristics of single molecules/atoms. For these large systems, quantum mechanics methods including ab initio calculation fail to provide detailed structural, dynamic, and thermodynamic information.

To facilitate the study of large molecular systems beyond the limits of ab initio calculations, nonquantum mechanics simulation methods should be adopted. The key essence of these methods is utilizing an empirical force field to describe the atomic interactions. Empirical force fields are developed based on the Born–Oppenheimer approximation, which states that in a system of atoms, the presence of electrons is neglected, and the system energy is solely determined by the positions of the nuclei.

Typical atomistic simulation approaches that can apply empirical force fields include the Monte Carlo simulation, MD simulation, lattice energy calculation, and lattice dynamics simulation methods. Monte Carlo simulation and MD simulation are the most widely applied. Note that in modern atomistic simulation methods, ab initio calculation is often combined with the Monte Carlo simulation and the MD simulation methods, leading to the ab initio Monte Carlo simulation method and the ab initio MD simulation method. In these methods, ab initio calculation is adopted to directly provide a detailed description of atomic interactions that would otherwise be described by the empirical force fields.

In the Monte Carlo method, atoms are treated as single point masses, and their movements are regarded to be random. The statistical thermodynamic properties of a system are obtained from probability distributions based on the principles of statistical mechanics. This method is frequently employed nowadays as it is powerful in terms of dealing with phase change processes and properties of complex systems. However, employing the Monte Carlo method can only accurately predict the equilibrium properties of a system and fails to provide dynamic properties associated with nonequilibrium processes.

The MD simulation method applies classical Newtonian mechanics to describe the movement of atoms and obtain the dynamic properties of the modeled system. An important characteristic of MD simulation is that the calculation of atomic movements is based on a solid physical foundation and is highly accurate. Hence, MD simulation can simultaneously provide the statistical and dynamic properties of a system and is applicable to a broad range of material systems. Compared to quantum mechanics methods, the MD simulation method is simple and computationally efficient. In certain circumstances, the results produced by the MD simulation method exhibit good agreement with those from quantum mechanics methods. Due to the rapid development of the MD simulation method, different types of force fields have been formulated systematically for various molecular systems, such as biomolecules, polymers, metals, and semiconductors. With the increase in the processing speed of computing resources, MD simulations will be able to model larger and more complicated systems.

1.1.2 Molecular dynamics simulation

As discussed above, MD simulation is a technique to investigate the structure, dynamics, and transport properties of many-body systems by calculating the motion of atoms/molecules in a molecular assembly. The first MD calculation was reported by Alder and Wainwright (1957) who formulated an MD method to investigate the phase diagram of a hard-sphere system. The spheres were modeled to interact via instantaneous collisions while moving freely between collisions. Since then, the complexity of MD simulation systems has increased gradually. Rahman (1964) from Argonne National Laboratory conducted a pioneering MD simulation of liquid argon using the Lennard–Jones potential. Rahman's computer code provided a solid framework for many succeeding MD programs. A few years later, Rahman and Stillinger (1971) modeled liquid water using MD simulation, while McCammon et al. (1977) conducted the first MD simulation of protein.

In an MD simulation, the system obeys the classical Newton's laws of motion. Consider a molecular system containing N particles with their position vectors and momentum vectors denoted by $\boldsymbol{r}_i = (x_i, \ y_i, \ z_i)$ and $\boldsymbol{p}_i = (p_{i,x}, \ p_{i,y}, \ p_{i,z})$, respectively. Note that in this book, symbols denoting scalar quantities or variables are presented in italics, and symbols denoting vectors, matrices, and higher-order tensors are in both bold and italics. The Hamiltonian H of the system is written as

$$H(\boldsymbol{R}^N, \boldsymbol{P}^N) = \sum_i^N \sum_\alpha \frac{p_{i,\alpha}^2}{2m_i} + U(\boldsymbol{R}^N). \tag{1.1}$$

Here, $\boldsymbol{R}^N = \{\boldsymbol{r}_1, \boldsymbol{r}_2, \ldots, \boldsymbol{r}_N\}$ and $\boldsymbol{P}^N = \{\boldsymbol{p}_1, \boldsymbol{p}_2, \ldots, \boldsymbol{p}_N\}$, which represent the generalized spatial coordinates and momenta of all the particles in the system, respectively. The first and second terms on the right-hand side represent the kinetic and potential energy of the particles, respectively; α represents each of the three directions x, y, and z; m_i denotes the mass of the ith particle.

The force field acting on the system is conservative in nature, which means that the force acting on each particle in the system can be obtained based on the derivative of the potential energy of the system with respect to the position of the particle. The direction of the first-order derivative of the potential energy (i.e., gradient) indicates where the minimum lies, and the magnitude of the gradient indicates the steepness of the local slope. The energy of the system can be lowered by moving each particle in response to the force acting on it. Hence, the force is given by the negative of the gradient and is written as (Leach, 1996)

$$\boldsymbol{F}_i(\boldsymbol{R}^N) = -\frac{\partial U(\boldsymbol{R}^N)}{\partial \boldsymbol{r}_i}. \tag{1.2}$$

Hence, the movement of a particle can be described by Newton's second law as

$$m_i \ddot{\boldsymbol{r}}_i = \boldsymbol{F}_i(\boldsymbol{R}^N), \tag{1.3}$$

where $\ddot{\boldsymbol{r}}_i$ denotes the second-order derivative of $\boldsymbol{r}_i$ with respect to time. Given the initial positions $\boldsymbol{R}^N$ and momenta $\boldsymbol{P}^N$ of the particles, the integration of the expressions in Eq. (1.3) with respect to time produces the trajectories of the particles. As a result, MD simulation is a deterministic method whereby with predefined initial conditions, the subsequent time evolution is determined in principle.

Herein, the solution of the above initial value problem is demonstrated by examining the case of a simple one-dimensional (1D) harmonic oscillator. For such an oscillator, its classical Hamiltonian H is written as

$$H = \frac{p(t)^2}{2m} + \frac{1}{2}k_s x(t)^2, \tag{1.4}$$

where $p(t)$ and $x(t)$ denote the instantaneous momentum and position of the oscillator at time t, respectively; m denotes its mass, and k_s denotes its spring constant. The Hamiltonian equations of this oscillator are a set of first-order differential equations with respect to time:

$$\frac{\mathrm{d}x}{\mathrm{d}t} = \frac{\partial H}{\partial p} = \frac{p}{m}, \quad \frac{\mathrm{d}p}{\mathrm{d}t} = -\frac{\partial H}{\partial x} = -k_s x. \tag{1.5}$$

To calculate the trajectory of the oscillator numerically, the finite difference method is adopted to convert the differential equations into algebraic equations. By regarding the integration timestep as Δt and using the forward difference scheme, Eq. (1.5) can be expressed as

$$\frac{\mathrm{d}x}{\mathrm{d}t} = \frac{x(t+\Delta t) - x(t)}{\Delta t} = \frac{p(t)}{m}, \quad \frac{\mathrm{d}p}{\mathrm{d}t} = \frac{p(t+\Delta t) - p(t)}{\Delta t} = -k_s x(t). \tag{1.6}$$

A simple manipulation of Eq. (1.6) leads to the solution of Eq. (1.4) as

$$x(t+\Delta t) = x(t) + \frac{p(t)\Delta t}{m}, \quad p(t+\Delta t) = p(t) - k_s x(t)\Delta t. \tag{1.7}$$

Given the initial position x^0 and initial momentum p^0 of the oscillator, Eq. (1.7) forms a couple of recurrence equations, and the time evolution of $x(t)$ and $p(t)$ can be calculated in a stepwise manner.

In MD simulation, the solution of Eq. (1.1) is more difficult to obtain than that of Eq. (1.4) because the movement of each atom is under the influence of its interactions with all the other atoms in the system. Hence, MD simulation is, in essence, handling a many-body problem. For a system containing N particles, there are $6N$ variables ($3N$ positions and $3N$ momenta) to solve for, and an analytical solution is virtually impossible to obtain since these variables are closely coupled. Using the finite difference method discussed above, the $6N$ variables can be decoupled, and the key to resolving the Hamiltonian in Eq. (1.1) is to obtain an expression for $\boldsymbol{F}_i(\boldsymbol{R}^N)$ in Eq. (1.3). In MD simulation, a conservative force field with a concomitant potential energy function $U(\boldsymbol{R}^N)$ is introduced to describe the interatomic interactions, and $\boldsymbol{F}_i(\boldsymbol{R}^N)$ is calculated using Eq. (1.2).

In most cases, the atomic trajectories, i.e., the evolution of $\boldsymbol{R}^N$ and $\boldsymbol{P}^N$, are not the end results of an MD simulation. One important question is:

how shall one deduce physical quantities of the system from these trajectories? The answer lies in the fact that MD simulation is a statistical mechanical approach, although the movement of the atoms is described by classical Newtonian mechanics. The atomic trajectories can be regarded as a set of configurations following a certain statistical distribution. For example, for a canonical ensemble (see Section 1.2.8) in which the temperature is kept constant, the probability density distribution of the system configurations is proportional to the Boltzmann function $\exp\left[-H(\boldsymbol{R}^N, \boldsymbol{P}^N)/(k_B T)\right]$, where H denotes the Hamiltonian, k_B is the Boltzmann constant (1.380649×10^{-23} J/K), and T denotes the temperature. As a result, physical quantities of the system are obtained by summing over all possible configurations, each weighted by a corresponding probability. Mathematically, a physical quantity $Q(\boldsymbol{R}^N, \boldsymbol{P}^N)$ can be written as

$$\left\langle Q(\boldsymbol{R}^N, \boldsymbol{P}^N)\right\rangle = \int \mathrm{d}\,\boldsymbol{R}^N \int \mathrm{d}\,\boldsymbol{P}^N \rho(\boldsymbol{R}^N, \boldsymbol{P}^N) Q(\boldsymbol{R}^N, \boldsymbol{P}^N), \tag{1.8}$$

with $\rho(\boldsymbol{R}^N, \boldsymbol{P}^N)$ denoting the probability density distribution of the system phases. Here, the angle bracket $\langle\rangle$ indicates that the physical quantity is averaged over all the possible phases of the system that are described by both $\boldsymbol{R}^N$ and $\boldsymbol{P}^N$. As mentioned previously, the probability density distribution for configurations that follow the Boltzmann function is a function of the Hamiltonian, and it is calculated simply by normalizing the Boltzmann function:

$$\rho(\boldsymbol{R}^N, \boldsymbol{P}^N) = \frac{\exp\left[-H(\boldsymbol{R}^N, \boldsymbol{P}^N)/(k_B T)\right]}{\int \mathrm{d}\boldsymbol{R}^N \int \mathrm{d}\boldsymbol{P}^N \exp\left[-H(\boldsymbol{R}^N, \boldsymbol{P}^N)/(k_B T)\right]}. \tag{1.9}$$

A full sampling of this phase space is virtually impossible to achieve. According to the ergodic hypothesis, given a sufficiently long time, the system can experience all the possible states in the phase space, and thus quantities averaged across the phase space can be approximated by their counterparts averaged over time as

$$\left\langle Q(\boldsymbol{R}^N, \boldsymbol{P}^N)\right\rangle = \lim_{\tau \to \infty} \frac{1}{\tau} \int_0^{\tau} Q(\boldsymbol{R}^N, \boldsymbol{P}^N)\mathrm{d}t. \tag{1.10}$$

In practice, the simulation cannot last indefinitely, but one should always exercise caution to ensure that a proper sampling of the phase space is achieved.

The fact that MD simulation models materials on the atomistic level determines its spatial and temporal scales. Consider a crystal with a simple cubic lattice structure and a lattice constant of 1 nm. A cubic nanoparticle of such a crystal with a side length of 1 μm comprises approximately 10^9 atoms. Modeling a system containing such a huge number of atoms is still beyond the capabilities of MD simulation due to the high requirements of computing resources and data storage. To obtain a better resolution of the modeled system and reduce statistical errors, the simulation timestep should be kept as small as possible. The timestep of an MD simulation is usually set at a tenth of the shortest vibration period of the atoms in the simulated system. The vibration frequency of an atom in bonding is inversely proportional to its atomic mass. The vibration period of the lightest H atom is in the order of 10 fs when bonded with other elements. Hence, a timestep of 1 fs is usually required when simulating a system containing H atoms. For a simulation time of 1 ns, a timestep of 1 fs corresponds to 10^6 simulation steps, which may require a very long computing time, particularly for molecular systems with a large number of atoms.

1.1.3 Applications of molecular dynamics simulation

After over six decades of development, MD simulation has found broad applications in the fields of materials science, biophysics and biochemistry, chemistry, and medicine. In the field of materials science, MD simulation can be conducted to investigate equilibrium thermodynamics, defects, phase transitions, nucleation and surface growth, heat and mass transfer, and a wide range of other nanoscale properties and phenomena. In bioscience, MD simulation can be applied to understand protein folding, predict molecular structures, and investigate the biocompatibility between different materials. In chemistry, MD simulation can procure information on intramolecular and intermolecular interactions, simulate chemical reactions, and calculate free energy.

For instance, MD simulation is widely adopted for the study of defects in materials. A variety of defects play a significant role in determining the physical and mechanical properties of materials and are hence of great technological interest. Point defects (e.g., vacancies), line defects (e.g., dislocations), and planar defects (e.g., grain boundaries) are common defects occurring in materials, particularly crystals. Direct experimental observations of these defects and the interactions among them are often challenging and costly due to their generally small sizes and high localization. Furthermore, although theoretical models have been developed to describe the relationships between defects and material properties, difficulties are encountered in describing the dynamic behaviors of those defects subjected to mechanical and/or thermal loading. In contrast, MD simulation can provide detailed information of the movement trajectories of all the atoms in a system, based on which the dynamic behaviors of defects in the system and their effects on its material properties can be well understood.

MD simulation is also an efficient research approach for investigating surface-related problems encountered in various aspects of materials science. Using MD simulation, complicated physical processes associated with surfaces can be studied, such as surface reconstruction, surface melting, surface diffusion, surface roughening and friction. Atomic trajectories obtained in the simulation can provide a direct observation of the surface structure, morphology and evolution, thereby serving as a high-resolution in situ microscopy tool.

1.1.4 Limitations of molecular dynamics simulation

MD simulation is a powerful and efficient simulation technique but has its limitations. The most significant limitation is that MD simulation highly relies on the accuracy of potential energy functions in describing a simulated system (see more discussions in Chapter 2: Potential Energy Functions). These functions are introduced to accurately depict the interactions between the atoms or molecules in the system, which in turn determine the atomic trajectories. Hence, a potential energy function that inaccurately describes such interactions will certainly result in a false description of the dynamic behaviors of the simulated system.

Potential energy functions are typically semiempirical formulae obtained by the mathematical fitting of experimental data. Hence, the specific conditions under which experimental data are obtained determine the applicability of the corresponding potential energy function. For instance, if a potential energy function is developed to model a particular phase of a material, it may not necessarily be applicable for modeling other phases of the same material. Therefore, selecting a suitable potential energy function and testing it are crucial in obtaining accurate simulation results.

In recent years, new materials have been discovered at an unprecedented rate due to the rapid growth of knowledge in materials science and the development of material synthesis and characterization techniques. However, an in-depth understanding of the properties of these materials via MD simulation requires the development of suitable potential energy functions, which is not only scientifically challenging but also time-consuming. Unfortunately, there is still a lack of highly efficient procedures for developing such potential energy functions. The most widely adopted procedure is to first perform ab initio calculations on a simulated system to obtain the necessary parameters, based on which mathematical fittings can be performed to establish the potential energy function. Hence, the accuracy of the potential energy function largely depends on the developer's knowledge of materials science, experience in conducting ab initio calculations, and mathematical proficiency. Normally, only potential energy functions that have been intensively tested by researchers for years are qualified for the studies of material properties. As a result, the application of MD simulation in the investigation of new materials is usually limited by the unavailability of a suitable potential energy function.

Another limitation of MD simulation lies in its time and length scales. MD simulation is usually adopted to handle a molecular system with a length scale of less than 100 nm and a timescale of less than 100 ns. In some cases, 100 ns is much longer than the relaxation time of certain properties of interest and is thus sufficient to ensure that a well-converged value is obtained. However, this is not always the case as different properties exhibit different relaxation times. The relaxation times of some physical properties have orders of magnitude larger than 100 ns, which is particularly true in cases where molecular systems tend to demonstrate slow dynamic behaviors in the vicinity of an equilibrium or phase transition point.

Besides the timescale limit, the length scale limit is also a problem commonly encountered in MD simulations. For instance, a simple protein molecule may contain 10^3–10^5 atoms, and a tiny biological cell typically consists of hundreds or thousands of protein molecules. Hence, a short simulation of even a simple biological cell system will involve approximately 10^6 atoms. However, such a simulation is far from satisfactory for understanding the microscopic world. For ideal crystals, the length scale limitation may be bypassed by introducing periodic boundary conditions (PBCs) to eliminate surface effects (see more discussions on PBCs in Section 1.2.3). However, for materials lacking periodicity, such as organic materials, biomaterials, and materials containing defects, the application of PBCs is usually difficult. In this case, the problem of length scale limitation can sometimes be partially resolved by a method known as finite-size scaling. For this method, a physical quantity Q is calculated for different system lengths L, and the following mathematical fitting is performed on Q and L:

$$Q(L) = Q_0 + \frac{c}{L^n}, \tag{1.11}$$

where Q_0, c, and n are fitting parameters. For an infinitely large L, Q_0 corresponds to an estimate of the quantity Q for a bulk material.

1.2 Fundamentals of molecular dynamics simulation

The first MD simulation was reported in the 1960s. However, it took nearly half a century for such a method to be established as a solid, well-accepted, and widely adopted tool. In the past decades, researchers have developed and accumulated a number of techniques to improve MD simulation. Some of these techniques have become indispensable foundations for most of the popular contemporary MD software packages. In this section, further details on the basics of MD simulation are introduced with the crucial techniques highlighted, such as the PBCs, time integration algorithms, and neighbor list.

1.2.1 Procedure

The workflow of a classical MD simulation generally contains the following steps shown in the flowchart below (Fig. 1.1).

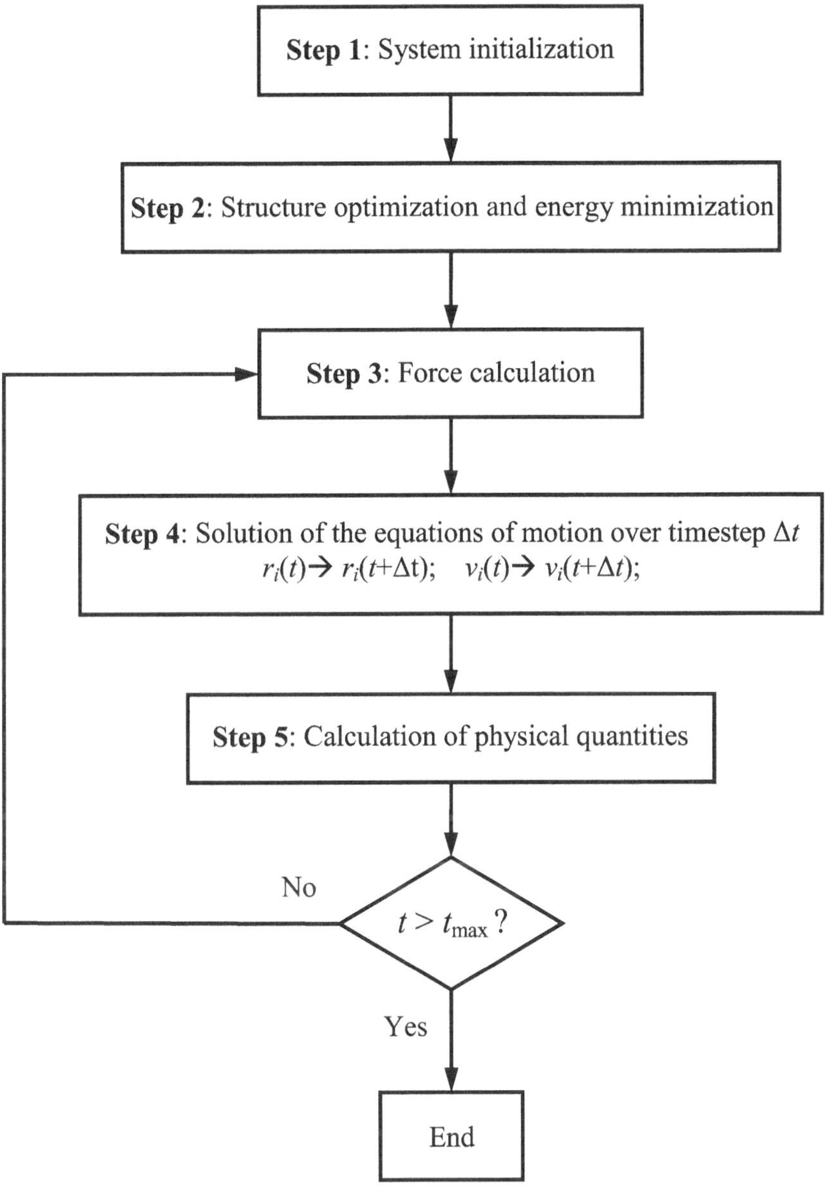

FIGURE 1.1 Flowchart of a classical MD simulation.

In step 1, the initial system conditions are established, including the initial atomic positions, velocities, boundary conditions and constraints. The potential energy function should also be specified in this step.

In step 2, structural optimization and energy minimization are performed to readjust the initial structure to better resemble an equilibrium structure based on the selected potential energy function. This step produces a superior "starting point" for the system and is always beneficial for the smooth execution of an MD simulation.

In step 3, the forces acting on each atom/molecule are calculated according to the selected potential energy function. Afterward, the equation of motion in Eq. (1.3) is solved for each particle over a timestep Δt.

In step 4, the atomic positions and velocities are updated accordingly.

In step 5, the physical quantities of interest are calculated based on the information collected in step 4. When the current simulation time reaches the preset maximum simulation time or if some other conditions are met, the simulation will end. Otherwise, the simulation continues by repeating steps 3–5.

It should be noted that step 2 is not always necessary and can be omitted in certain cases. For example, if an MD simulation is continued from a prior simulation whose system has achieved a well-equilibrated state, then structural optimization or energy minimization would not be repeated. An MD simulation could also run smoothly and produce reasonably accurate results without structure optimization and energy minimization even it is started from scratch, provided that its initial atomic structure happens to be close to the equilibrium structure corresponding to the potential energy function selected. However, this situation is unlikely because the initial atomic structures are usually constructed based on information obtained from experiments, ab initio calculations, or previous MD simulations using different potential functions or subjected to different constraints. In the following sections, details on the crucial techniques involved in each of these steps are introduced in sequence.

1.2.2 System initialization

To perform a simulation, an MD box should first be defined, and the positions and velocities of atoms should be initiated. If system initialization is not properly performed, it will usually take a long time for the system to reach thermal equilibrium. There are two types of MD simulation initialization: (1) starting a simulation from scratch and (2) continuing from a previous simulation.

If the simulation is started from scratch, the positions of the atoms will usually be initiated based on a lattice structure. The lattice constants are set according to actual reported values from experiments or theoretical calculations. Consider a simple example of an orthogonal simulation box containing a face-centered cubic (FCC) lattice with the lattice constant a. In its unit cell, there are four atoms, and their relative positions with respect to the origin of the unit cell can be expressed as $r_1 = (0, 0, 0)$, $r_2 = (0, 0.5a, 0.5a)$, $r_3 = (0.5a, 0, 0.5a)$, and $r_4 = (0.5a, 0.5a, 0)$. The simulation box is then filled with atoms by duplicating the unit cell and stacking the copies. Hence, the atomic positions in the simulation box can be defined by $r_n + a\,(N_x,\ N_y,\ N_z)$, where n takes on values from 1 to 4, and N_x, N_y, and N_z are integers denoting the shifts of the original unit cell in the x-, y- and z-directions. This is the standard procedure for treating crystalline materials such as metals and semiconductors. For amorphous materials and liquid systems that are not crystalline, a suitable structure can be constructed according to X-ray diffraction experimental results of the corresponding system. If such results are unavailable, a crystalline structure can be assumed, but further treatment is required.

Take the modeling of liquid water as an example; water molecules can be initially arranged to occupy the lattice sites of an FCC lattice structure. The lattice constant is set at a value corresponding to the actual density of water. The dipole orientation of the H_2O molecules can be arranged to be either uniform or random. Afterward, a quenching simulation is performed to allow the water molecules to move freely and subsequently reach a state of thermodynamic equilibrium. During quenching, the system temperature is first raised rapidly and then decreased gradually. Its volume or pressure can be controlled depending on its expected final density or structure. The advantage of the quenching method is that poorly initiated coordinates of atoms or molecules, which result in overlapping and overly small separation, can be avoided. In addition, the density of the structure can be adjusted, and the orientation of the molecules can be randomized. Some complex systems (e.g., polymeric systems) consist of large molecules that contain a substantial number of atoms, and the spatial conformations of these molecules play a crucial role in determining the final properties of the entire system. In such cases, the initial simulation box should be large enough to provide adequate space for the polymer molecules to rotate and fold during the quenching process. To obtain a density close to experimental measurements, the volume of the simulation box is decreased simultaneously with the temperature of the system.

In principle, the initial velocities of the atoms can take any arbitrary values, provided that system stability is retained. After the simulation ensues for a certain amount of time, the atomic velocities will be automatically refined to approach reasonable values in accordance with the assigned simulation conditions and constraints. However, to accelerate the simulation and ensure its smooth performance, the initial atomic velocities should follow certain distributions. Since the simulation is typically carried out at a temperature above 0 K, a good choice is to set initial velocities that meet this temperature requirement, which is usually based on the Maxwell–Boltzmann distribution that can guarantee a quick relaxation of the initial system. The Maxwell–Boltzmann distribution is a Gaussian distribution of a random variable x and is given by

$$P(x)_{\mu,\sigma} = \frac{1}{\sigma\sqrt{2\pi}} \exp\left[-\frac{(x-\mu)^2}{2\sigma^2}\right], \tag{1.12}$$

with μ denoting the mean and σ^2 the variance. According to statistical mechanics, the temperature of a molecular system and the average atomic velocity are related by

$$\left\langle v_{i,\alpha}^2 \right\rangle = \frac{k_B T}{m_i}, \tag{1.13}$$

where $v_{i,\alpha}$ denotes a component of the velocity of the ith atom, with α representing the x-, y-, or z-direction, k_B is the Boltzmann constant, T denotes the temperature, and m_i is the atomic mass. A random variable x following a Gaussian distribution with $\mu = 0$ and $\sigma = 1$ can be transformed into another random variable x' that follows a Gaussian distribution with μ' and σ' by the transformation $x' = \mu' + \sigma' x$. Hence, an initial atomic velocity distribution (which is Gaussian) can be generated by setting $\sigma' = \sqrt{k_B T/m_i}$, $\mu' = 0$, and $x' = v_{i,\alpha}$. The resulting velocity distribution is then written as

$$P(v_{i,\alpha}) = \left(\frac{m_i}{2\pi k_B T}\right)^{1/2} \exp\left(-\frac{1}{2}\frac{m_i v_{i,\alpha}^2}{k_B T}\right). \tag{1.14}$$

After the initialization of the velocities, they are usually further adjusted to ensure that the net momentum of the atoms is zero, and the system does not drift during the simulation. If the system possesses a nonzero momentum for which $P_\alpha = \sum_i m_i v_{i,\alpha}$, the initial atomic velocities should be adjusted by $v'_{i,\alpha} = v_{i,\alpha} - \langle v_{i,\alpha} \rangle$, where $\langle v_{i,\alpha} \rangle = P_\alpha / \sum_i m_i$.

For the initialization method that continues from a previous MD simulation, the initial atomic positions and velocities are taken from the final positions and velocities of that simulation. This practice is common in MD simulations, particularly for those involving multiple stages. The final system configuration for each stage can be taken as the initial configuration for the next stage.

1.2.3 Periodic boundary conditions

During the initialization of an MD simulation, a simulation box with a finite size is defined first. One common question about the simulation box is: how should its boundary conditions be defined? Due to the surface effect, atoms or molecules located at the boundaries have fewer neighbors than their counterparts inside the box; thus, the dynamic behaviors of the former deviate dramatically from those of the latter. Unless one intends to simulate a cluster of atoms, the existence of surfaces will most likely be undesirable. These surface effects cause the simulation results to deviate from those obtained on the macroscopic scale, and the extent of such a deviation depends on the ratio of the surface atoms to the total atoms. For example, if one takes a cubic simulation box with a side length of a and assumes an even distribution of atoms inside the box, then the total atom number N_t should be proportional to the box volume, i.e., $N_t \propto a^3$. Similarly, the number of atoms on the box surface, N_s, is proportional to the box surface area, i.e., $N_s \propto 6a^2$. Hence, the ratio of the surface atoms to the total atoms is proportional to $1/a$ or $1/\sqrt[3]{N_t}$. For an MD simulation system containing 1000 atoms, such a ratio is in the order of tenths. Increasing the total atom number to 10^6 that is sufficiently large for common MD simulations will still result in this ratio being in the order of hundredths. In fact, no matter how large the simulation system is, its number of atoms will always be negligible compared with the number of atoms contained in a macroscopic chunk of material (in the order of 10^{23}). As a result, the ratio of the surface atoms to the total atoms will always be much larger in MD simulation than in reality.

A solution to this surface problem is to apply PBCs, which are some of the most frequently used boundary conditions in MD simulations. To apply the PBCs, a primary simulation cell is extracted from a realistic atomic system, as shown in Fig. 1.2A. Atoms selected to form the primary cell are far from the surface and are free from boundary

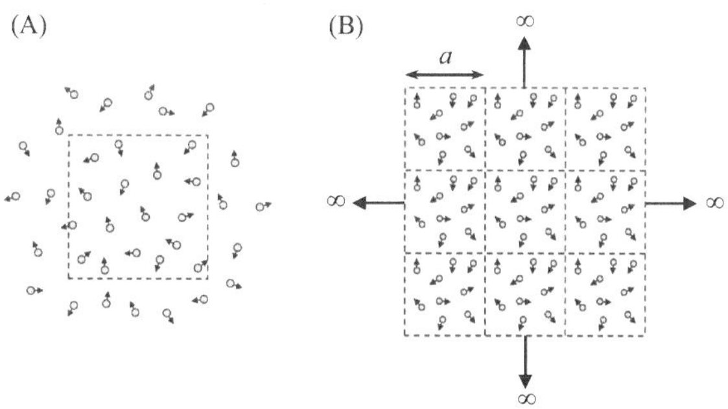

FIGURE 1.2 (A) Schematic of a realistic atomic system. Black arrows indicate the movement direction of the atoms. Atoms enclosed by the dotted lines are selected to form a simulation cell. (B) Schematic of an artificial atomic system constructed by duplicating the simulation cell in all three dimensions and extending them to infinity.

effects. By duplicating and packing the primary cell periodically in all three directions, the simulation space is completely filled with atoms to model macroscopic materials, as shown in Fig. 1.2B. Hence, an atom in the primary cell with the position vector $\boldsymbol{r}_i$ represents an infinite set of image atoms located at $\boldsymbol{r}_i + l\boldsymbol{a} + m\boldsymbol{b} + n\boldsymbol{c}$, where l, m, and n are integers, and $\boldsymbol{a}$, $\boldsymbol{b}$, and $\boldsymbol{c}$ are the edge vectors of the primary simulation cell. In practical MD simulations, the duplication of the primary cell is not physically performed, and the effects of the image atoms are accounted for by calculating their interactions with the representative atoms inside the primary cell. In each timestep, the positions of the representative atoms are first updated, and those of the image atoms are then updated accordingly by simply imposing the translation periodicity.

The application of PBCs brings about two main advantages. First, each atom in the primary simulation cell interacts not only with other atoms inside the cell but also with their images in nearby virtual cells. Hence, the atoms near the boundaries of the simulation cell with PBCs have a similar number of neighbors as those inside the cell, and thus the surface effect is virtually eliminated. Secondly, when an atom in the primary simulation cell moves out from one side, an image atom in the neighboring cell enters the simulation cell from the other side, as illustrated in Fig. 1.3. Therefore, the number of atoms inside the simulation cell will remain constant, which is a common requirement for most MD simulations.

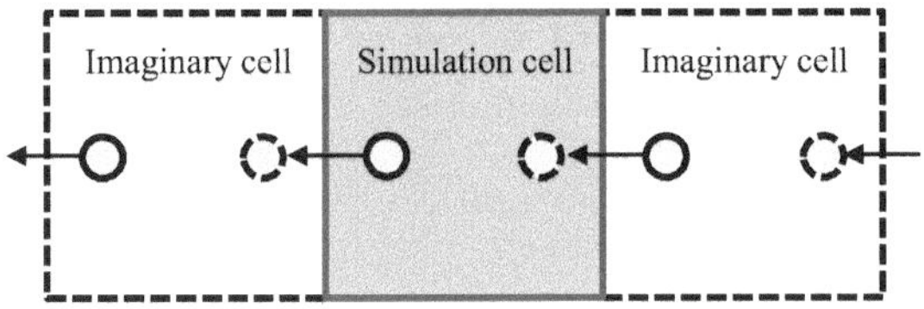

FIGURE 1.3 Schematic showing the synchronous movement of an atom in the primary simulation cell and that of its images. An atom (denoted by the circle with the solid line) moving out of the simulation cell is accompanied by one image atom (denoted by the circle with the dash line) moving inside.

To apply PBCs, the shape of the simulation cell should enable all space to be filled after applying periodicity in three dimensions. There are five choices for the cell shape: cubic (orthogonal or orthorhombic), truncated octahedron, hexagonal prism, rhombic dodecahedron, and "elongated" dodecahedron. Among these five shapes, cubic and hexagonal prism are much more frequently used due to their easy implementation. The others are preferable in cases where liquid solutions are modeled, and the number of solvent molecules needs to be minimized. For example, due to their close-to-spherical shapes, the truncated octahedron and rhombic dodecahedron simulation cells are preferred choices for modeling spherical molecules or proteins.

A problem resulting from PBCs is that the number of interaction pairs increases to be arbitrarily large as there are infinite image atoms after imposing PBCs. Apparently, the calculation of all these interactions is impossible for practical MD simulations. Fortunately, most atomic interactions diminish rapidly with distance. Beyond a certain critical distance R_c, the interaction between two atoms is sufficiently weak, and it is thus usually truncated in practical simulations.

One possible situation is that for the *i*th atom in the primary simulation cell, both atom *j* and several of its images are within the distance R_c of it and thus contribute to the net force acting on it. This situation increases the complexity of MD simulations and may lead to unrealistic results. To solve the above problem, the minimum image criterion is applied. This criterion states that the *i*th atom of the simulation cell interacts with either atom *j* or one of its images nearest to atom *i*. To demonstrate this, consider a rectangular simulation cell containing four atoms denoted by A (yellow), B (red), C (cyan), and D (green), as illustrated in Fig. 1.4. The primary simulation cell is surrounded by its 8 image cells. All of the neighbors of atom A within the critical distance R_c do not remain inside the primary cell. Hence, when calculating the

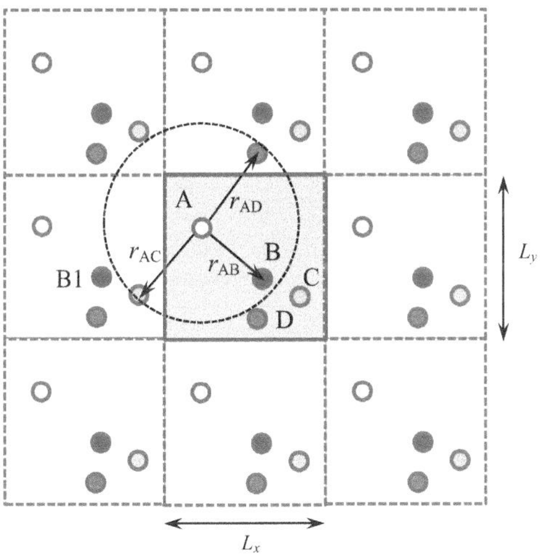

FIGURE 1.4 Schematic showing the closest neighbors of atom A in the primary simulation cell as determined by the minimum image criterion. The dimensions of each simulation cell are $L_x \times L_y$. The atoms located within a distance of R_c of atom A are listed as its closest neighbors and can interact with it.

force on atom A, atom B and the images of atoms C and D in the neighboring "virtual" cells are taken into account.

The minimum image criterion can be naturally satisfied when the dimensions of the MD simulation cell are no less than $2R_c$. For atom A to interact with both atom B and one of its images B1, atom A should remain within a distance R_c of both atoms B and B1. However, this situation is impossible as it will result in a cell size smaller than $2R_c$. Usually, R_c is not large but stays around 10 Å. A primary simulation cell with dimensions of approximately 20 Å can usually be successfully simulated even with regular PCs. Since the simulation cell dimensions are generally much larger than $2R_c$, the minimum image criterion is commonly imposed in conjunction with PBCs for most MD simulations.

1.2.4 Energy minimization and structure optimization

As discussed in Section 1.2.2, the initial structure constructed from scratch usually deviates from the equilibrium structure determined by the potential energy function adopted. Structural relaxation is usually required before the main simulation run. In addition, for some MD simulations, the ultimate goal is to obtain the most energetically stable atomic structure. In both cases, an energy minimization process is performed, during which the structure is optimized to exhibit the most stable configuration or the lowest energy.

During energy minimization, the ground state energy, or the potential energy of the system at 0 K, is the only concern, and its kinetic energy is not included. For a molecular system, its potential energy $U(\boldsymbol{R}^N)$ is a function of the position vector $\boldsymbol{R}^N$ of its N constituent atoms. A general and popular method of obtaining the minimum of a function is to utilize its derivatives.

A univariable function $f(x)$ possesses a minimum value under the conditions $f'(x) = 0$ and $f''(x) > 0$. For a multivariable function $f(x_1, x_2, \ldots, x_n)$, this condition can be similarly written as

$$\frac{\partial f}{\partial x_i} = 0; \quad \frac{\partial^2 f}{\partial x_i^2} > 0, \quad i = 1, 2, \ldots, n\,. \tag{1.15}$$

Hence, the energy minimization process of an MD system containing N atoms can be implemented by seeking a set of position vectors $\boldsymbol{R}^N_{\min} = \{\boldsymbol{r}_1^{\min}, \boldsymbol{r}_2^{\min}, \boldsymbol{r}_3^{\min}, \ldots, \boldsymbol{r}_N^{\min}\}$ for which

$$\left.\frac{\partial U(\boldsymbol{R}^N)}{\partial r_{i,\alpha}}\right|_{\boldsymbol{R}^N = \boldsymbol{R}^N_{\min}} = 0, \quad \left.\frac{\partial^2 U(\boldsymbol{R}^N)}{\partial r_{i,\alpha}^2}\right|_{\boldsymbol{R}^N = \boldsymbol{R}^N_{\min}} > 0,\ i = 1,\ 2,\ 3,\ \ldots,\ N;\ a = x,\ y,\ z. \tag{1.16}$$

Here, the subscript/superscript "min" indicates the minimum energy state of the MD system, and $r_{i,\alpha}$ is the component of $\boldsymbol{r}_i$ in the α-direction. The calculation of the derivatives of a function can involve either analytical or numerical methods. The analytical method allows for the deduction of a mathematical expression of the derivatives from the initial function and can thus lead to absolute accuracy. However, for multivariable functions, the mathematical deduction process is usually tedious, and the analytical expression is difficult to obtain. Hence, the numerical method based on the finite difference approach is usually applied.

For MD simulations, the steepest descent method and the conjugate gradient method are the most frequently applied methods that use the first-order derivative of $U(\boldsymbol{R}^N)$ to obtain the minimum energy. The basic scheme behind these two methods is summarized as follows. First, the potential energy $U(\boldsymbol{R}^N)$ is calculated for the initial structure. Afterward, each atom is moved along certain directions in small incremental steps such that the potential energy decreases by the maximum possible amount. Such movements continue until a minimum value of $U(\boldsymbol{R}^N)$ is found. The main difference between the steepest descent method and the conjugate gradient method is the manner of determining the movement direction.

Consider that after k steps of movement, the position vectors of the atoms are denoted by $\boldsymbol{R}^N_{(k)} = \left\{\boldsymbol{r}_1^{(k)}, \boldsymbol{r}_2^{(k)}, \ldots, \boldsymbol{r}_N^{(k)}\right\}$. After an atomic movement $\Delta\boldsymbol{R}^N$ in the $(k+1)$th step, the potential energy $U(\boldsymbol{R}^N + \Delta\boldsymbol{R}^N)$ can be expressed in terms of $U(\boldsymbol{R}^N)$ via the Taylor expansion as

$$U(\boldsymbol{R}^N_{(k)} + \Delta\boldsymbol{R}^N) = U(\boldsymbol{R}^N_{(k)}) + \Delta\boldsymbol{R}^N \cdot U'(\boldsymbol{R}^N_{(k)}) + \frac{1}{2}\Delta\boldsymbol{R}^N U''(\boldsymbol{R}^N_{(k)})(\Delta\boldsymbol{R}^N)^{\mathrm{T}} + \cdots,$$

$$U'(\boldsymbol{R}^N_{(k)}) = \left(\frac{\partial U(\boldsymbol{R}^N)}{\partial \boldsymbol{r}_1}, \frac{\partial U(\boldsymbol{R}^N)}{\partial \boldsymbol{r}_2}, \cdots, \frac{\partial U(\boldsymbol{R}^N)}{\partial \boldsymbol{r}_N}\right)\Bigg|_{\boldsymbol{R}^N = \boldsymbol{R}^N_{(k)}},$$

$$U''(\boldsymbol{R}^N_{(k)}) = \begin{bmatrix} \frac{\partial^2 U(\boldsymbol{R}^N)}{\partial \boldsymbol{r}_1^2} & \frac{\partial^2 U(\boldsymbol{R}^N)}{\partial \boldsymbol{r}_1 \partial \boldsymbol{r}_2} & \cdots & \frac{\partial^2 U(\boldsymbol{R}^N)}{\partial \boldsymbol{r}_1 \partial \boldsymbol{r}_N} \\ \frac{\partial^2 U(\boldsymbol{R}^N)}{\partial \boldsymbol{r}_2 \partial \boldsymbol{r}_1} & \frac{\partial^2 U(\boldsymbol{R}^N)}{\partial \boldsymbol{r}_2^2} & \cdots & \frac{\partial^2 U(\boldsymbol{R}^N)}{\partial \boldsymbol{r}_2 \partial \boldsymbol{r}_N} \\ \cdots & \cdots & \cdots & \cdots \\ \frac{\partial^2 U(\boldsymbol{R}^N)}{\partial \boldsymbol{r}_N \partial \boldsymbol{r}_1} & \frac{\partial^2 U(\boldsymbol{R}^N)}{\partial \boldsymbol{r}_N \partial \boldsymbol{r}_2} & \cdots & \frac{\partial^2 U(\boldsymbol{R}^N)}{\partial \boldsymbol{r}_N^2} \end{bmatrix}\Bigg|_{\boldsymbol{R}^N = \boldsymbol{R}^N_{(k)}}. \tag{1.17}$$

Here, $U'(\boldsymbol{R}^N_{(k)})$ is the gradient of $U(\boldsymbol{R}^N_{(k)})$ at $\boldsymbol{R}^N_{(k)}$ and is simply denoted as the vector $\boldsymbol{g}_{(k)}$ hereafter; $U''(\boldsymbol{R}^N_{(k)})$ is the Hessian matrix of $U(\boldsymbol{R}^N)$ at $\boldsymbol{R}^N_{(k)}$ and is simply denoted as $\boldsymbol{H}_{(k)}$; and $(\Delta\boldsymbol{R}^N)^{\mathrm{T}}$ denotes the transpose of the vector $\Delta\boldsymbol{R}^N$.

For the steepest descent method, the movement direction of the $(k+1)$th step is equivalent to the direction of the gradient $g_{(k)}$ of step k. If the vector $s_{(k)}$ is defined as

$$s_{(k)} = -\frac{g_{(k)}}{\left|g_{(k)}\right|}, \tag{1.18}$$

the position $R^N_{(k+1)}$ will then be obtained by

$$R^N_{(k+1)} = R^N_{(k)} + \lambda_{(k)} s_{(k)}. \tag{1.19}$$

Here, $\lambda_{(k)}$ is a positive coefficient that determines the magnitude of the movement. Since the negative gradient of a function determines its direction of fastest decrease, Eq. (1.18) determines the "steepest" descending direction of the potential energy.

The coefficient $\lambda_{(k)}$ can be determined in various ways, and the line search method is one of the most frequently adopted ways. Assume that the search starts from point A and $\lambda_{(k)}$ is set at a predefined value λ_0. In each subsequent step of the line search process, $\lambda_{(k)}$ will be increased by a constant value c if the previous step leads to a decreased value of $U(R^N_{(k+1)})$. This process continues until the function values of three successive steps follow $U(R^N_{(k)}) > U(R^N_{(k+1)})$ and $U(R^N_{(k+2)}) > U(R^N_{(k+1)})$. In this case, there must be a value of λ that falls between $\lambda_0 + ck$ and $\lambda_0 + c(k+1)$, which leads to the minimum of $U(R^N_{(k+1)})$.

To find such a value of λ, the three points $U(R^N_{(k)})$, $U(R^N_{(k+1)})$, and $U(R^N_{(k+2)})$ are fitted to a quadratic curve, and the minimum energy U_{min} can thus be determined according to the mathematical form of the fitting. Afterward, the quadratic fitting process continues with $U(R^N_{(k)})$, $U(R^N_{(k+1)})$ and U_{min} (if $\lambda < \lambda_{(k+1)}$) or $U(R^N_{(k+1)})$, $U(R^N_{(k+2)})$, and U_{min} (if $\lambda > \lambda_{(k+1)}$) until a well converged $\lambda_{(k)}$ is obtained. To better illustrate how the line search method works, a bivariable function $f(x,y)$ is taken as an example. Consider the contour plot of $f(x,y)$ that is shown in Fig. 1.5A. The line search starts from point A along the gradient direction indicated by the solid line. With an initial value of λ_0, points 1, 2, and 3 along the gradient direction are obtained with $f_1 > f_2$ and $f_3 > f_2$. Hence, the minimum should be somewhere between points 1 and 3, as shown in Fig. 1.5B. The quadratic fitting of points 1, 2, and 3 gives the minimum point 4. By conducting the search again, the quadratic fitting of points 1, 2, and 4 gives a new minimum point 5. Repeating this process produces a value that eventually converges to the minimum of the function $f(x,y)$.

Furthermore, the arbitrary step method can be used to determine $\lambda_{(k)}$. The value of $\lambda_{(k)}$ is initially set at 1. If $\lambda_{(k)}$ leads to a reduced $U(R^N_{(k+1)})$,

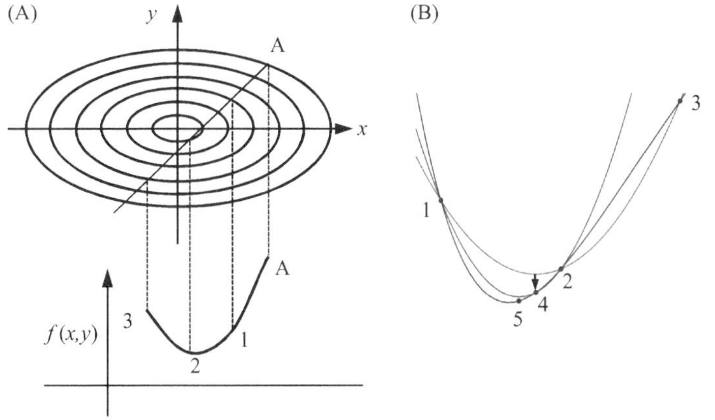

FIGURE 1.5 (A) Schematic of the steepest descent method in searching for the minimum of a bivariable function $f(x, y)$. (B) Illustration of the line search approach for searching a minimum value. The minimum point 4 is located by fitting a quadratic curve to the initial set of three points 1, 2, and 3. A better estimation of the minimum can then be found by fitting a new quadratic to points 1, 2, and 4 to locate a new minimum point 5.

$\lambda_{(k)}$ will be iteratively increased by a multiplicative factor to calculate $U(\boldsymbol{R}^N_{(k+1)})$ until an optimal value of $\lambda_{(k)}$ is determined that leads to a reduced and well-converged value of $U(\boldsymbol{R}^N_{(k+1)})$. If the initial value of $\lambda_{(k)} = 1$ leads to an increased $U(\boldsymbol{R}^N_{(k+1)})$ at the first step, $\lambda_{(k)}$ will be directly reduced by a multiplicative factor in order to obtain a reduced $U(\boldsymbol{R}^N_{(k+1)})$ before the aforementioned iterative process starts.

For the steepest descent method in conjunction with the line search method, the gradients of two successive steps follow the relation $g_{(k)} \cdot g_{(k+1)} = 0$, i.e., the gradient directions of two successive steps are orthogonal with respect to each other. This characteristic may result in the failure of the steepest descent at points on the surfaces of long narrow valleys of the potential energy landscape. This failure is because the algorithm may lead to an oscillatory searching process between the sides of the valley instead of approaching its bottom.

The conjugate gradient method does not lead to such oscillatory behavior in narrow valleys. For this method, the direction vector $\boldsymbol{s}_{(k)}$ is given by

$$\boldsymbol{s}_{(k)} = -g_{(k)} + \gamma_{(k)}\boldsymbol{s}_{(k-1)}, \quad \gamma_{(k)} = \frac{g_{(k)} \cdot g_{(k)}}{g_{(k-1)} \cdot g_{(k-1)}}. \tag{1.20}$$

The position vector $\boldsymbol{R}^N_{(k+1)}$ is still obtained by Eq. (1.19), and both the line search and arbitrary step methods can be used. For the conjugate gradient method, the following relationships are fulfilled:

$$g_{(i)} \cdot g_{(j)} = 0, \quad i \neq j, \tag{1.21}$$

$$s_{(i)} H_{(j)} s_{(j)}^{\mathrm{T}} = 0, \quad i \neq j, \tag{1.22}$$

$$g_{(i)} \cdot s_{(j)} = 0, \quad i \neq j. \tag{1.23}$$

Conjugate directions do not interfere with one another in the sense that when a new direction is conjugate to a previous direction, the changes of the variables along the new direction do not lead to an increase in the function value of $U(\boldsymbol{R}^N)$ with respect to the previous direction.

Besides first-order derivative methods (such as the steepest descent method and the conjugate gradient method discussed above) that search the minimum using the information of the slope, there are several second-order derivative methods that utilize information of the curvature and thus result in higher accuracy. One frequently employed second-order derivative method is the Newton–Raphson method. Provided that in Eq. (1.17), $U(\boldsymbol{R}^N)$ has a minimum at the point $\boldsymbol{R}_{(k)}^N + \Delta \boldsymbol{R}^N$, its first-order derivative with respect to $\Delta \boldsymbol{R}^N$ is given by

$$\nabla U \Big|_{\boldsymbol{R}^N = \boldsymbol{R}_{(k)}^N + \Delta \boldsymbol{R}^N} = \boldsymbol{H}_{(k)} \cdot \Delta \boldsymbol{R}^N + g_{(k)}. \tag{1.24}$$

Setting $\nabla U \Big|_{\boldsymbol{R}^N = \boldsymbol{R}_{(k)}^N + \Delta \boldsymbol{R}^N} = 0$, one has $\Delta \boldsymbol{R}^N = -(\boldsymbol{H}_{(k)})^{-1} g_{(k)}$. In the Newton–Raphson method, one simply sets $\boldsymbol{R}_{(k+1)}^N = \boldsymbol{R}_{(k)}^N - (\boldsymbol{H}_{(k)})^{-1} g_{(k)}$, and the iteration of $\boldsymbol{R}_{(k+1)}^N$ will eventually lead to the solution of the minimum point. The closer the initial guess $\boldsymbol{R}_{(0)}^N$ is to the final minimum point $\boldsymbol{R}^N$, the better the performance of the Newton–Raphson method will be.

However, in practical implementation, the Newton–Raphson method requires all the eigenvalues of the Hessian matrix to be positive. In addition, the calculation of the inverse of the Hessian matrix requires intensive computing resources and storage, which renders the Newton–Raphson method only suitable for MD simulation systems containing fewer than hundreds of atoms. Furthermore, analytical second-order derivatives of $U(\boldsymbol{R}^N)$ are not always available. To increase the efficiency and applicability of the Newton–Raphson method, quasi-Newton methods were developed, in which the inverse of $\boldsymbol{H}_{(k)}$ is built upon successive iterations instead of the direct calculation of the second-order derivatives. Some of the most frequently employed quasi-Newton methods are the Davidon–Fletcher–Powell method, the Broyden–Fletcher–Goldfarb–Shanno method, and the Murtagh–Sargent method.

In practical MD simulations, a combination of first-order and second-order derivative methods is usually implemented. The energy minimization process can be divided into two stages. In the first stage, the initial system configuration usually has a much higher energy than the minimum potential energy. Therefore, first-order derivative methods, such as the steepest descent method and the conjugate gradient method, are used to rapidly reduce the energy difference. In the second stage, second-order derivative methods are used to further refine the atomic configuration to seek the final minimum energy state.

Achieving the true minimum energy state is impossible, and one can only obtain estimations in MD simulations. There are several criteria to determine whether an energy minimization process should stop. The first criterion involves the energy difference between two successive iteration steps. If such an energy difference is smaller than a preset value, the program will consider that the energy minimum has been obtained. The second criterion involves the magnitudes of atomic displacements during two successive steps. The energy minimization process ends when the condition $|\boldsymbol{R}_{(k+1)}^N - \boldsymbol{R}_{(k)}^N|^2 < \varepsilon$ is met. Here, ε is a preset constant. The third criterion entails the root-mean-square (RMS) deviation of the gradient $g_{(k)}$. Here, the RMS of $g_{(k)}$ is defined as $\sqrt{g_{(k)} \cdot (g_{(k)})^{\mathrm{T}}/(3N)}$. The energy minimization process ends when the magnitude of RMS deviation meets the condition $|\Delta(\mathrm{RMS})| < \varepsilon$.

1.2.5 Force calculation

Given a potential energy function, the force acting on an atom is calculated by obtaining the first-order derivative of $U(\boldsymbol{R}^N)$ as indicated by Eq. (1.2). For many common potential energy functions employed in MD simulations, analytical expressions of the first-order derivative of $U(\boldsymbol{R}^N)$ are available and have been implemented in most MD codes. For example, in the book *"Computational simulation of liquids"* authored by Allen and Tildesley (1987), the analytical expressions for the forces involved in three representative examples are listed: (1) a polymer chain with constrained bond lengths but realistic bond angle bending and torsional potentials, (2) a molecular fluid of linear molecules among which the interactions include permanent electrostatic interactions, and (3) a fluid of atoms with three-body interactions modeled by the AxilrodTeller triple−dipole potential.

Here, a simple chain molecule is taken as an example to demonstrate the force calculation process in MD simulations. The geometry of this molecule is illustrated in Fig. 1.6. Its potential energy mainly involves

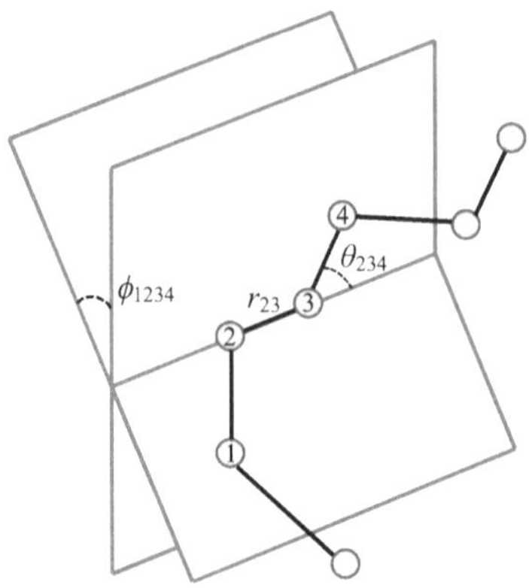

FIGURE 1.6 Schematic of a simple chain molecule with the interatomic distance r_{23}, bond angle θ_{234}, and torsion angle ϕ_{1234} labeled.

contributions from its bond stretching, bond angle bending and torsion, and interactions with other molecules.

Suppose that the bond bending potential component $U_{\text{bond-bending}}$ can be simply written as

$$U_{\text{bond-bending}} = \sum_{i \neq j \neq k} k_{ijk}^{\theta} \cos\theta_{ijk}, \tag{1.25}$$

where k_{ijk}^{θ} is a constant. The angle θ_{ijk} is between two successive bond vectors $\boldsymbol{r}_{ij}$ and $\boldsymbol{r}_{jk}$ and can be represented by

$$\cos\theta_{ijk} = (\boldsymbol{r}_{ij} \cdot \boldsymbol{r}_{ij})^{-1/2} (\boldsymbol{r}_{jk} \cdot \boldsymbol{r}_{jk})^{-1/2} (\boldsymbol{r}_{ij} \cdot \boldsymbol{r}_{jk}). \tag{1.26}$$

Note that in this book, the bond vector $\boldsymbol{r}_{ij}$ is defined by $\boldsymbol{r}_i - \boldsymbol{r}_j$ rather than $\boldsymbol{r}_j - \boldsymbol{r}_i$.

The chain component consisting of atoms 2, 3, and 4 is taken as an example. To calculate the forces caused by the bending of the angle θ_{234} acting on these atoms, the following derivatives are required according to Eq. (1.2):

$$\begin{aligned}
&\frac{\partial}{\partial \boldsymbol{r}_2}(\boldsymbol{r}_{23} \cdot \boldsymbol{r}_{34}) = \boldsymbol{r}_{34}, \ \frac{\partial}{\partial \boldsymbol{r}_3}(\boldsymbol{r}_{23} \cdot \boldsymbol{r}_{34}) = \boldsymbol{r}_{23} - \boldsymbol{r}_{34}, \ \frac{\partial}{\partial \boldsymbol{r}_4}(\boldsymbol{r}_{23} \cdot \boldsymbol{r}_{34}) = -\boldsymbol{r}_{23}, \\
&\frac{\partial}{\partial \boldsymbol{r}_2}(\boldsymbol{r}_{23} \cdot \boldsymbol{r}_{23}) = 2\boldsymbol{r}_{23}, \ \frac{\partial}{\partial \boldsymbol{r}_3}(\boldsymbol{r}_{23} \cdot \boldsymbol{r}_{23}) = -2\boldsymbol{r}_{23}, \ \frac{\partial}{\partial \boldsymbol{r}_4}(\boldsymbol{r}_{23} \cdot \boldsymbol{r}_{23}) = 0, \\
&\frac{\partial}{\partial \boldsymbol{r}_2}(\boldsymbol{r}_{34} \cdot \boldsymbol{r}_{34}) = 0, \ \frac{\partial}{\partial \boldsymbol{r}_3}(\boldsymbol{r}_{34} \cdot \boldsymbol{r}_{34}) = 2\boldsymbol{r}_{34}, \ \frac{\partial}{\partial \boldsymbol{r}_4}(\boldsymbol{r}_{34} \cdot \boldsymbol{r}_{34}) = -2\boldsymbol{r}_{34}.
\end{aligned} \tag{1.27}$$

The forces acting on atoms 2, 3, and 4 due to the bending of θ_{234} are then obtained as

$$f_2 = \frac{\partial}{\partial \boldsymbol{r}_2} k_{234}^{\theta} \cos\theta_{234} = k_{234}^{\theta} r_{23}^{-1} r_{34}^{-1} (\boldsymbol{r}_{34} - \frac{\boldsymbol{r}_{23} \cdot \boldsymbol{r}_{34}}{r_{23}^2} \boldsymbol{r}_{23}),$$

$$f_3 = \frac{\partial}{\partial \boldsymbol{r}_3} k_{234}^{\theta} \cos\theta_{234} = k_{234}^{\theta} r_{23}^{-1} r_{34}^{-1} (\frac{\boldsymbol{r}_{23} \cdot \boldsymbol{r}_{34}}{r_{23}^2} \boldsymbol{r}_{23} - \frac{\boldsymbol{r}_{23} \cdot \boldsymbol{r}_{34}}{r_{34}^2} \boldsymbol{r}_{34} + \boldsymbol{r}_{23} - \boldsymbol{r}_{34}), \quad (1.28)$$

$$f_4 = \frac{\partial}{\partial \boldsymbol{r}_4} k_{234}^{\theta} \cos\theta_{234} = k_{234}^{\theta} r_{23}^{-1} r_{34}^{-1} (\frac{\boldsymbol{r}_{23} \cdot \boldsymbol{r}_{34}}{r_{34}^2} \boldsymbol{r}_{34} - \boldsymbol{r}_{23}).$$

A similar approach can be used to calculate the forces caused by the torsion of the bond formed by atoms 2 and 3 for all the four atoms (1, 2, 3, and 4) involved.

The example shown above deals with the covalent interaction of bonded atoms, which is a short-range interaction. For such short-range interactions, the total force acting on an atom can be represented by a direct summation over all the forces exerted by a certain number of neighboring atoms. In principle, for long-range interactions, the force calculation should consider the interactions of all the atom pairs. For a system containing N particles, there are $N(N-1)/2$ pair interactions, and hence $N(N-1)/2$ force calculation operations are required. Therefore, the computing task increases dramatically with the system size. As discussed in Section 1.2.3, most atomic interactions diminish rapidly with increasing interatomic distances. Thus, the interaction between two atoms is usually truncated at a certain critical distance, which reduces the computational cost of the force calculation.

To safely truncate a potential function at a cutoff distance, the potential should decay with distance at a rate no lower than $U(r) \propto r^{-3}$. However, in ionic systems, the Columbic interaction decays slowly with distance according to the relationship $U(r) \propto r^{-1}$, and only approaches zero at very large distances. Hence, the truncation of the Columbic interaction at a small cutoff distance will inevitably lead to inaccuracies in the simulation results. However, a complete consideration of the long-range Columbic interactions will greatly increase the system size and computational cost, which significantly prohibits the simulation of many ionic systems. To circumvent this problem, different approaches have been developed to handle the long-range Columbic interaction, of which the Ewald summation method is the most widely used.

For the Ewald method, an atom in the simulation cell interacts not only with other atoms in the same cell but also with all the atoms in the image cells. Provided that the origin positions of the image cells are denoted by the vector $\boldsymbol{n} = l\boldsymbol{a} + m\boldsymbol{b} + n\boldsymbol{c}$, the Columbic energy of an ionic system can be described as

$$U(\boldsymbol{R}^N) = \frac{1}{2}\sum_{n}\left[\sum_{i}\sum_{j}\frac{q_i q_j}{4\pi\varepsilon_0|\boldsymbol{r}_{ij} - \boldsymbol{n}|}\right]. \tag{1.29}$$

Here, q_i and q_j are the charges of particle i and j, respectively; ε_0 is the vacuum dielectric constant (8.854×10^{-12} F/m); n spans over all the image cells. The most prominent problem is that the Columbic energy of the system does not necessarily converge. To overcome this, the single slowly and conditionally convergent series in Eq. (1.29) is replaced by a sum of two rapidly converging series as

$$U_{Ewald} = U^{(r)} + U^{(m)}. \tag{1.30}$$

The two terms on the right-hand side of Eq. (1.30) represent the real space sum and the imaginary (or Fourier) sum, respectively.

Physically, the decomposition of the Ewald sum can be interpreted as follows. Each point charge in the system is considered to be surrounded by a Gaussian-shaped charge distribution of equal magnitude and opposite sign. The artificially introduced charge distribution screens the interactions between the point-charge and neighboring point-charges, thus limiting their interactions to only short-range interactions. Consequently, the sum $U^{(r)}$ converges rapidly. To cancel the artificial screening effect introduced by the Gaussian charge distribution, a second Gaussian charge distribution of the same sign and magnitude as the point charge is added. The summation of the second charge distribution is in the Fourier space, which ensures rapid convergence. The decomposition of the Ewald sum for a 1D point charge system is illustrated in Fig. 1.7.

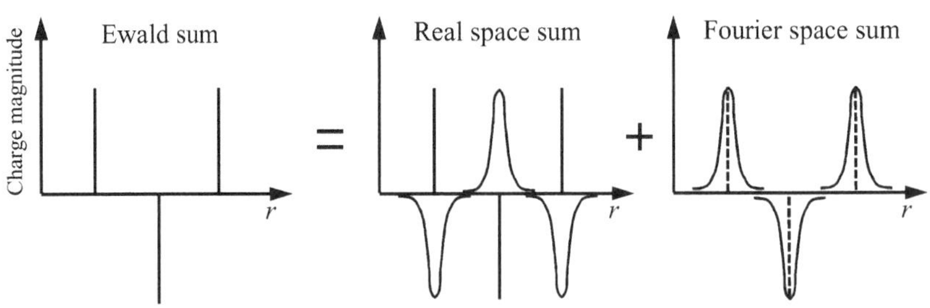

FIGURE 1.7 Decomposition of the Ewald sum of a 1D point-charge system. The vertical lines represent unit charges, and the Gaussian distributions are normalized to unity.

1.2.6 Time integration algorithms

The time integration algorithm serves as the engine of an MD simulation program. The time integration of the equations of motion for interacting atoms produces their trajectories. Time integration algorithms are developed based on the finite difference method, in which time is discretized on a finite grid, and the distance between consecutive points on this grid is denoted by the timestep Δt. Given the atomic positions, velocities, and accelerations at time t, the integration scheme returns the same quantities at a later time $t+\Delta t$. For common MD simulations, the Verlet algorithm and the predictor–corrector algorithm are the most popular schemes.

For the earliest Verlet method, the atomic position is expressed in the form of a second-order Taylor polynomial (Verlet, 1967):

$$\boldsymbol{r}(t+\Delta t)=\boldsymbol{r}(t)+\frac{d\boldsymbol{r}(t)}{dt}\Delta t+\frac{1}{2!}\frac{d^2\boldsymbol{r}(t)}{dt^2}\Delta t^2+\cdots. \tag{1.31}$$

Replacing Δt with $-\Delta t$ leads to

$$\boldsymbol{r}(t-\Delta t)=\boldsymbol{r}(t)-\frac{d\boldsymbol{r}(t)}{dt}\Delta t+\frac{1}{2!}\frac{d^2\boldsymbol{r}(t)}{dt^2}\Delta t^2+\cdots. \tag{1.32}$$

Combining Eqs. (1.31) and (1.32) and discarding their higher-order terms result in the following equation:

$$\boldsymbol{r}(t+\Delta t)=-\boldsymbol{r}(t-\Delta t)+2\boldsymbol{r}(t)+\frac{\mathrm{d}^2}{\mathrm{d}t^2}\boldsymbol{r}(t)\Delta t^2. \tag{1.33}$$

Since $d^2\boldsymbol{r}(t)/dt^2$, which defines the acceleration, can be obtained by force calculation, Eq. (1.33) should return the atomic position at $t+\Delta t$ if the atomic positions at $t-\Delta t$ and t are known. Similarly, subtracting Eq. (1.32) from Eq. (1.31) gives

$$\boldsymbol{v}(t)=\frac{\mathrm{d}\,\boldsymbol{r}(t)}{\mathrm{d}\,t}=\frac{1}{2\Delta t}[\boldsymbol{r}(t+\Delta t)-\boldsymbol{r}(t-\Delta t)]. \tag{1.34}$$

Eq. (1.34) indicates that the velocity vector $\boldsymbol{v}(t)$ can be determined by the atomic position vectors $\boldsymbol{r}(t)$ at the time $t-\Delta t$ and $t+\Delta t$. The advantage of the Verlet method is that in each timestep, the force calculation only needs to be performed once and the calculation of the velocity and position is time-reversible. However, in Eq. (1.34), there is a term $1/\Delta t$ on the right-hand side, which can introduce a large error to the velocity when the timestep Δt is very small. In addition, the atomic positions are independent of the velocities, and thus the MD system cannot be coupled with an external thermal bath.

To circumvent these issues, Verlet developed another integration algorithm that was later named the leap-frog method. In this method, the atomic velocities and positions are calculated as follows:

$$v(t+\frac{1}{2}\Delta t) = v(t-\frac{1}{2}\Delta t) + a(t)\Delta t, \tag{1.35}$$

$$r(t+\Delta t) = r(t) + v(t+\frac{1}{2}\Delta t)\Delta t. \tag{1.36}$$

During the integration process, the forces acting on the atoms are first calculated based on the atomic positions at time t and the atomic acceleration vectors $a(t)$ can thus be obtained. Given the velocity of an atom at time $t-\Delta t/2$, its velocity at $t+\Delta t/2$ is obtained according to Eq. (1.35). Afterward, the atomic velocity at time t is calculated by

$$v(t) = \frac{1}{2}v(t+\frac{1}{2}\Delta t) + \frac{1}{2}v(t-\frac{1}{2}\Delta t). \tag{1.37}$$

Compared with the earlier Verlet method, the leap-frog method boosts calculation accuracy, and the atomic trajectory is now dependent on the velocity, thereby allowing the MD system to be coupled with an external thermal bath. However, the leap-frog method corresponds to higher computing costs and storage. Moreover, the velocity update always lags behind the position update by one timestep.

An algorithm superior to the leap-frog method is called the velocity Verlet method, in which atomic positions and velocities at time $t+\Delta t/2$ are obtained simultaneously from their values at time t in the following manner:

$$r(t+\Delta t) = r(t) + v(t)\Delta t + \frac{1}{2}a(t)\Delta t^2, \tag{1.38}$$

$$v(t+\frac{1}{2}\Delta t) = v(t) + \frac{1}{2}a(t)\Delta t, \tag{1.39}$$

$$v(t+\Delta t) = v(t+\frac{1}{2}\Delta t) + \frac{1}{2}a(t+\frac{1}{2}\Delta t)\Delta t. \tag{1.40}$$

The predictor–corrector algorithm is another commonly used time integration algorithm for equations of motion. Since the movement trajectories of particles in MD simulation are continuous, their positions, velocities, and accelerations at time $t+\Delta t$ can be predicted from the atomic positions at time t. Meanwhile, their time derivatives can also be predicted by invoking Taylor expansions up to a certain order, which are given by

$$r_{\text{pre}}(t+\Delta t) = r(t) + v(t)\Delta t + \frac{1}{2}a(t)\Delta t^2 + \frac{1}{6}b(t)\Delta t^3 + \cdots, \tag{1.41}$$

$$v_{\text{pre}}(t+\Delta t) = v(t) + a(t)\Delta t + \frac{1}{2}b(t)\Delta t^2 + \cdots, \tag{1.42}$$

$$a_{\text{pre}}(t+\Delta t) = a(t) + b(t)\Delta t + \cdots, \tag{1.43}$$

$$b_{\text{pre}}(t+\Delta t) = b(t) + \cdots. \tag{1.44}$$

Here, $v(t)$, $a(t)$, and $b(t)$ are the first-, second-, and third-order time derivatives of the position vector $r(t)$, respectively, and the subscript "pre" stands for "prediction." Based on the position prediction at time $t+\Delta t$, one can calculate the atomic forces by computing the gradient of the potential energy function at $r_{\text{pre}}(t+\Delta t)$. The resulting acceleration will generally differ from that predicted by utilizing the Taylor expansion shown in Eq. (1.43). The difference $\Delta a(t+\Delta t)$ between the predicted acceleration $a_{\text{pre}}(t+\Delta t)$ and corrected acceleration $a_{\text{cor}}(t+\Delta t)$ constitutes an "error signal" as

$$\Delta a(t+\Delta t) = a_{\text{cor}}(t+\Delta t) - a_{\text{pre}}(t+\Delta t). \tag{1.45}$$

This error signal is then used to "correct" atomic positions and their time derivatives. All the corrections are proportional to $\Delta a(t+\Delta t)$, and they are given by

$$r_{\text{cor}}(t+\Delta t) = r_{\text{pre}}(t+\Delta t) + c_0\Delta a(t+\Delta t), \tag{1.46}$$

$$v_{\text{cor}}(t+\Delta t) = v_{\text{pre}}(t+\Delta t) + c_1\Delta a(t+\Delta t), \tag{1.47}$$

$$a_{\text{cor}}(t+\Delta t) = a_{\text{pre}}(t+\Delta t) + c_2\Delta a(t+\Delta t), \tag{1.48}$$

$$b_{\text{cor}}(t+\Delta t) = b_{\text{pre}}(t+\Delta t) + c_3\Delta a(t+\Delta t), \tag{1.49}$$

where c_0, c_1, c_2, and c_3 are constants, and the subscript "cor" stands for "correction." The predictor–corrector method results in higher accuracy than the Verlet method for a small timestep Δt.

1.2.7 Neighbor list

As discussed previously, most nonbonded atomic interactions decay rapidly with distance and can thus be truncated at a certain cutoff distance to reduce the computing cost. Hence, the interactions of two atoms with a separation distance larger than the cutoff distance can be ignored. This simplification is a great benefit for large MD simulation systems. However, one remaining problem is that distance calculations for all atomic pairs are still required. The time needed for calculating all pair separation distances is proportional to the number of atomic pairs $N(N-1)/2$ for an N-particle system. Performing $N(N-1)/2$ distance calculations is almost as time-consuming as evaluating the forces. Fortunately, in MD simulations, the timestep is usually small and in the order of femtoseconds. Hence, the displacement of an atom within a timestep is normally small. Consequently, the neighbors of an atom seldom change in a short period. By taking advantage of such a situation, a neighbor list can be generated, stored, and updated for each atom. For each timestep, distance and force calculations only need to be performed between an atom and its neighbors on this list.

There are mainly two popular methods to generate the neighbor list. The first one is the Verlet neighbor list method proposed by Verlet in the 1960s. In this method, a neighbor-list distance r_{nei} is defined to be equal to the potential cutoff distance r_{cut} in addition to a skin distance, as shown in Fig. 1.8. All the atoms j are stored in the neighbor list of atom i for which $r_{ij} < r_{\text{nei}}$. Here, r_{ij} is the distance between atoms i and j. In each MD step, only the distances between atom i and its neighbors in the list are calculated, and their interactions with atom i are considered only when they remain in the inner core region, i.e., $r_{ij} < r_{\text{cut}}$.

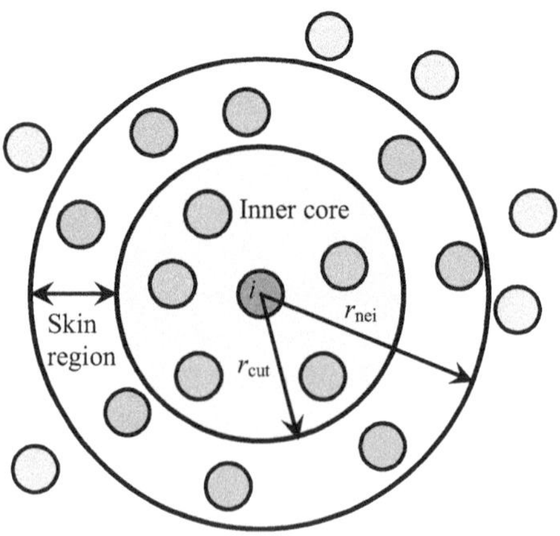

FIGURE 1.8 Schematic showing the neighbors of atom i according to the Verlet neighbor list method. The circle with the radius of r_{cut} defines the inner core region; the region outside the inner core but enclosed by the outer circle with the radius of r_{nei} defines the skin region. The small red, orange, blue, and green circles represent atom i, atoms in the inner core, atoms in the skin region, and atoms excluded from the neighbor list, respectively.

The choice of r_{nei} is based on a compromise: the larger the neighbor-list distance is, the less frequently the neighbor list has to be updated, but more atomic pairs must be examined for possible force calculations. Provided that the neighbor list is updated every N_t timesteps, the skin distance must meet the following requirement:

$$r_{nei} - r_{cut} > N_t \langle v_{max} \rangle \Delta t, \tag{1.50}$$

where $\langle v_{max} \rangle$ is the time average of the highest atomic velocity in the MD system. The Verlet method is probably the easiest way of generating a neighbor list. For a practical implementation of this method, a row-vector is used to store the neighbors of each atom, as shown in Fig. 1.9. Here, $N_{i,nei}$ is the number of neighbors of atom i, $j_1 \cdots j_{N_{i,nei}}$ are their indices, and N is the total number of atoms.

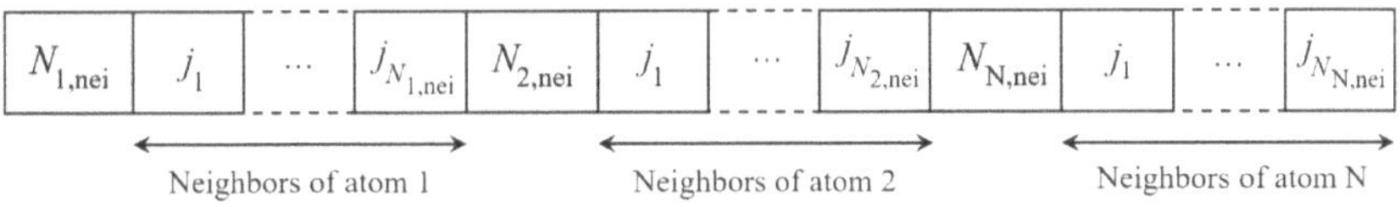

FIGURE 1.9 Practical implementation of the Verlet neighbor list by using a row-vector to store the neighbors of each atom.

The efficiency of the Verlet neighbor list method depends on the update frequency N_m, which is usually limited to a small value (typically 10–20) to ensure high simulation accuracy. Hence, the number of distance calculation operations is still in the order of N^2. The cell index method described below is a more efficient method and is more suitable for large systems with dimensions much larger than r_c.

To illustrate the cell index method, a two-dimensional (2D) square simulation box with a side length L and a total number of atoms N is taken as an example. The simulation box is divided into a regular lattice of $M \times M$ cells and the number of atoms N_c in each cell equals N/M^2. The value of M is chosen so that the side length L/M meets the requirement $L/M > r_{cut}$. For an atom in the central cell 13 shown in Fig. 1.10, all of its neighbors within a distance smaller than r_{cut} must remain in either cell 13 or 1 of the 8 adjacent cells 7, 8, 9, 12, 14, 17, 18, and 19. Hence, only $9N_c$ distance calculations will be required for this atom when its force is calculated. Similarly, for a three-dimensional (3D) cubic simulation box, the number of atoms N_c in each cell equals N/M^3, and only 27 N_c distance calculations are required for each atom during the force calculation. For the whole MD system, the number of distance calculations for each timestep is $27N_cN$. Consequently, the total number of distance calculations is reduced to the order of N. At the beginning of an MD simulation, an initial list is set up for each cell, and this list is updated during the simulation every 10–20 timesteps. During the update, atoms moving out of this cell are deleted from the list. Correspondingly, atoms entering into this cell are inserted into the list.

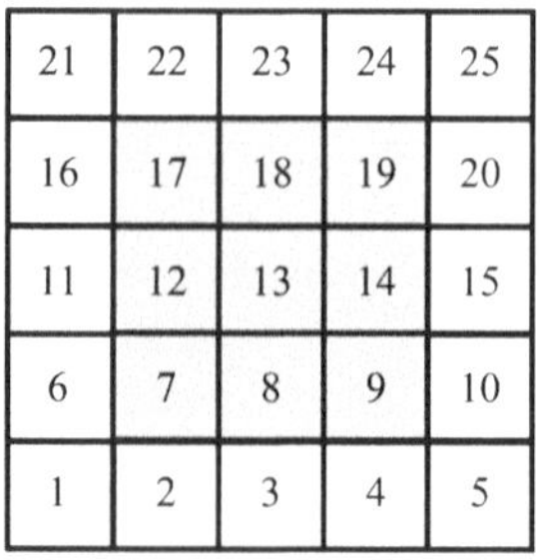

FIGURE 1.10 Schematic of the cell index method for a 2D simulation box.

1.2.8 Ensemble and statistical observables

Based on the previous discussions, one can understand that MD simulation is a statistical method. In classical statistical thermodynamics, the inquiry is not about the individual particles that constitute a macroscopic system but about their average behaviors or properties. The common practice is to make many copies of the system and calculate the average values of these replicates. Note that all these replicates are not required to be exactly the same on the atomic level; otherwise, the averaged values obtained based on these replicates become meaningless. A key requirement is that all these replicates share the same values of certain thermodynamic properties, for example, the number of atoms, temperature, pressure, volume, chemical potential, and total energy. A collection of these independent replicates forms a statistical ensemble.

Four important statistical ensembles are frequently applied in MD simulation. The first is the canonical ensemble, in which the number of atoms N, system volume V, and temperature T are the same for all the replicates of the system. The second is the microcanonical ensemble, in which N, V, and the total energy E are kept constant. The third is the isothermal–isobaric ensemble, in which N, T, and the pressure P are kept constant. The last is the grand canonical ensemble, in which V, T, and the chemical potential are kept constant.

According to the ergodic hypothesis, given a sufficiently long simulation time, an MD simulation system will experience all the possible states in its phase space. Hence, in MD simulations, replicates of a molecular system are generated by recording the microscopic states of the same system at selected time intervals over a long period of time. Setting the MD system under different conditions allows for the measurement of its properties in different ensembles. The most commonly measured physical properties are discussed as follows.

The potential energy of an MD system is directly given by the potential energy function and is usually calculated in tandem with the force calculation. The instantaneous kinetic energy at time t is given by

$$K(t) = \frac{1}{2}\sum_i m_i[v_i(t)]^2. \quad (1.51)$$

The total energy is then obtained as the sum of the potential and kinetic energy.

Based on the kinetic energy, the instantaneous system temperature at time t can be evaluated by

$$T(t) = \frac{2K(t)}{f}, \quad (1.52)$$

where f denotes the degrees of freedom of the system. For an MD system containing N atoms, the degrees of freedom can generally be written as $f = 3N - 3$. If there are additional constraints imposed on the system, the final degrees of freedom should equal $(3N - 3)$ minus the total number of such constraints. For example, in a dual-atom molecular system containing N_m molecules, if the bond length of each molecule is constrained, the degrees of freedom of the system should be $f = N_m \times 2 \times 3 - 3 - N_m$.

The calculation of pressure P in an MD simulation is based on the virial equation, which is given by

$$PV = Nk_BT + \frac{2}{3}W, \quad (1.53)$$

where V is the system volume and the virial term W is defined as

$$W = \frac{1}{2}\sum_{i=1}^{N} \boldsymbol{r}_i \cdot \boldsymbol{F}_i, \quad (1.54)$$

where $\boldsymbol{F}_i$ denotes the total force imposed on atom i. For an MD system with PBCs, to avoid unphysical self-interactions between an atom and its images, Eq. (1.54) can be rewritten as

$$W = \frac{1}{2}\sum_{i<j} \boldsymbol{r}_{ij} \cdot \boldsymbol{F}_{ij}, \quad (1.55)$$

where $\boldsymbol{F}_{ij}$ is the force imposed on atom i due to its interaction with atom j. Since the system temperature can be obtained by Eqs. (1.51) and (1.52), the final expression for pressure can be written as $P = 2(K + W)/(3V)$.

Besides the total energy, temperature, and pressure, the radial distribution function (RDF) is another static quantity that is usually used to characterize the structural property of an MD system. The physical meaning of the RDF can be illustrated by Fig. 1.11. The pink sphere located in the center represents the "target atom," and the number of atoms (gray spheres) with a distance from the target atom that is larger than r and smaller than $r + dr$ is denoted by dN. The RDF $g(r)$ is then defined by

$$g(r) = \frac{\mathrm{d}N}{4\rho_{\mathrm{N}}\pi r^2 \mathrm{d}r}, \tag{1.56}$$

where ρ_{N} is the number density of the system.

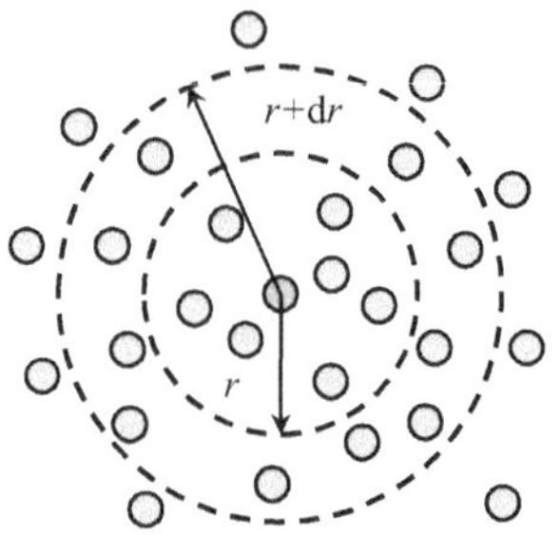

FIGURE 1.11 Illustration of the radial distribution function.

Physically, the RDF can be understood as the ratio of the local density to the average bulk density. The local density near the target atom is different from the bulk density. As the distance r increases, the local density approaches the bulk density, and the value of $g(r)$ also approaches 1. The plot of $g(r)$ with respect to r gives information about the density fluctuations from which structural properties can be deduced.

Mean square displacement (MSD) R_{MSD} is a quantity that is usually calculated to characterize the dynamic properties of an MD system. By definition, the MSD of a group of atoms in a time period of t can be written as the average of the square of their displacements over this period:

$$R_{\mathrm{MSD}}(t) = \left\langle \left| \boldsymbol{r}_i(t) - \boldsymbol{r}_i(0) \right|^2 \right\rangle. \tag{1.57}$$

It is not required for $\boldsymbol{r}_i(0)$ to be recorded at the beginning of the MD simulation. According to statistical mechanics, any arbitrary point in time during the simulation can be adopted as the starting point for calculating the MSD, as long as the simulation system contains a sufficiently large number of atoms and lasts for a sufficiently long time. Suppose that the trajectories of a group of atoms are recorded for n timesteps, and the position vectors are $\boldsymbol{r}_i(\Delta t), \boldsymbol{r}_i(2\Delta t), \cdots, \boldsymbol{r}_i(n\Delta t)$. For the calculation of the MSD, these position vectors are divided into two groups and an averaging process is performed on $n/2$ datasets. For instance, $R(\Delta t)$ can be calculated as

$$R(\Delta t) = \left\langle \frac{\left|r_i(2\Delta t)-r_i(\Delta t)\right|^2 + \left|r_i(3\Delta t)-r_i(2\Delta t)\right|^2 + \cdots + \left|r_i((n\Delta t/2)+\Delta t)-r_i(n\Delta t/2)\right|^2}{n/2} \right\rangle. \tag{1.58}$$

For a gaseous or liquid system, the magnitude of the MSD increases linearly with t after an initial short nonlinear stage. Based on Einstein's diffusion law, the diffusion constant D of a gaseous or liquid system can thus be related to the MSD via the relationship:

$$\lim_{t\to\infty} \left\langle \left|r_i(t)-r_i(0)\right|^2 \right\rangle = 6Dt. \tag{1.59}$$

The correlation function is another important quantity for the investigation of the dynamic properties of an MD system and describes the time relativity of a physical quantity with itself or other quantities. The autocorrelation functions of two physical quantities A and B are written as

$$C_A(t) = \langle A(t)A(0) \rangle \text{ and } C_B(t) = \langle B(t)B(0) \rangle. \tag{1.60}$$

As with the calculation of the MSD, any point of time of the MD simulation can be adopted as the starting point of time for the calculation of $A(0)$ and $B(0)$. The cross-correlation function of A and B is defined as

$$C_{AB}(t) = \langle A(t)B(0) \rangle. \tag{1.61}$$

The calculation of the correlation function is performed by averaging $A(t)A(0)$ across a group of datasets as

$$C_A(m\Delta t) = \langle A(m\Delta t)A(0) \rangle = \frac{1}{n}\sum_{i=0}^{n-1} [A(m\Delta t+i\Delta t)A(i\Delta t)], \tag{1.62}$$

where m is an integer that determines the time interval for which the self-correlation of A is calculated and n is the number of measurements of the quantity A. In practical MD simulations, the calculation of the correlation function can be expedited by fast Fourier transformation.

1.2.9 Accuracy of MD simulation

There are mainly two types of errors in computer simulations (including MD simulation): systematic errors and statistical errors. Systematic errors are usually difficult to detect, while statistical errors are traceable and should always be estimated and minimized.

Systematic errors do not occur by chance but are caused by inaccuracies inherent to the system. They cause the measurement of a property to diverge from its true value. Possible reasons for systematic errors in computer simulations include faults in the algorithm or model, inappropriate approximations, and round-off errors. A simple approach for identifying systematic errors is to obtain multiple measurements of simple thermodynamic properties of a simulated system and check their distributions, which should usually be an unbiased normal distribution if no systematic error is present. Alternatively, systematic errors can be detected by performing simulations on different hardware or software platforms, or by employing different models or algorithms.

Statistical errors in MD simulations are typically caused by insufficient sampling of the phase space. Although the ergodic hypothesis states that the simulated system can experience all possible states in the phase space given a sufficiently long simulation time, such an indefinite time interval cannot be realistically achieved for MD systems. Hence, for any physical quantity A, its time-averaged value $\langle A \rangle_{\mathrm{sim}}$ obtained during a simulation is usually different from its true mean value $\langle A \rangle$. To estimate the statistical error, the standard deviation is usually calculated. Suppose that one obtains M values for the quantity A in each measurement, and the standard deviation of the average value $\langle A \rangle_{\mathrm{sim}}$ (not A) for all these measurements is defined as

$$\sigma[\langle A \rangle_{\mathrm{sim}}] = \frac{\sigma[\langle A \rangle]}{\sqrt{M}} = \frac{\sqrt{\sum_{i=1}^{M} [A_i - \langle A \rangle_{\mathrm{sim}}]^2}}{M}, \tag{1.63}$$

where $\sigma[\langle A \rangle]$ is the standard deviation of the true mean value $\langle A \rangle$. To apply this equation, M observations are required to be independent of one another. However, in MD simulations, there exists a correlation or relaxation time for each property under investigation. Two observations conducted within a time interval smaller than the relaxation time are not independent of each other. Hence, various approaches have been developed for evaluating the statistical errors of MD simulation.

One such approach is the block average method. In this method, the entire simulation is divided into n_{b} blocks, each lasting for τ_{b} timesteps. For each simulation block, the average value of A is calculated as

$$\langle A \rangle_{\mathrm{blk}} = \frac{1}{\tau_{\mathrm{b}}} \sum_{i=1}^{\tau_{\mathrm{b}}} A_i(t). \tag{1.64}$$

If the time length of each simulation block is sufficiently large, the values of $\langle A \rangle_{\mathrm{blk}}$ obtained from different blocks will be considered uncorrelated. The standard error $\sigma[\langle A\rangle_{\text{sim}}]$ for the measurement of $\langle A\rangle_{\text{sim}}$ can thus be estimated by

$$\sigma^2[\langle A\rangle_{\text{sim}}] = \frac{1}{n_\text{b}}\sum_{i=1}^{n_\text{b}}\left[\langle A(t)\rangle_{\text{blk}} - \langle A\rangle_{\text{sim}}\right]^2. \tag{1.65}$$

The limiting value s of τ_b that ensures the independence of $\langle A(t)\rangle_{\text{blk}}$for each simulation block can be calculated by

$$s = \lim_{\tau_\text{b}\to\infty}\frac{\tau_\text{b}\sigma^2\left[\langle A(t)\rangle_{\text{blk}}\right]}{\sigma^2[\langle A\rangle]}. \tag{1.66}$$

In MD simulations, evaluating the limiting value s at an arbitrarily large value of τ_b is impractical. Alternatively, s is obtained by plotting $\tau_\text{b}\sigma^2[\langle A(t)\rangle_{\text{blk}}]/\sigma^2[\langle A\rangle]$ against an increasing finite value of τ_b. For small values of τ_b, this quantity increases sharply, but it eventually reaches a plateau as τ_b reaches a critical value. By approximating s with this plateau value, the standard error $\sigma[\langle A\rangle_{\text{sim}}]$ is estimated by

$$\sigma[\langle A\rangle_{\text{sim}}] \approx \sigma[\langle A\rangle]\sqrt{\frac{s}{M}}. \tag{1.67}$$

1.3 Hardware and software for MD simulation

Compared with other computer simulation techniques in computational materials science, MD simulation is not particularly demanding in terms of computing resources due to its usage of classical potential energy functions. When applying a simple potential energy function for a gaseous molecular system containing fewer than 10,000 atoms, a common PC will be adequate to perform the simulation, and the execution time generally lasts for minutes or hours. For simple liquid systems with few interface problems, the number of atoms that a common PC can handle usually does not exceed 1000 due to the possible calculation of long-range Columbic interactions. The simulation execution time may last for hours or days in such cases. For large complex molecular systems, the demand for computing resources significantly increases with the system size, especially when a complicated potential energy function is used. For MD simulations commonly conducted in contemporary research of materials science or condensed matter physics, parallel computing is generally preferred. For a biological system containing approximately 50,000 atoms, an MD simulation with a timestep of 1 ns requires around two weeks to process for a single central processing unit (CPU) core, but the execution time can be reduced to around a day through parallel computing with 16 CPU cores. Owing to the development of the fast and efficient message passing interface (MPI) standard, the efficiency of parallel computing involving multiple CPU cores is tremendously boosted.

However, computing efficiency does not increase linearly with the number of CPU cores employed in parallel computing. Since the processing speed of each CPU core is much higher than the data transfer speed between different cores, the main limitation in parallel computing is the rate of data exchange among these cores. To increase the efficiency of parallel computing, more advanced hardware for data transfer or more efficient data transfer algorithms must be developed.

The performance of a parallel MD simulation is not only determined by the system size but also the execution platform. The machines, operating systems, code compilers, and development tools employed are all important factors. Hence, selecting the optimal hardware platform for MD simulations proves to be a challenge. A common strategy is using benchmarks, which are computer programs developed for testing purposes under controlled conditions and are seldom modified. By running these benchmark programs on different hardware platforms, the corresponding execution time and outputs are compared, and the optimal platform is selected. However, this approach is usually impractical because of difficulties in accessing different types of hardware. Fortunately, an MD benchmark program named MDBNCH (molecular dynamics benchmark) has been developed (Ercolessi, 1988). A database containing the execution time of this benchmark program on different machines has been established, and this database is presently growing because of the contributions of researchers in the field. Although it may differ greatly from MD simulations in practical scientific research, the MDBNCH program can provide a more accurate performance estimation of these simulations than other general benchmark programs such as the well-known Linpack and SPEC (standard performance evaluation corporation).

There are several popular MD simulation packages that enable parallel computing, including NAMD, AMBER, CHARMM, LAMMPS, and GROMACS, among which NAMD, LAMMPS, and GROMACS are the most widely used due to their free accessibility to the public and excellent updates and maintenance.

NAMD, or Nanoscale Molecular Dynamics, is a parallel MD code designed for high-performance simulations of large biomolecular systems. Typical MD simulations with NAMD may involve hundreds of CPU cores. According to its official description, over 500,000 cores can be used for the largest NAMD simulations. NAMD has a high level of code design that grants high simulation efficiency and is compatible with various types of input and output file formats, thereby facilitating great versatility between different material modeling packages. Hence, NAMD is an excellent choice for modeling large and complex biomolecules.

LAMMPS (large-scale atomic/molecular massively parallel simulator) might be the most popular MD simulation code for the modeling of condensed matter. Besides its high efficiency, LAMMPS permits the usage of a wide range of potential energy functions for solid-state materials (metals and semiconductors), soft matter (biomolecules and polymers), and coarse-grained or mesoscopic systems. LAMMPS can not only model atomic systems but also be applied to simulate particles on the mesoscopic or continuum scale. Being an open-source software, it provides a high degree of flexibility for users to develop problem-oriented codes.

GROMACS is primarily designed for modeling biochemical molecules, such as proteins and lipids, which possess a large number of complicated bonded interactions. Since it is also capable of computing nonbonded interactions, GROMACS has been widely applied to model nonbiological systems, including polymers and liquid solutions. It provides a high computational efficiency due to the multitude of advanced optimization algorithms implemented in its code. One of the most prominent features of GROMACS is its simple user-interface with command-line options for the input and output files. During the simulation process, the instantaneous execution status is also present on the user interface. Handy tools for trajectory analysis that greatly assist data postprocessing are also included in GROMACS.

AMBER and CHARMM are also widely used MD simulation packages, although the user will be charged for their usage. The main reason for their popularity is the great versatility of their unique built-in force fields for the simulation of different biomolecular systems. The term "AMBER" (assisted model building with energy refinement) here can mean two different things. First, it refers to a set of molecular mechanics force fields for the simulation of biomolecules. Second, it indicates the corresponding package of MD simulation programs. The development of AMBER began in the 1970s and is currently maintained by an active development community. It provides high flexibility in constructing novel molecules, but its simulation efficiency pales next to that of GROMACS. Meanwhile, "CHARMM" (Chemistry at Harvard Macromolecular Mechanics) was also developed primarily for modeling biological systems, including peptides, proteins, prosthetic groups, small molecule ligands, nucleic acids, lipids, and carbohydrates. A comprehensive set of analysis and model building tools is included in CHARMM, which results in its wide application in the study of inorganic material systems. However, compared with the other MD codes, such as GROMACS, AMBER, and LAMMPS, CHARMM has mostly poor simulation efficiency.

References

Alder, B.J., Wainwright, T.E., 1957. Phase transition for a hard sphere system. J. Chem. Phys. 27, 1208–1209.

Allen, M.P., Tildesley, D.J., 1987. Computer Simulation of Liquids. Clarendon Press, Oxford.

Ercolessi, F. 1988. http://www.sissa.it/furio/mdbnch.html.

Leach, A.R., 1996. Molecular Modelling: Principles and Applications. Longman, Edinburgh.

McCammon, J.A., Gelin, B.R., Karplus, M., 1977. Dynamics of folded proteins. Nature 267, 585–590.

Rahman, A., 1964. Correlations in the motion of atoms in liquid argon. Phys. Rev. 136, A405–A411.

Rahman, A., Stillinger, F.H., 1971. Molecular dynamics simulation of liquid water. J. Chem. Phys. 55, 3336–3359.

Verlet, L., 1967. Computer experiments on classical fluids. I. thermodynamical perperties of Lennard–Jones molecules. Phys. Rev. 159, 98–103.

CHAPTER

2

Potential energy functions

This chapter presents the basics of the potential energy functions for molecular dynamics simulations. It begins with the Born–Oppenheimer assumption and then describes the general procedure of constructing such potential energy functions. Typical two-body (or pairwise) and many-body potentials are first introduced, followed by specific potential energy functions for different materials, including ionic materials, metals, covalent materials, and molecular systems. The most commonly employed potential functions are also discussed with respect to their detailed functional forms.

2.1 The Born–Oppenheimer assumption

In molecular dynamics (MD) simulations the force acting on each atom is derived from the gradient of the potential energy function with respect to the atomic position. Therefore, constructing an appropriate potential energy function becomes a core task of MD simulation.

A system of interacting atoms consists of N_n nuclei and N_e electrons that interact with one another. The true Hamiltonian for such a system can be written as

$$H = \sum_i \frac{\boldsymbol{p}_i^{\mathrm{nul}} \cdot \boldsymbol{p}_i^{\mathrm{nul}}}{2m_i} + \sum_n \frac{\boldsymbol{p}_n^{\mathrm{elc}} \cdot \boldsymbol{p}_n^{\mathrm{elc}}}{2m_\mathrm{e}} + \frac{1}{2}\sum_{ij} \frac{Z_i Z_j e^2}{\left|\boldsymbol{r}_i^{\mathrm{nul}} - \boldsymbol{r}_j^{\mathrm{nul}}\right|} + \frac{1}{2}\sum_{n,n'} \frac{e^2}{\left|\boldsymbol{r}_n^{\mathrm{elc}} - \boldsymbol{r}_{n'}^{\mathrm{elc}}\right|} - \sum_{i,n} \frac{Z_i e^2}{\left|\boldsymbol{r}_i^{\mathrm{nul}} - \boldsymbol{r}_n^{\mathrm{elc}}\right|}, \tag{2.1}$$

where i and j index the nuclei, and n and n' the electrons; $\boldsymbol{r}_i^{\mathrm{nul}}$ and $\boldsymbol{p}_i^{\mathrm{nul}}$ denote the position and momentum vectors of the nuclei, respectively; $\boldsymbol{r}_i^{\mathrm{elc}}$ and $\boldsymbol{p}_n^{\mathrm{elc}}$ denote those of the electrons; m_e represents the mass of an electron (9.109383×10^{-31} kg); e denotes the elementary electron charge ($1.60217662 \times 10^{-19}$ C); Z_i and m_i denote the atomic number and mass of nucleus i, respectively. The first two terms represent the kinetic energies of the nuclei and electrons, respectively, and the last three terms describe the Coulombic interactions among them. The Schrödinger equation must be solved to obtain the full information of the electron

DOI: https://doi.org/10.1016/B978-0-12-816419-8.00007-6

distribution or wave function $\Psi(\boldsymbol{R}^{N_n}, \boldsymbol{r}^{N_e})$, based on which the interactions among the nuclei and electrons can be fully considered. Here, $\boldsymbol{R}^{N_n} = \{\boldsymbol{r}_1^{\text{nul}}, \ldots, \boldsymbol{r}_{N_n}^{\text{nul}}\}$ and $\boldsymbol{r}^{N_e} = \{\boldsymbol{r}_1^{\text{elc}}, \ldots, \boldsymbol{r}_{N_e}^{\text{elc}}\}$.

Solving the Schrödinger equation is usually challenging, both analytically and numerically. Exact solutions exist only for a very limited number of systems, and thus approximation schemes are required. One of these schemes is the Born–Oppenheimer assumption. Since the mass M of a nucleus is much greater than the mass m of an electron, nuclei move on a timescale about two orders of magnitude greater than that of electrons, according to the relation:

$$\frac{\omega_e}{\omega_n} \sim \sqrt{\frac{M}{m}} \sim 100, \tag{2.2}$$

where ω_e and ω_n denote the characteristic vibration frequency of the electrons and nuclei, respectively; M and m represent the masses of a nucleus and an electron, respectively. Based on Eq. (2.2), Born and Oppenheimer assumed that the nuclei could be regarded as fixed when the movement of electrons is concerned. Based on this assumption, the wave function $\Psi(\boldsymbol{R}^{N_n}, \boldsymbol{r}^{N_e})$ of the system is simply the product of the wave functions of the electrons and nuclei:

$$\Psi(\boldsymbol{R}^{N_n}, \boldsymbol{r}^{N_e}) = \Xi(\boldsymbol{R}^{N_n})\Phi(\boldsymbol{r}^{N_e}, \boldsymbol{R}^{N_n}), \tag{2.3}$$

where $\Xi(\boldsymbol{R}^{N_n})$ and $\Phi(\boldsymbol{r}^{N_e}, \boldsymbol{R}^{N_n})$ denote the wave functions of the nuclei and electrons, respectively. Consequently, the wave function $\Psi(\boldsymbol{R}^{N_n}, \boldsymbol{r}^{N_e})$ can be obtained by solving two separate Schrödinger equations:

$$H_e\Phi(\boldsymbol{r}^{N_e}, \boldsymbol{R}^{N_n}) = U(\boldsymbol{R}^{N_n})\Phi(\boldsymbol{r}^{N_e}, \boldsymbol{R}^{N_n}), \tag{2.4}$$

$$\left[\sum_i \frac{\boldsymbol{p}_i^{\text{nul}} \cdot \boldsymbol{p}_i^{\text{nul}}}{2m_i} + U\left(\boldsymbol{R}^{N_n}\right)\right]\Xi\left(\boldsymbol{R}^{N_n}\right) = E\Xi\left(\boldsymbol{R}^{N_n}\right), \tag{2.5}$$

where

$$H_e = \sum_n \frac{\boldsymbol{p}_n^{\text{elc}} \cdot \boldsymbol{p}_n^{\text{elc}}}{2m_e} + \frac{1}{2}\sum_{ij} \frac{Z_i Z_j e^2}{\left|\boldsymbol{r}_i^{\text{nul}} - \boldsymbol{r}_j^{\text{nul}}\right|} + \frac{1}{2}\sum_{n,n'} \frac{e^2}{\left|\boldsymbol{r}_n^{\text{elc}} - \boldsymbol{r}_{n'}^{\text{elc}}\right|} - \sum_{i,n} \frac{Z_i e^2}{\left|\boldsymbol{r}_i^{\text{nul}} - \boldsymbol{r}_n^{\text{elc}}\right|}. \tag{2.6}$$

Eq. (2.4) is the Schrödinger equation for the system of electrons subjected to the Born–Oppenheimer approximation, of which the nuclei are considered to be fixed in place. The eigenvalue $U(\boldsymbol{R}^{N_n})$ depends parametrically on the coordinates of the nuclei and is known as the interatomic potential, based on which the motion of the nuclei can be obtained via solving Eq. (2.5). By incorporating the electronic effects into $U(\boldsymbol{R}^{N_n})$, the Schrödinger equation pertaining to the nuclei (Eq. 2.5) can be replaced by the classical Newtonian equations of motion, and the movement of the nuclei is then described classically.

The validity of simplifying Eq. (2.5) by the classical Newtonian equations of motion can be assessed by calculating the de Broglie thermal wavelength Λ, which is defined by

$$\Lambda = \sqrt{\frac{2\pi\hbar^2}{Mk_B T}}, \tag{2.7}$$

where $\hbar = h/(2\pi)$ is the reduced Planck constant (1.054572×10^{-34} J s), k_B is the Boltzmann constant (1.380649×10^{-23} J/K), M denotes the atomic mass, and T denotes the temperature. When Λ satisfies the condition $\Lambda \ll a$ (a denotes the mean nearest-neighbor separation), the simplification of describing the movement of the nuclei classically is validated. For cases in which an atomic system is under a sufficiently low temperature or contains relatively light atoms, the thermal wavelength could be comparable to the separation distance a, and the simulation results should be interpreted with caution.

2.1.1 Construction of potential energy functions

Solving the Schrödinger equation presented in Eq. (2.4) poses a daunting challenge even for modern researchers. A more practical strategy for constructing a potential energy function is to adopt and modify certain preexisting potential energy functions with specific functional forms and free parameters.

According to such a strategy, constructing a potential energy function can be divided into two steps. In the first step, an analytical form is chosen for the potential function in question. The selected functional form should be capable of well representing the "true" potential in terms of describing the physical properties and behaviors of specific materials. In the past, choices of the functional form were very limited, and it was typically expressed as the sum of two-body (or pairwise) terms, with the potential energy of an atomic pair depending solely on their separation. Presently, novel many-body forms are established to capture as much information as possible regarding the physical behaviors and chemical bonds of the system. A typical approach for constructing a potential energy function U^{tot} is to split it into parts:

$$U^{tot} = \sum_i U_1(r_i) + \sum_{i,j} U_2(r_i, r_j) + \sum_{i,j,k} U_3(r_i, r_j, r_k) + \cdots. \tag{2.8}$$

The first term represents an external potential, the second term represents two-body interactions such as chemical bonding or van der Waals interactions, and the third term represents three-body interactions such as bond bending or bond torsion. Four-body and higher terms can also be introduced to increase the complexity and accuracy of the potential energy function.

The functional form constructed in the first step usually contains several undetermined parameters that exhibit varying values for different materials. Hence, the second step of constructing a potential is parameterizing the functions that constitute the functional form. Several methods have been developed over the years to achieve this goal. At one extreme, the parameterization is based solely upon ab initio calculations of which the electronic effects are fully considered; at the other extreme, the parameterization is realized by fitting experimental data, and the highest priority is assigned to complementing these results instead of merely adhering to ab initio calculations. Regardless of the construction method, potential energy functions are not universally applicable and should each be limited to a certain range of materials. Electronic structures of atoms of the same element may differ significantly depending on their interactions with surrounding atoms (regardless of element types). For instance, it will be virtually impossible to construct a potential energy function capable of modeling both a bulk solid and a diatomic molecule of the same element. Therefore, when a potential energy function is chosen for an MD simulation, one should always be familiar with its applicable range and be cautious in interpreting the simulation results.

2.1.2 Two-body potentials

The simplest potential energy functions are two-body (or pairwise) potentials, which were commonly employed in early MD simulations. Two-body potentials are usually idealized and can be parameterized to effectively reproduce many-body interactions after performing appropriate combinations. The hard-sphere and square well potentials are examples of such potentials. The hard-sphere potential is defined as

$$U(r) = \begin{cases} \infty, & r < \sigma \\ 0, & r \geq \sigma \end{cases}, \tag{2.9}$$

and the square well potential is defined as

$$U(r) = \begin{cases} \infty, & r < \sigma_1 \\ -\varepsilon, & \sigma_1 \leq r < \sigma_2 \\ 0, & r \geq \sigma_2 \end{cases}, \tag{2.10}$$

where r denotes the distance between two atoms, ε denotes the depth of the potential energy well, and σ, σ_1, and σ_2 denote the zero-crossing distances. These potentials are the earliest MD potentials and perform well for problems related to molecule packing. Fig. 2.1 shows the schematics of these two potentials.

Compared with the hard-sphere and square well potentials, the soft sphere potential introduces an exponential function to describe the two-

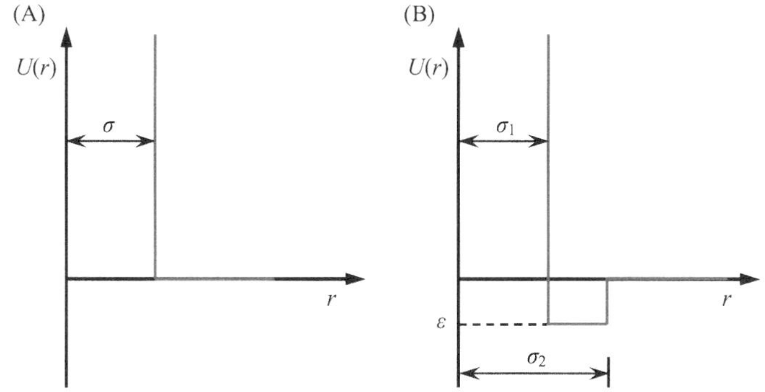

FIGURE 2.1 Schematics of (A) the hard-sphere potential and (B) the square well potential.

body interaction, but it only considers the repulsive component. The soft sphere potential possesses the analytical form:

$$U(r) = \varepsilon\left(\frac{\sigma}{r}\right)^n, \tag{2.11}$$

where ε and σ have the same physical meanings as those for the hard-sphere and square well potentials. By increasing the exponential index n, a harder potential is obtained.

For modern MD simulations, the Lennard–Jones (LJ) potential (Jones and Chapman, 1924a,b) is probably the most frequently applied two-body potential that describes van der Waals interactions accurately. The common form of the LJ potential is written as

$$U^{\mathrm{LJ}}(r) = 4\varepsilon\left[\left(\frac{\sigma}{r}\right)^{12} - \left(\frac{\sigma}{r}\right)^{6}\right]. \tag{2.12}$$

The first component with $(\sigma/r)^{12}$ represents a repulsive interaction, while the second component with $-(\sigma/r)^6$ describes an attractive interaction. The attractive component dominates at large distances and provides cohesion to the system. When atoms interact with one another, dipole moments are induced and result in an attractive interaction, which is essentially a van der Waals interaction. Although the attractive interactions are rather weak, they dominate the bonding of closed-shell systems, namely, noble gases such as Ar and Kr. On the other hand, the repulsive component is dominant at short distances and models the interaction between two atoms when they are brought close together. With a decrease in the separation distance of the two atoms, their wave functions will begin to overlap at some point, which will result in a positive potential energy of such a diatomic system according to Pauli's exclusion principle. The critical separation distance for a negative potential energy is denoted by σ, a value below which repulsive interactions are induced. In Fig. 2.2, the value of σ can be interpreted as the coordinate of the point at which the potential

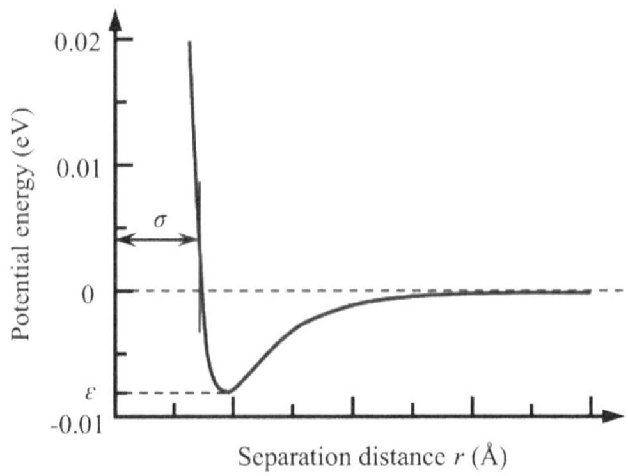

FIGURE 2.2 Potential energy profile of two Ar atoms engaging in a weak van der Waals interaction, which is described by the Lennard–Jones (LJ) potential energy function.

energy is zero. The parameter ε is the well depth and defines the potential energy minimum at which two atoms are separated by an equilibrium distance that can be calculated by $\sqrt[6]{2}\sigma$. As the separation distance approaches an arbitrarily large value, the potential energy approaches zero.

There are also alternative functional forms for the LJ potential, such as

$$U^{\mathrm{LJ}}(r) = \varepsilon\left[\left(\frac{r_{\mathrm{m}}}{r}\right)^{12} - 2\left(\frac{r_{\mathrm{m}}}{r}\right)^{6}\right], \tag{2.13}$$

and

$$U^{\mathrm{LJ}}(r) = \left(\frac{A}{r^{12}}\right) - \left(\frac{C}{r^{6}}\right). \tag{2.14}$$

The relationships between the empirical parameters r_{m}, A, and C in the above two functional forms are straightforward, of which $A = \varepsilon r_{\mathrm{m}}^{12} = 4\varepsilon\sigma^{12}$ and $C = 2\varepsilon r_{\mathrm{m}}^{6} = 4\varepsilon\sigma^{6}$. Generally, r_{m}, A, and C are obtained by fitting experimental data for a specific atom type. When a system contains different atom types, the LJ interaction parameters for each atomic pair can be obtained by applying certain mixing rules. The Lorentz–Berthelot rule is one of the most commonly implemented mixing rules, of which the zero-crossing distance σ_{AB} and well depth $\varepsilon_{\mathrm{AB}}$ for two interacting atom types, A and B, are obtained, respectively, by

$$\sigma_{\mathrm{AB}} = \frac{1}{2}(\sigma_{\mathrm{AA}} + \sigma_{\mathrm{BB}}), \tag{2.15}$$

$$\varepsilon_{\mathrm{AB}} = \sqrt{\varepsilon_{\mathrm{AA}}\varepsilon_{\mathrm{BB}}}. \tag{2.16}$$

The application of this rule is highly successful in modeling interactions between similar atomic species but may overestimate the well depth.

The arithmetic mean is often replaced by the geometric mean for the zero-crossing distance σ_{AB}, which is given by

$$\sigma_{AB} = \sqrt{\sigma_{AA}\sigma_{BB}}. \tag{2.17}$$

A limitation of the LJ potential is that it approaches a singularity when two atoms exhibit a very small interatomic separation, as the term $(\sigma/r)^{12}$ renders the potential energy unphysically large. The Buckingham potential (Buckingham, 1938) resolves this issue by replacing the term $(\sigma/r)^{12}$ with an exponential term:

$$U^{\text{Buck}}(r) = A\exp\left(-\frac{r}{\rho}\right) - \frac{C}{r^6}, \tag{2.18}$$

where A, C, and ρ are all material-dependent constants. For the Buckingham potential, the potential energy can remain finite even at small interatomic distances.

In most cases, the main form of interaction between two atoms is chemical bonding instead of the weak van der Waals interaction. A typical potential energy function for a chemical bond is the Morse potential function (Morse, 1929), which is given by

$$U^{\text{Morse}}(r) = D_e\exp(-2a(r-r_0)) - 2D_e\exp(-a(r-r_0)), \tag{2.19}$$

where D_e denotes the well depth, r_0 the minimum energy distance, and $a = \omega\sqrt{\mu/(2D_e)}$ with ω denoting the bond vibration frequency and μ the corresponding reduced mass of the atom. The Morse potential is an ideal anharmonic potential function that permits exact mathematical treatment. In particular, it will be reduced to a harmonic oscillator expression in the harmonic limit, which allows for analytical solutions of the Schrödinger equation.

2.1.3 Many-body potentials

The LJ and Morse potentials are probably the most widely applied two-body potentials. However, materials that can be realistically modeled by a two-body potential are limited to noble gases and diatomic molecules. Systems of greater research interest, such as ionic materials, covalent materials, metals, and molecular systems, cannot be accurately described by two-body potentials alone as many-body potentials are required to consider the interactions among three or more bodies. For instance, the cohesive energy between two metal atoms in a diatomic system is much smaller than that of the bulk phase of the same metal. Similarly, the vacancy formation energy for the bulk phase of a metal is also much lower than the cohesive energy of a diatomic system. If one considers semiconductors or complex molecular systems, such discrepancies could be even greater. In addition to two-body terms (e.g., covalent bonds for semiconductors), a many-body potential energy function for metals or semiconductors should also take into account bond angle bending (three-body), bond torsion (four-body), and other possible many-body interactions.

A general form of a many-body potential function can be written as

$$U = \sum\nolimits_{\text{bonds}} \frac{k_i^{\text{b}}}{2}(l_i - l_{i,0})^2 + \sum\nolimits_{\text{angles}} \frac{k_i^{\theta}}{2}(\theta_i - \theta_{i,0})^2 + \sum\nolimits_{\text{torsion}} \frac{V_n}{2}(1 + \cos(n\phi - \phi_0))$$
$$+ \sum_{i=1}^{N} \sum_{j=i+1}^{N} \left(\frac{q_i q_j}{4\pi\varepsilon_0 r_{ij}} + 4\varepsilon_{ij} \left[\left(\frac{\sigma_{ij}}{r_{ij}} \right)^{12} - \left(\frac{\sigma_{ij}}{r_{ij}} \right)^{6} \right] \right). \tag{2.20}$$

The five terms on the right-hand side of Eq. (2.20) correspond to the five types of interactions illustrated in Fig. 2.3. Here, bond stretching and bond angle bending interactions are modeled by harmonic potentials that increase the potential energy as the bond length l_i and the bond angle θ_i deviate from their reference values $l_{i,0}$ and $\theta_{i,0}$, respectively. The parameters k_i^{b} and k_i^{θ} are material-dependent constants that determine the interaction strength. The third term represents the contribution of the bond torsion to the total potential energy of the system. The variable V_n describes the energy barriers for the bond torsion, n denotes the number of minimum points in the potential energy function as the bond is rotated 360°, ϕ denotes the torsion angles, and ϕ_0 denotes the values of ϕ at the points of minimum potential energy. The fourth term represents Coulombic interactions, and the last term, modeled by the LJ potential, describes van der Waals interactions.

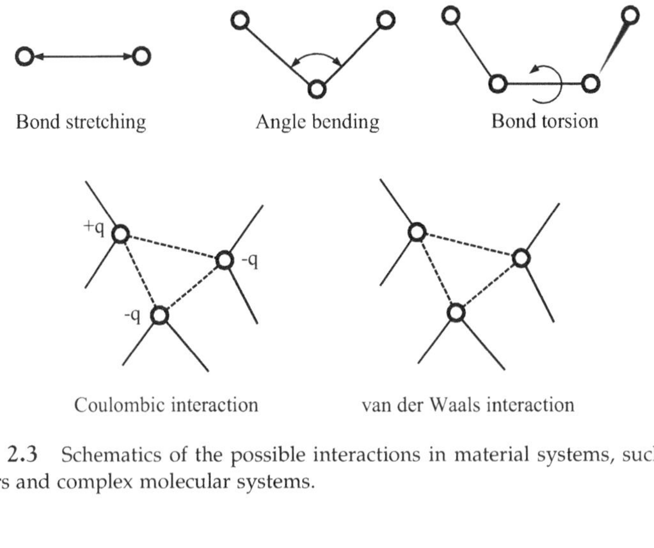

FIGURE 2.3 Schematics of the possible interactions in material systems, such as semiconductors and complex molecular systems.

More sophisticated potential functions may have additional terms, for example, the term regarding the out-of-plane bending interaction (illustrated in Fig. 2.4). Such a term is commonly adopted in the so-called united atom force fields that treat the H and C atoms in methyl groups or in methylene bridges as a united atom to maintain stereochemistry at chiral centers. However, it should be noted that the out-of-plane bending interaction term may not always be necessary, and its inclusion may have a deleterious effect on the performance of a potential function, particularly when the atomic vibrational frequencies of a modeled molecular system are of interest since they are rather sensitive to the presence of the out-of-plane bending interaction terms.

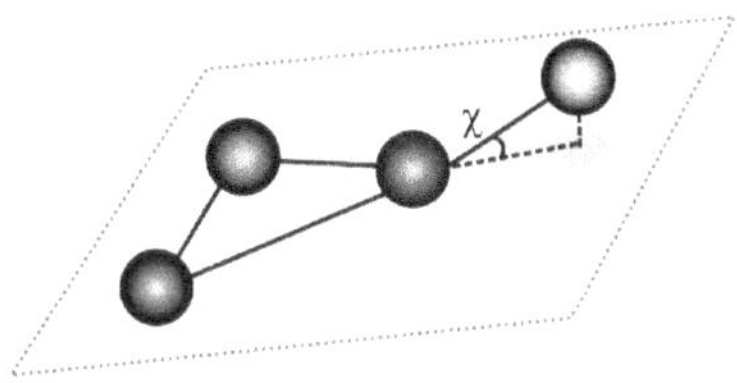

FIGURE 2.4 Illustration of out-of-plane bending of a four-atom system.

Examples of potential functions that follow the above functional forms with different levels of modification include the AMBER (assisted model building with energy refinement), CHARMM (Chemistry at Harvard Macromolecular Mechanics) and MM (molecular mechanics) force fields, CVFF (consistent-valence force field), and PCFF (polymer consistent force field). It should be noted that in the subject of chemistry, it is a common practice to directly describe a force acting on a particle without defining its corresponding potential. Hence, interaction potential functions are often termed "force fields." In the following discussions, the terms "potential functions" and "force fields" shall be used interchangeably. These force fields have been designed for a wide range of material systems and are usually considered as generalized force fields. For instance, the AMBER force field (Cornell et al., 1995) is highly applicable to most proteins and nucleic acids, and the CHARMM force field (Brooks et al., 1983) can be employed to model a wide variety of organic molecules, solutions, polymers, and biomolecules.

Although these force fields have a wide range of applicability, they only cover a portion of the elements in the periodic table and are not fully generalized. In addition, most of the parameters in these force fields were obtained through fitting experimental data, and thus they lack a rigorous foundation. To address these issues, other force fields, such as the ESFF (extensible and systematic force field), UFF (universal force field), and Dreiding force field, were developed based on quantum calculations of fundamental elements. The ESFF force field (Shi et al., 2003) can precisely predict the behaviors of gases, condensed organic and inorganic molecules, and organic metallic molecules, but it can only provide rough estimations of configuration energies and vibrating frequencies. The Dreiding force field (Mayo et al., 1990) can be employed to calculate various properties of molecular clusters, but its applicability does not extend across the entire periodic table. The UFF (Rappe et al., 1992) covers all the elements and applies to almost any molecular or atomic system while providing better predictions than the Dreiding force field in terms of molecular structures, but it will result in significant errors when modeling intermolecular interactions.

Besides these generalized force fields, many specialized force fields were developed to model specific classes of materials. For instance, the Vashishta potential and the Shell model were designed for modeling ionic materials, the embedded atom method (EAM)-based potential was for metals, and the Stillinger–Weber (SW) potential and the Tersoff potential were for covalent materials. These specialized force fields may not be as versatile as the generalized force fields, but they provide more accurate predictions. In the following sections, these force fields are briefly introduced based on their applicable range of materials.

2.2 Potential energy functions for different materials

Atoms interact with one another by forming different types of bonds. Common bond types in solids include ionic bonds, covalent bonds, and metallic bonds. For molecular systems, nonbonded interactions including van der Waals and Coulombic forces are also common.

The formation of ionic bonds occurs because of charge transfer between two atoms that exhibit a large difference in their electronegativity values. When an ionic bond forms, one of the atoms (e.g., a halogen atom) rips off one or more electrons from the other (e.g., an alkali metal atom). Upon the transfer of the electron(s), the atom that has lost the electron(s) is now positively charged, while the atom that has gained the extra electron(s) is negatively charged. For ions with fixed charges, their interactions are mainly Coulombic in nature and depend on their atomic separation distances, whereas their bond angles play a negligible role.

When two atoms exhibit similar electronegativity values, no charge transfer occurs, and these two atoms share some of their valence electrons, forming covalent bonds. Since the shared electrons are spatially localized, the covalent bond energetics depend not only on the bond lengths but also on the bond angles.

In contrast to ionic and covalent bonds, metallic bonds cannot exist in isolation. Instead, it is collective in nature and always involves a group of atoms. Metals are good electrical conductors as their electrons are highly delocalized and can move freely within the metallic crystal. Hence, metallic bonds are, in essence, the result of the attraction between positive nuclei and delocalized electrons.

Due to the distinguishing characteristics of ionic, covalent, and metallic bonds, potential energy functions for materials with different bond types exhibit different functional forms. In the following section, these functional forms are briefly described.

2.2.1 Ionic materials

Ionic systems are formed by oppositely charged ions interacting with one another via ionic bonding (or electrostatic interactions), which can be modeled by Coulombic interactions and are expressed by

$$U^{\mathrm{Coul}}(r) = \frac{e^2}{4\pi\varepsilon_0}\frac{q_i q_j}{r_{ij}}, \tag{2.21}$$

where e denotes the elementary electron charge (1.602176×10^{-19} C), q_i denotes the partial charge of ith atom, ε_0 is the vacuum dielectric constant (8.854×10^{-12} F/m), and r_{ij} is the distance between atoms i and j. However, Eq. (2.21) does not include terms that describe electric repulsion at small interatomic distances. By adding a short-range repulsive term $U^{\mathrm{SR}}(r_{ij})$ that accounts for the repulsion of the electron shells to Eq. (2.21), the total potential energy function of an ionic system can be rewritten as

$$U(r_{ij}) = \frac{e^2}{4\pi\varepsilon_0}\frac{q_i q_j}{r_{ij}} + U^{\mathrm{SR}}(r_{ij}). \tag{2.22}$$

Several forms have been proposed for the term $U^{\mathrm{SR}}(r_{ij})$, and commonly adopted ones include the Buckingham potential:

$$U^{\mathrm{SR}}(r_{ij}) = A\exp\left(-\frac{r_{ij}}{\rho}\right) - \frac{C}{{r_{ij}}^6}, \tag{2.23}$$

the Born–Huggins–Mayer potential (Huggins and Mayer, 1933):

$$U^{\mathrm{SR}}(r_{ij}) = A\exp\left[-B(r_{ij} - \sigma)\right] - \frac{C}{{r_{ij}}^6} - \frac{D}{{r_{ij}}^8}, \tag{2.24}$$

and the Morse potential given by Eq. (2.19). In Eqs. (2.23) and (2.24), A, B, C, D, and ρ are all material-dependent constants, and σ represents the zero-crossing distance similar to that of the LJ potential. Note that only the exponential terms in the above equations account for short-range repulsive interactions. The terms $-C/r_{ij}^6$ and $-D/r_{ij}^8$ represent the attractive components of the van der Waals interactions.

For purely ionic bonding systems, the functional form of Eq. (2.22) is usually sufficient to model interatomic interactions. However, in reality, all ionic compounds exhibit certain characteristics of covalent bonds, and thus additional terms are required in certain cases. As covalent bonds usually involve many-body effects, a three-body term is typically introduced to consider the covalent-bonding effects. One good example would be the ionic interaction model for SiO_2 developed by Vashishta et al. (1990). The Vashishta potential comprises two parts and reads as

$$U = \sum_{i<j} U_2(r_{ij}) + \sum_{i<j<k} U_3\left(r_{ij}, r_{jk}, r_{ik}\right). \tag{2.25}$$

The two-body term $U_2(r_{ij})$ accounts for the steric repulsion due to atomic sizes, the Coulombic interactions due to charge transfer, and the charge-dipole interaction due to the effects of electronic polarizability of anions. The three-body term $U_3(r_{ij}, r_{jk}, r_{ik})$ represents the covalent O–Si–O and Si–O–Si bonding interactions that are angle-dependent and functions of the Si–O distance. A detailed functional expression of the two-body term $U_2(r_{ij})$ consists of three parts and is written as

$$U_2(r_{ij}) = \frac{H_{ij}}{r_{ij}^{\eta_{ij}}} + \frac{Z_{i,\text{eff}} Z_{j,\text{eff}}}{r_{ij}} - \frac{\alpha_i Z_{j,\text{eff}}^2 + \alpha_j Z_{i,\text{eff}}^2}{2 r_{ij}^4} \exp\left(-\frac{r_{ij}}{r_{4s}}\right). \tag{2.26}$$

Here, H_{ij} and η_{ij} denote the strength and exponent of the steric repulsion due to the spatial arrangement of the atoms, respectively; $Z_{i,\text{eff}}$ and α_i denote the effective nuclear charge and electronic polarizability of the ith ion, respectively. The last term of Eq. (2.26) is used to describe the charge-dipole interactions, of which the exponential screening expression $\exp(-r_{ij}/r_{4s})$ provides a reasonable cutoff for such interactions. The decay length r_{4s} is taken to be a few angstroms so that the magnitude of the third term at the boundaries of the simulation cells is reduced to a small fraction of the magnitude of this term in the absence of $\exp(-r_{ij}/r_{4s})$.

The three-body term accounting for covalent bonds is written as

$$U_3 = B_{jik} f\left(r_{ij}, r_{ik}\right) p\left(\theta_{jik}, \theta_{jik}^0\right), \tag{2.27}$$

where B_{ijk} is a constant to denote the interaction strength, the function $f(r_{ij}, r_{ik})$ represents the effects of bond stretching, and $p(\theta_{jik}, \theta_{jik}^0)$ accounts for the bending of the bond angle. The expressions for $f(r_{ij}, r_{ik})$ and $p(\theta_{jik}, \theta^0_{jik})$ read as

$$f(r_{ij}, r_{ik}) = \begin{cases} \exp\left(\dfrac{l}{r_{ij} - r_c} + \dfrac{l}{r_{ik} - r_c}\right) & \text{for} \quad r_{ij} < r_c, \ r_{ik} < r_c \\ 0 & \text{for} \quad r_{ij} \geq r_c, \ r_{ik} \geq r_c \end{cases}, \tag{2.28}$$

$$p\left(\theta_{jik}, \theta^0_{jik}\right) = \left(\cos\theta_{jik} - \cos\theta^0_{jik}\right)^2, \tag{2.29}$$

where r_c is the cutoff distance for three-body interactions, and θ^0_{jik} is the equilibrium bond angle. The functional form of $f(r_{ij}, r_{ik})$ is chosen so that it cuts off at r_c without any discontinuity.

The Vashishta potential performs well for modeling pure SiO_2 and can provide an accurate prediction of its structure and properties, which are in good agreement with experimental results. However, the Vashishta potential and other potential functions with similar forms do not consider charge transfer among atoms when their local environments change. For instance, Si atoms possess no net charge in bulk Si but are positively charged in SiO_2. When bulk Si and SiO_2 are brought together to form an interface, a charge transfer occurs between Si atoms on both sides of the interface. Such a situation is not considered by the Vashishta potential.

One approach for considering such charge transfers was established by Streitz and Mintmire (1994). They formulated the electrostatic energy U^{ES} (a portion of the total interaction energy of an ionic system) of a group of interacting atoms as the sum of the atomic energy and the electrostatic interaction energy of all the atomic pairs:

$$U^{ES} = \sum_i E_i(q_i) + \frac{1}{2}\sum_{i \neq j} U_{ij}\left(\boldsymbol{r}_{ij}; q_i, q_j\right). \tag{2.30}$$

The first term $\sum_i E_i(q_i)$ accounts for the atomic energy, and the second term $U_{ij}(\boldsymbol{r}_{ij}; q_i, q_j)$ accounts for Coulombic interactions and is given by

$$U_{ij}\left(\boldsymbol{r}_{ij}; q_i, q_j\right) = \int d^3r_1 \int d^3r_2 \rho_i(\boldsymbol{r}_1, \boldsymbol{r}_i; q_i) \rho_j\left(\boldsymbol{r}_2, \boldsymbol{r}_j; q_j\right) / r_{12}, \tag{2.31}$$

where the bond vector $\boldsymbol{r}_{ij} = \boldsymbol{r}_i - \boldsymbol{r}_j$, and $\rho_i(\boldsymbol{r}, \boldsymbol{r}_i; q_i)$ represents the charge distribution (including the nuclear charge) of atom i with a partial charge q_i at position $\boldsymbol{r}_i$ and can be written as

$$\rho_i\left(\boldsymbol{r}, \boldsymbol{r}_i; q_i\right) = Z_{i,\text{eff}}\delta(\boldsymbol{r} - \boldsymbol{r}_i) + \left(q_i - Z_{i,\text{eff}}\right) f_i(\boldsymbol{r} - \boldsymbol{r}_i), \tag{2.32}$$

where $Z_{i,\text{eff}}$ is the effective nuclear charge of the ith atom, which satisfies the condition $0 < Z_{i,\text{eff}} < T_i$, with T_i being the total nuclear charge of the atom, whereas the function $f_i(\boldsymbol{r} - \boldsymbol{r}_i)$ describes the radial distribution of the remaining charge outside the core. For simplicity, $f_i(\boldsymbol{r}-\boldsymbol{r}_i)$ can adopt the functional form of the spherically symmetric Slater-type orbitals, which is given by

$$f_i(|\boldsymbol{r}-\boldsymbol{r}_i|) = \left(\xi_i^3/\pi\right)\exp\left(-2\xi_i|\boldsymbol{r}-\boldsymbol{r}_i|\right), \tag{2.33}$$

where ξ_i is a material-dependent constant. The exact partial charge q_i on atom i is solved at each site for each MD simulation step by performing an energy minimization of the electrostatic energy U^{ES}. Upon reaching the state of minimum energy, an electronegativity equilibration can be established in the modeled system. The residual atomic interactions, other than the electrostatic interactions, can be modeled in various ways, for example, by the EAM (embedded atom method) adopted by Streitz and Mintmire (1994).

The Shell model is also widely applied to model atomic interactions within an ionic system and has proved to be an efficient model for calculating its electrostatic interaction energy. In the Shell model, ions are split into a core and a massless shell, of which the latter is expected to react to the electrostatic environment. The ith core possesses a charge Q, and the shell possesses a charge q; they are connected by a harmonic spring. The charges of the core and shell add up to the total ion charge, and their values are calibrated to reproduce the mechanical and physical properties of the modeled system. The separation distance between the core and shell charges represents the electronic polarization. The electrostatic interaction between different ions is thus composed of four parts, namely, one shell–shell term, two shell–core terms, and one core–core term, and they are given by

$$\begin{aligned} U^{ES} = &\sum_{i<j}^{N}\frac{q_{i,s}q_{j,s}}{4\pi\varepsilon_0|\boldsymbol{r}_{i,s}-\boldsymbol{r}_{j,s}|} + \sum_{i<j}^{N}\frac{Q_{i,c}q_{j,s}}{4\pi\varepsilon_0|\boldsymbol{r}_{i,c}-\boldsymbol{r}_{j,s}|} \\ &+ \sum_{i<j}^{N}\frac{q_{i,s}Q_{j,c}}{4\pi\varepsilon_0|\boldsymbol{r}_{i,s}-\boldsymbol{r}_{j,c}|} + \sum_{i<j}^{N}\frac{Q_{i,c}Q_{j,c}}{4\pi\varepsilon_0|\boldsymbol{r}_{i,c}-\boldsymbol{r}_{j,c}|}, \end{aligned} \tag{2.34}$$

where $\boldsymbol{r}_{i,c}$ and $\boldsymbol{r}_{i,s}$ denote the positions of the ith core and shell, respectively. The total potential energy of the system is then obtained as the sum of the potential energy from the short-range repulsive interactions, the potential energy from the van der Waals interactions, and the electrostatic energy.

2.2.2 Metals

In metals, atoms participate in metallic bonding, which is essentially the attractive interactions between positively charged nuclei and the electrons that are loosely bound to them. A widely applied method of describing the interactions among metal atoms is to model the near-free electrons as an electron gas. The effective medium theory (EMT) assumes that the electron density of the electron gas is nonhomogeneous, and the interactions among the atoms are dependent on their local electronic environment. The functional form of the EMT potential is written as

$$U^{\text{tot}} = \sum_i F[n_i(\boldsymbol{r}_i)] + \frac{1}{2}\sum_{i,j} \Phi(\boldsymbol{r}_i - \boldsymbol{r}_j), \tag{2.35}$$

where F is a function of the local electron density $n_i(\boldsymbol{r}_i)$ at the atomic site $\boldsymbol{r}_i$, and $\Phi(\boldsymbol{r}_i - \boldsymbol{r}_j)$ is typically a pure two-body potential to model the nuclei. The local electron densities can be obtained by ab initio calculations.

As with the EMT method, the notion of the EAM potential is based on the idea of an electron gas (Daw and Baskes, 1984). The EAM potential is currently the most widely adopted potential for modeling both pure metals and alloys. The difference between the EMT potential and the EAM potential is that the latter has been developed semiempirically through fitting experimental data. The functional form of the EAM potential can be written as

$$U^{\text{tot}} = \sum_i F[\rho_i] + \frac{1}{2}\sum_{ij} V(r_{ij}), \tag{2.36}$$

which is highly similar to the functional form of the EMT potential. Here, F is the embedding function that represents the amount of energy required to place atom i within the electron gas; ρ_i is the electron density at the site of atom i and is obtained through the relation $\rho_i = \sum_{j,j\neq i} \rho_{ij}(r_{ij})$, with $\rho_{ij}(r_{ij})$ being the electron density of atom j at the site of atom i. $V(r_{ij})$ is a two-body potential that accounts for the interactions among the nuclei.

Through fitting experimental data, Daw et al. (1993) obtained the functional form of $V(r_{ij})$, which is written as

$$V(r_{ij}) = \frac{1}{4\pi\varepsilon_0}\frac{Z_{i,\text{eff}}Z_{j,\text{eff}}}{r_{ij}}. \tag{2.37}$$

The electron density ρ_i used in the EAM potential is obtained as a set of data points from ab initio calculations.

2.2.3 Covalent materials

Covalent bonds are formed by electrons localized between nuclei. The localized electrons tend to repel one another, leading to the formation of atomic structures that maximizes the distance between the bonds by orientating them to reach optimal angles. For instance, in the case of a material formed by atoms with four equal bonds, a diamond structure can be expected. Each atom remains in the center of a regular tetrahedron with a bond angle of $\cos^{-1}(1/3) = 109.47^\circ$. Due to the structural preference for certain bond angles, the total potential energy of a covalent system depends not only on the interatomic distances and bond lengths, but also on the bond angles.

Semiconductors and insulators are typical examples of covalent materials that compose a large portion of engineering materials. The SW potential is one of the pioneering attempts to model a semiconductor, specifically bulk Si, with a classical model. Initially, the SW potential was constructed to predict the melting point of bulk Si, but it turned out to be capable of providing accurate descriptions of its point-defect energies and surface properties (Stillinger and Weber, 1985). After years of development, the SW potential has become one of the most widely applied potentials for modeling covalent materials. The SW potential consists of a two-body term and a three-body term and is expressed as

$$U = \sum_{i \neq j} V_2(r_{ij}) + \sum_{i \neq j \neq k} V_3\left(r_{ij}, r_{jk}, r_{ik}\right), \tag{2.38}$$

$$V_2(r_{ij}) = \varepsilon f_2\left(r_{ij}/\sigma\right), \tag{2.39}$$

$$V_3(r_{ij}, r_{jk}, r_{ik}) = \varepsilon f_3(r_i/\sigma, r_j/\sigma, r_k/\sigma), \tag{2.40}$$

where ε and σ are scaling factors in units of energy and distance, respectively, and the functions $f_2(r)$ and $f_3(r_i, r_j, r_k)$ are defined as

$$f_2(r) = \begin{cases} A(Br^{-p} - r^{-q})\exp\left[(r-a)^{-1}\right] & , \quad r < a \\ 0 & , \quad r \geq a \end{cases}, \tag{2.41}$$

$$f_3\left(r_i, r_j, r_k\right) = h\left(r_{ij}, r_{ik}, \theta_{jik}\right) + h\left(r_{ji}, r_{jk}, \theta_{ijk}\right) + h\left(r_{ki}, r_{kj}, \theta_{ikj}\right), \tag{2.42}$$

with

$$h(r_{ij}, r_{ik}, \theta_{jik}) = \begin{cases} \lambda \exp\left[\gamma(r_{ij}-a)^{-1} + \gamma(r_{ik}-a)^{-1}\right](\cos\theta_{jik} + \frac{1}{3})^2, \\ \qquad r_{ij} < a \quad \text{and} \quad r_{ik} < a \\ 0 \quad , \quad \text{else} \end{cases}, \tag{2.43}$$

where θ_{jik} is the angle formed by the *ij* bond and the *ik* bond. It is apparent that the SW potential prefers an atomic configuration for which $\cos\theta_{jik} = -1/3$ and the bond angles are as close as possible to those in a diamond-like tetrahedral structure, regardless of the material modeled. In addition, the constants A, B, p, q, a, λ, and γ are obtained via empirical fitting so that the diamond structure is the most stable one. As a result, bulk Si modeled by the SW potential exhibits the diamond structure instead of a compact structure.

The SW potential provides a fairly accurate description of crystalline Si. However, the inherent bias toward tetrahedral structures leads to severe transferability issues. When the actual conditions deviate from those that the SW potential was designed for, it instantly loses credibility. For instance, the SW potential cannot predict the properties of bulk Si when the phase transformation occurs under high pressure or temperature. This undesirable situation has led to the development of new potential functions of which the local environment of covalent bonds is taken into account. Nevertheless, potentials developed based on the functional form of the SW potential (using only geometrical quantities such as bond lengths and bond angles) are still widely used today for modeling organic systems. Although these potentials have limited transferability and are unsuitable for modeling changes in molecular topology or simulating materials under high pressures or temperatures, they are still capable of providing precise information regarding the structural and dynamic properties of a wide range of materials.

The Tersoff potential is a well-known and widely applied potential for modeling covalent systems, and various versions of this potential have been reported so far (Tersoff, 1986, 1988a,b,c, 1989). These variants consider the local environment of bonds or, more precisely, the effect of the bond order that approximates the number of chemical bonds between a pair of atoms. This number can be regarded as a measurement of the strength of the bond and varies as the bond length and bond angle change. The analytical form of the Tersoff potential is written as

$$U = \frac{1}{2} f_c(r_{ij}) \left[\sum_{i \neq j} f_R(r_{ij}) + \sum_{i \neq j \neq k} b_{ij} f_A(r_{ij}) \right], \tag{2.44}$$

$$f_c(r_{ij}) = \begin{cases} 1, & r_{ij} < R - D \\ \frac{1}{2}\left\{1 - \sin\left[\frac{\pi}{2}\frac{(r_{ij} - R)}{D}\right]\right\}, & R - D < r_{ij} < R + D, \\ 0, & r_{ij} > R + D \end{cases} \tag{2.45}$$

$$f_R(r_{ij}) = A \exp\left(-\lambda_1 r_{ij}\right), \tag{2.46}$$

$$f_A(r_{ij}) = -B \exp\left(-\lambda_2 r_{ij}\right), \tag{2.47}$$

$$b_{ij} = \left(1 + \beta^n \zeta_{ij}^n\right)^{-\frac{1}{2n}}, \tag{2.48}$$

$$\zeta_{ij} = \sum_{k \neq i,j} f_c(r_{ik}) g(\theta_{ijk}) \exp\left[\lambda_3^3 (r_{ij} - r_{ik})^3\right], \tag{2.49}$$

$$g(\theta_{ijk}) = 1 + \frac{c^2}{d^2} - \frac{c^2}{\left[d^2 + (\cos\theta_{ijk} - \cos\theta_0)^2\right]}, \tag{2.50}$$

where $f_c(r_{ij})$ is the cutoff function ensuring that only the nearest-neighbor interactions are considered, with both R and D being the cutoff distance parameters. The functions $f_R(r_{ij})$ and $f_A(r_{ij})$ represent repulsive and attractive atomic interactions, respectively. The parameters A and B denote the cohesive energies for the repulsive and attractive interactions, respectively, and λ_1 and λ_2 denote lattice parameters. The other symbols in Eqs. (2.44)–(2.50) that are unaccounted for are material-dependent constants.

The variable b_{ij} represents the bond order and is determined by the local coordination number of the ith atom. The reason for introducing b_{ij} is that in actual systems, the bond ij is weakened by the presence of other bonds involving atom i (e.g., bond ik). The degree of weakening depends on the positions and orientations of these bonds. In the expression of b_{ij}, ζ_{ij} denotes the effective coordination number of the atoms, i.e., the number of the nearest neighbors, and considers the bond angle θ_{ijk} and the relative distances $r_{ij} - r_{ik}$; β and n are the adjusting parameters selected to describe the lattice structure and physical properties of a specific material; λ_3 is a cutoff parameter that determines the range for the calculation of ζ_{ij}. In the expression of the bond angle distribution function $g(\theta_{ijk})$, θ_0 is the equilibrium bond angle and c and d are dimensionless parameters that determine the extent of the bond angle effect on the covalent bond.

The Tersoff potential was initially calibrated for Si systems and later for C systems and was tested by calculating the cohesive energy, elastic constants, phonon frequencies, defect energy, and migration barriers. The results obtained with this potential are in good agreement with the results of experiments and ab initio calculations. So far, the Tersoff potential has been developed to accommodate various systems, including Si–C hybrids, Si–O oxides, and several other semiconductor systems such as bulk Ge and bulk Ga. However, the Tersoff potential also has several limitations, with the most severe limitation being the difficulty encountered in fitting the parameters for all seven functions in Eqs. (2.44)–(2.50).

Brenner et al. (2002) refined the Tersoff potential to model hydrocarbons, and this refined potential is now frequently referred to as the reactive empirical bond order (REBO) potential. The REBO potential can model interatomic interactions within a variety of small hydrocarbon molecules as well as graphite and diamond. It is a highly parameterized expression of the Tersoff potential with additional terms that counteract an inherent overbinding of radicals and include nonlocal effects. In the REBO potential, the bond order term b_{ij} incorporates the angular function $g(\theta_{ijk})$ and is given by

$$b_{ij} = \left(1+\chi_{ij}\right)^{-\frac{1}{2}}, \tag{2.51}$$

$$\chi_{ij} = \sum_{k\neq i,j} f_{ik}(r_{ik})g_{ik}(\theta_{ijk})\exp\left[2\mu_{ik}\left(r_{ij}-r_{ik}\right)\right], \tag{2.52}$$

$$g(\theta_{ijk}) = \gamma\left(1+\frac{c^2}{d^2}-\frac{c^2}{\left[d^2+\left(1+\cos\theta_{ijk}\right)^2\right]}\right). \tag{2.53}$$

The repulsive and attractive components of the potential are given respectively by

$$f_{\mathrm{R}}(r_{ij}) = \frac{D_0}{S-1}\exp\left[-\beta\sqrt{2S}(r_{ij}-r_0)\right], \tag{2.54}$$

$$f_{\mathrm{A}}(r_{ij}) = \frac{SD_0}{S-1}\exp\left[-\beta\sqrt{2/S}(r_{ij}-r_0)\right], \tag{2.55}$$

where D_0 denotes the dimer binding energy and r_0 is the equilibrium distance. Other symbols in Eqs. (2.51)–(2.55) that are unaccounted for are material-dependent constants. The cutoff function $f_c(r_{ij})$ that restricts the two-body potential to the nearest neighbors is given by

$$f_c(r_{ij}) = \begin{cases} 1, & r_{ij} < R^{(1)} \\ \frac{1}{2}\left\{1+\cos\left[\frac{\pi\left(r_{ij}-R^{(1)}\right)}{R^{(2)}-R^{(1)}}\right]\right\}, & R^{(1)} < r_{ij} < R^{(2)}, \\ 0, & R^{(2)} < r_{ij} \end{cases} \tag{2.56}$$

where both $R^{(1)}$ and $R^{(2)}$ are cutoff distance parameters. When S equals 2, the two-body terms $f_{\mathrm{R}}(r_{ij})$ and $f_{\mathrm{A}}(r_{ij})$ are reduced to the usual Morse potential that explicitly includes the effect of bond breaking and accounts for the anharmonicity of actual bonds.

One limitation of the REBO potential is that it is highly short-ranged and excludes the covalent-bonding interactions between non-neighboring atoms, causing difficulties in modeling intermolecular interactions and interactions between surfaces. The adaptive inter-atomic REBO (AIREBO) potential (Stuart et al., 2000) is an optimized version of the REBO potential, and it adds torsion, dispersion, and nonbonded repulsion interaction terms to the original REBO potential via an adaptive method. Having overcome the limitations of its predecessor in studying systems involving intermolecular interactions, the AIREBO potential enables the study of reactivity in molecular condensed phases.

The AIREBO potential can be expressed as

$$U = \frac{1}{2}\sum_{i}\sum_{j\neq i}\left[V_{ij}^{\mathrm{REBO}} + V_{ij}^{\mathrm{LJ}} + \sum_{k\neq i,j}\sum_{l\neq i,j,k} V_{ijkl}^{\mathrm{Torsion}}\right]. \tag{2.57}$$

The first term V_{ij}^{REBO} is the basic term of the entire potential, with an expression similar to that of the original REBO potential. It describes the short-range interactions between two covalently bonded atoms. The second term V_{ij}^{LJ} describes the long-range interactions between nonbonded atoms and shares the same expression as the LJ potential. The third term $V_{ij}^{\mathrm{Torsion}}$ describes the effects of bond rotations caused by changes in the bond angles.

Another potential, the modified EAM (MEAM) potential, is also frequently adopted to model covalent materials. The MEAM potential was developed by Baskes (1987), of which more distant neighbors and angle-dependent electron densities are included in the functional form of the EAM potential. For the MEAM potential, the total energy of a solid is given by the same function in Eq. (2.36), but the local electron density is given by

$$\rho_i = \sum_{j,j\neq i}\rho_{ij}(r_{ij}) - a\sum_{j\neq i,k\neq i}\left(1 - 3\cos^2\theta_{jik}\right)\rho_{ij}(r_{ij})\rho_{ik}(r_{ik}), \tag{2.58}$$

where a is a constant determined by fitting measurements of the shear modulus of the solid of interest, and θ_{jik} denotes the angle between atomic bonds ij and ik. The electron density is described by two terms: the linear superposition of the electron densities from the neighboring atoms (similar to the first term of the EAM potential) and an angular correction term. The specific angular form for the electron density is chosen to minimize the total energy of the system under a tetrahedral distortion of the electron density. Such an angular form vanishes in cases of ideal cubic structures. The initial MEAM potential was targeted at modeling Si, and its later versions (Baskes, 1992; Baskes et al., 1989) were extended to model Si, Ge, and their alloys.

2.2.4 Molecular systems

Compared with ionic, covalent, and metallic materials that usually exhibit well-defined crystalline structures, molecular systems, such as polymers, proteins, and nucleic acids, generally contain complex molecules, and their relaxed geometrical configurations are difficult to determine. Such systems are typically modeled by classical mechanics, and several popular force fields have been developed, each possessing a unique advantage and applicability. A common characteristic of these force fields is their wide applicability to a large range of materials.

According to the complexities of the force fields for modeling molecular systems, they can be classified into three classes. For the first-class force fields, harmonic functions are used to describe the bond stretching, bond angle bending, and bond torsion, and their coupling effects are not counted. Nonbonding interactions are described by the LJ potential. For the second-class force fields, cubic and biquadratic functions are added to describe the bond stretching, bond angle bending, and bond torsion, with their coupling effects being considered. Exponential-type functions are added to describe nonbonding interactions. In the third-class force fields, hyperconjugation and polarization effects are further included.

Compared with the first-class force field, the second-class force fields (e.g., consistent force field, PCFF, and MM), and the third-class force fields [e.g., AMOEBA (atomic multipole optimized energetics for biomolecular applications) and polarizable intermolecular potential function] are complex and require numerous force constants. Obtaining these force constants requires not only large quantities of experimental data but also accurate quantum calculation results. The purpose of developing second-class and third-class force fields is to obtain more accurate simulation results of certain molecular properties, including dynamic spectra, thermodynamic characteristics, and crystal information.

The AMBER force field is a widely applied first-class force field for modeling biomolecules such as small proteins, nucleic acids, and polysaccharose. The functional form of the AMBER force field is given by

$$
\begin{aligned}
U = {} & \sum\nolimits_b k_b(l-l_0)^2 + \sum\nolimits_\theta k_\theta(\theta-\theta_0)^2 + \sum\nolimits_\phi \frac{1}{2} V_0\left[1+\cos(n\phi-\phi_0)\right] \\
& + \sum \varepsilon\left[\left(\frac{r^*}{r_{ij}}\right)^{12} - 2\left(\frac{r^*}{r_{ij}}\right)^{6}\right] + \sum \frac{q_i q_j}{\varepsilon_{ij} r_{ij}} + \sum\left[\frac{C_{ij}}{r_{ij}^{12}} - \frac{D_{ij}}{r_{ij}^{6}}\right].
\end{aligned}
\tag{2.59}
$$

One can clearly identify the first five terms on the right side as the bond stretching, bond angle bending, bond angle torsion, van der Waals interaction, and electrostatic interaction terms. The last term accounts for hydrogen bond interactions. No out-of-plane or cross terms are included in the AMBER force field, and thus it has a higher efficiency than the MM force field.

The CHARMM force field is another widely applied first-class force field for the simulation of molecular systems, including organic molecules, solutions, polymers, and biomolecules. Compared with the AMBER force field, the CHARMM force field consists of out-of-plane terms and a relatively sophisticated term describing hydrogen bonding. Other force fields similar to AMBER and CHARMM include the CVFF force field. The CVFF force field was initially developed to model biomolecules, particularly, amino acids, water, and various functional groups (Dauber-Osguthorpe et al., 1988). After further improvements, it can now be adopted to model a wide range of polypeptides, proteins, and various organic molecules. Similarly to the MM force field, the CVFF force field considers the coupling effects among bond stretching, bond angle bending, and bond torsion; it is thus advantageous for providing reasonable predictions of system configurations, cohesive energies, and vibration frequencies.

The MM force field is a typical second-class force field designated for modeling molecular systems. Different versions of the MM force field (e.g., MM2, MM3, MM4, and MM +) were successively developed to extend its applicability to a wider range of organic compounds (Allinger and Flanagan, 1983; Allinger and Sprague, 1973; Allinger et al., 1990a,b, 1994, 1996). The MM force field is capable of providing accurate predictions of the structural configurations, configuration energies, dynamic properties, and vibration spectra of a molecular system. The functional form of the MM force field reads as

$$U = U_{nb} + U_b + U_\theta + U_\phi + U_\chi + U_{el} + U_{cross}. \quad (2.60)$$

The first term on the right-hand side represents the nonbonded van der Waals interactions and is expressed in the form of the Buckingham potential as

$$U_{nb} = ac \exp\left(-\frac{c\sigma}{r}\right) - b\varepsilon\left(\frac{\sigma}{r}\right)^6. \quad (2.61)$$

The bond stretching, bond angle bending, and bond torsion terms are given, respectively, by

$$U_b = \frac{k_b}{2}(r-r_0)^2\left\{1 - k'_b(r-r_0) - k''_b(r-r_0)^2 - k'''_b(r-r_0)^3\right\}, \quad (2.62)$$

$$U_\theta = \frac{k_\theta}{2}(\theta-\theta_0)^2\left\{1 - k'_\theta(\theta-\theta_0) - k''_\theta(\theta-\theta_0)^2 - k'''_\theta(\theta-\theta_0)^3\right\}, \quad (2.63)$$

$$U_\phi = \sum \frac{V_n}{2}\left(1 + \cos(n\phi - \phi_0)\right). \quad (2.64)$$

In Eqs. (2.61)–(2.63), the various terms k_b, k'_b, k''_b, k'''_b, k_θ, k'_θ, k''_θ, and k'''_θ are all force constants, r denotes the bond length, r_0 the equilibrium bond length, θ the bond angle, θ_0 the equilibrium bond angle; ϕ, ϕ_0, n, and V_n have the same physical meaning as those in Eq. (2.20). For a system of four atoms, there will be an energy loss if one of the atoms is moved out of the plane formed by the other three atoms, as shown in Fig. 2.4.

The term U_χ accounts for the out-of-plane bending interactions that are closely related to the bond torsion term and is expressed as

$$U_\chi = k_\chi(1 - \cos 2\chi), \tag{2.65}$$

where k_χ denotes the force constant and χ the dihedral angle formed by the out-of-plane bond with the plane. The term U_{el} accounts for the Coulombic interactions, and U_{cross} accounts for the coupling effects among the bond stretching, bond angle bending, and bond torsion terms since they are interdependent. For instance, the coupling effect of the stretching of two bonds in connection is described by

$$U(r_1, r_2) = \frac{1}{2}k_{12}(r_1 - r_{1,0})(r_1 - r_{2,0}), \tag{2.66}$$

where $r_{1,0}$ and $r_{2,0}$ denote the lengths of the two bonds in the equilibrium state and k_{12} is a constant. The coupling effect between the bending of a bond angle and the stretching of its two constitutive bonds is expressed by

$$U(r_1, r_2, \theta) = \frac{1}{2}k_{12\theta}\left[(r_1 - r_{1,0}) + (r_1 - r_{2,0})\right](\theta - \theta_0), \tag{2.67}$$

where θ_0 denotes the equilibrium bond angle and $k_{12\theta}$ is a constant. Finally, the coupling effect between the bond torsion angle and the stretching of one of its constitutive bonds is given by

$$U(r, \phi) = k(r - r_0)\cos(n\phi). \tag{2.68}$$

The MM force field fully considers the coupling effects among bond stretching, bond angle bending, and bond torsion terms, leading to better simulation results than other force fields. However, the complexity of the MM force field results in challenges with respect to its programming and computing efficiency.

References

Allinger, N.L., Flanagan, H.L., 1983. Isotope effects in molecular mechanics (MM2). Calculations on deuterium compounds. J. Comput. Chem. 4, 399–403.

Allinger, N.L., Sprague, J.T., 1973. Calculation of the structures of hydrocarbons containing delocalized electronic systems by the molecular mechanics method. J. Am. Chem. Soc. 95, 3893–3907.

Allinger, N.L., Chen, K., Lii, J.-H., 1996. An improved force field (MM4) for saturated hydrocarbons. J. Comput. Chem. 17, 642–668.
Allinger, N.L., Li, F., Yan, L., 1990a. Molecular mechanics. The MM3 force field for alkenes. J. Comput. Chem. 11, 848–867.
Allinger, N.L., Rahman, M., Lii, J.-H., 1990b. A molecular mechanics force field (MM3) for alcohols and ethers. J. Am. Chem. Soc. 112, 8293–8307.
Allinger, N.L., Zhou, X., Bergsma, J., 1994. Molecular mechanics parameters. J. Mol. Structure: THEOCHEM 312, 69–83.
Baskes, M.I., 1987. Application of the embedded-atom method to covalent materials: a semiempirical potential for silicon. Phys. Rev. Lett. 59, 2666–2669.
Baskes, M.I., 1992. Modified embedded-atom potentials for cubic materials and impurities. Phys. Rev. B 46, 2727–2742.
Baskes, M.I., Nelson, J.S., Wright, A.F., 1989. Semiempirical modified embedded-atom potentials for silicon and germanium. Phys. Rev. B 40, 6085–6100.
Brenner, D.W., Shenderova, O.A., Harrison, J.A., Stuart, S.J., Ni, B., Sinnott, S.B., 2002. A second-generation reactive empirical bond order (REBO) potential energy expression for hydrocarbons. J. Phys. Condens. Matter 14, 783–802.
Brooks, B.R., Bruccoleri, R.E., Olafson, B.D., States, D.J., Swaminathan, S., Karplus, M., 1983. CHARMM: a program for macromolecular energy, minimization, and dynamics calculations. J. Comput. Chem. 4, 187–217.
Buckingham, R.A., 1938. The classical equation of state of gaseous helium, neon and argon. Proc. R. Soc. London. Ser. A. Math. Phys. Sci. 168, 264–283.
Cornell, W.D., Cieplak, P., Bayly, C.I., Gould, I.R., Merz, K.M., Ferguson, D.M., et al., 1995. A second generation force field for the simulation of proteins, nucleic acids, and organic molecules. J. Am. Chem. Soc. 117, 5179–5197.
Dauber-Osguthorpe, P., Roberts, V.A., Osguthorpe, D.J., Wolff, J., Genest, M., Hagler, A.T., 1988. Structure and energetics of ligand binding to proteins: *Escherichia coli* dihydrofolate reductase-trimethoprim, a drug-receptor system. Proteins 4, 31–47.
Daw, M.S., Baskes, M.I., 1984. Embedded-atom method: derivation and application to impurities, surfaces, and other defects in metals. Phys. Rev. B 29, 6443–6453.
Daw, M.S., Foiles, S.M., Baskes, M.I., 1993. The embedded-atom method: a review of theory and applications. Mater. Sci. Rep. 9, 251–310.
Huggins, M.L., Mayer, J.E., 1933. Interatomic distances in crystals of the alkali halides. J. Chem. Phys. 1, 643–646.
Jones, J.E., Chapman, S., 1924a. On the determination of molecular fields-I. From the variation of the viscosity of a gas with temperature. Proc. R. Soc. London. Ser. A 106, 441–462.
Jones, J.E., Chapman, S., 1924b. On the determination of molecular fields-II. From the equation of state of a gas. Proc. R. Soc. London. Ser. A 106, 463–477.
Mayo, S.L., Olafson, B.D., Goddard, W.A., 1990. Dreiding: a generic force field for molecular simulations. J. Phys. Chem. 94, 8897–8909.
Morse, P.M., 1929. Diatomic molecules according to the wave mechanics. II. Vibrational levels. Phys. Rev. 34, 57–64.
Rappe, A.K., Casewit, C.J., Colwell, K.S., Goddard III, W.A., Skiff, W.M., 1992. UFF, a full periodic table force field for molecular mechanics and molecular dynamics simulations. J. Am. Chem. Soc. 114, 10024–10035.
Shi, S., Yan, L., Yang, Y., Fisher-Shaulsky, J., Thacher, T., 2003. An extensible and systematic force field, ESFF, for molecular modeling of organic, inorganic, and organometallic systems. J. Comput. Chem. 24, 1059–1076.
Stillinger, F.H., Weber, T.A., 1985. Computer simulation of local order in condensed phases of silicon. Phys. Rev. B 31, 5262–5271.

Streitz, F.H., Mintmire, J.W., 1994. Electrostatic potentials for metal-oxide surfaces and interfaces. Phys. Rev. B 50, 11996–12003.

Stuart, S.J., Tutein, A.B., Harrison, J.A., 2000. A reactive potential for hydrocarbons with intermolecular interactions. J. Chem. Phys. 112, 6472–6486.

Tersoff, J., 1988a. Empirical interatomic potential for silicon with improved elastic properties. Phys. Rev. B 38, 9902–9905.

Tersoff, J., 1988b. New empirical approach for the structure and energy of covalent systems. Phys. Rev. B 37, 6991–7000.

Tersoff, J., 1989. Modeling solid-state chemistry: interatomic potentials for multicomponent systems. Phys. Rev. B 39, 5566–5568.

Tersoff, J., 1986. New empirical model for the structural properties of silicon. Phys. Rev. Lett. 56, 632–635.

Tersoff, J., 1988c. Empirical interatomic potential for carbon, with applications to amorphous carbon. Phys. Rev. Lett. 61, 2879–2882.

Vashishta, P., Kalia, R.K., Rino, J.P., Ebbsjö, I., 1990. Interaction potential for SiO_2: a molecular-dynamics study of structural correlations. Phys. Rev. B 41, 12197–12209.

CHAPTER

3

Control techniques of molecular dynamics simulation

This chapter presents the main control techniques involved in a classical molecular dynamics (MD) simulation. It begins with the classification of three main types of constraints applied in MD simulations for proper dynamic process control. The thermodynamic constraint is first introduced by highlighting four statistical ensembles that are frequently adopted in MD simulations. A variety of thermostat and barostat algorithms are then presented, which are indispensable for realizing these ensembles. The thermostat algorithms include a thermostat based on simple velocity rescaling, the Gaussian thermostat, Berendsen thermostat, Andersen thermostat, Langevin thermostat, and Nosé–Hoover thermostat, while the barostat algorithms include the Berendsen barostat, Andersen barostat, Parrinello–Rahman barostat, and Martyna–Tuckerman–Tobias–Klein (MTTK) barostat. Afterward, general spatial boundary conditions, including the repulsive boundary condition, fixed boundary condition, and semi-rigid atomistic boundary condition, are illustrated. Finally, a typical geometrical constraint, i.e., the rigid bond constraint, which accelerates the simulation of molecules involving light atoms, is discussed.

3.1 Types of constraints in molecular dynamics simulation

Owing to the high resolution provided in describing space, time, and energy, MD simulation presents itself as a powerful complement to experimental techniques and can provide a fundamental understanding of experimentally observed phenomena. To realize the practical conditions under which experiments are conducted, certain constraints are imposed on the corresponding MD simulation systems. There are three main types of constraints applied in MD simulations for proper dynamic process control.

Molecular Dynamics Simulation
DOI: https://doi.org/10.1016/B978-0-12-816419-8.00008-8

The first type of simulation constraint is the *thermodynamic constraint*, which is limited to a set of three thermodynamic quantities characterizing the macroscopic state of an MD system. Generally, this set of quantities is constructed by choosing one quantity from each of the three following pairs that consists of one extensive and one intensive quantity: (1) the number of atoms N and chemical potential μ, (2) volume V and pressure P, (3) energy E and temperature T. Together with their reference values, the selected set of quantities define the thermodynamic ensemble that is sampled during an MD simulation. By default, the number of atoms N, volume V, and energy E of a system are kept constant in a standard MD simulation, i.e., the system can be described by the microcanonical ensemble. Applying specific modifications or constraints to the Hamiltonian or Newtonian equations of motion of the atoms renders it possible to keep other thermodynamic quantities of the system constant. This method allows the MD simulation program to generate and sample other kinds of thermodynamic ensembles for the system besides the conventional microcanonical ensemble.

The second type of simulation constraint is the *spatial boundary condition* that defines the shape of the simulated system and the nature of its environment. Typical spatial boundary conditions include periodic boundary conditions (PBCs), vacuum boundary conditions, fixed boundary conditions, repulsive boundary conditions, and semirigid atomic boundary conditions.

The third type of simulation constraint is the *geometrical constraint* imposed on the molecular level. A typical example would be fixing the bond lengths of a single molecule for a better approximation of the quantum-mechanical features of its high-frequency bond vibrations. In this chapter, the three types of constraints discussed above are introduced with a focus on temperature and pressure control techniques and their respective characteristics.

3.2 Thermodynamic ensembles

In mathematical physics and statistical mechanics, an ensemble is a collection of a large number of virtual copies of a system under the same macroscopic conditions. Each copy represents a possible state that the system may be in and is independent of one another. The ensemble can be understood to be analogous to a collection of numerous repetitions of an experiment under the same macroscopic conditions but with its microscopic details unconstrained. The ensemble is a concept introduced in statistical mechanics to better describe the statistical behavior of a thermodynamic system.

A possible microscopic state of a system corresponds to a point in its phase space. With time, the state of a system changes, and the system moves in its phase space accordingly, forming a trajectory. Over a sufficiently long time, the system will experience all the possible states in its phase space, as stated by the ergodic hypothesis. Although a practical system evolves, its corresponding ensemble does not necessarily have to evolve if it contains all the possible states of the system. Such an invariant statistical ensemble is deemed stationary and can be considered to be in statistical equilibrium. In many cases, it is possible to obtain the thermodynamic quantities of a system of interest by explicitly averaging these quantities across its entire thermodynamic ensemble.

Different macroscopic constraints correspond to different types of statistical ensembles with distinct statistical characteristics. Common statistical ensembles involved in MD simulations include the microcanonical (NVE) ensemble, canonical (NVT) ensemble, constant-pressure and constant-temperature (or isothermal-isobaric, NPT) ensemble, constant-pressure and constant-enthalpy (or isoenthalpic-isobaric, NPH) ensemble, and grand canonical (μVT) ensemble. The first three ensembles are the most frequently adopted.

In the NVE ensemble, the number of atoms N, system volume V, and total energy E are kept constant. Under these conditions, the system temperature T and pressure P may fluctuate around an average value. The system is isolated from its environment, and neither energy nor particle exchange occurs between them, thereby ensuring that statistical equilibrium is maintained.

In the NVT ensemble, the number of atoms N, system volume V, and temperature T are kept constant, while the total energy E and pressure P are not specified and hence fluctuate around their respective mean values. The NVT ensemble can be adopted to describe a closed system that is allowed to exchange thermal energy with its surrounding environment. To achieve statistical equilibrium for the NVT ensemble, the exchange of particles between the simulation system and its environment is not allowed.

In the NPT ensemble, the number of atoms N, system pressure P, and temperature T are kept constant. The simulated system is coupled with an external thermostat and barostat, but particle exchange with its environment is not allowed. The NPT ensemble closely corresponds to a system subjected to ambient temperature and pressure, for example, a chemical reaction occurring under typical laboratory conditions.

In the NPH ensemble, the number of atoms N, pressure P, and enthalpy H are kept constant. The NPH ensemble is an analog of the NVE ensemble, in which the parameter of the total energy E is replaced by enthalpy H. The pressure is kept constant without any temperature control. Since enthalpy H equals $E + PV$ and depends on both system pressure P and volume V, it is more complex to maintain a constant enthalpy. As a result, the NPH ensemble is seldom used in modern MD simulations except for several special cases.

In the μVT ensemble, the chemical potential μ, volume V, and temperature T of the system are invariant, while the total energy E, pressure P, and number of atoms N change over time. The system is open and can thus exchange energy and particles with its environment. Hence, contact between the system and an external thermostat, a barostat, chemical reservoir, or electrical reservoir is allowed, provided that the internal temperature and chemical potential are balanced with their external counterparts.

The selection of an appropriate ensemble for a specific MD simulation system (or one of its subsystems) is dictated by the conditions under which the simulation is performed. A basic rule of thumb is that the selected ensemble should ensure a representative sample of the microscopic states to make valid statistical arguments regarding the system of interest. The μVT ensemble, in principle, has the widest range of applicability due to its open-system characteristic. However, it possesses a larger number of variables or degrees of freedom than the other ensembles, which leads to a higher level of mathematical complexity. Most MD simulations are performed in the NVE, NVT, or NPT ensemble. Two or three ensembles can be adopted successively for one MD simulation system or be applied separately to its subsystems simultaneously. For instance, the NPT ensemble can be utilized first to relax an MD simulation system and reduce any initial stresses that are possibly induced by lattice mismatches or incorrect settings of the lattice parameters. Subsequently, the NPT ensemble is replaced by the NVE (or NVT) ensemble for a standard run, in which the total energy and system volume (or the system volume and temperature) must be conserved. When the thermal conductivity of a material is calculated using a nonequilibrium MD simulation (which is discussed in Chapter 6: Application of Molecular Dynamics Simulation in Thermal Problems), the NVT ensemble is adopted to control the temperatures of certain regions of the material, while the NVE is adopted for the other regions without temperature control.

3.3 Temperature control

In a standard MD simulation, the motion of N individual atoms enclosed in a specified volume V is determined by the classical Newton's laws of motion. Without any external constraints, the energy of the system is conserved, and the distribution of microscopic states corresponds to the NVE ensemble. Unfortunately, the microcanonical ensemble does not resemble the constant-temperature condition under which most actual experiments are conducted. As a result, the NVT, NPT, and μVT ensembles are more frequently adopted to keep the system temperature constant. Besides achieving a better representation of experimental conditions, there are other motivations for temperature control. For instance, the study of temperature-dependent properties or processes, such as thermal expansion coefficients, heat-induced conformational transformations of polymers, and phase transitions of metals, requires MD simulations to be conducted at specific temperatures. Moreover, to prevent the simulated system from unrealistic heating due to the energy drift caused by the accumulation of numerical errors over a long simulation period, proper temperature control is necessary. In some cases, particularly for which a statistical sampling in the full phase space is required, conducting an MD simulation at a specific temperature can improve the sampling efficiency.

Temperature control is achieved by coupling the simulated system with a so-called "thermostat" that adds or removes energy from the system by either directly altering the motion of atoms or modifying their Newtonian equations of motion. A variety of thermostat algorithms are currently available to achieve effective temperature control, and they can be categorized in several ways. For instance, the nature of these thermostat algorithms can be either stochastic or deterministic, depending on whether random processes/numbers are adopted to alter the atomic dynamics. They can also be classified as global or local, depending on whether they are applied to the entire system or each atom/molecule individually. In the following sections, various popular and historic thermostats are discussed.

3.3.1 Thermostat based on simple velocity rescaling

The simple velocity rescaling method is the earliest and most straightforward thermostat algorithm. However, it is also the most "unphysical" thermostat as it prohibits temperature fluctuation. This method directly alters the atomic dynamics described by Newton's laws of motion because velocity rescaling is conducted after the atomic positions and momenta are updated by a time-integrator.

The temperature of an MD simulation system is correlated with its kinetic energy by $< v_{i,\alpha}^2 > = k_B T/m_i$ [refer to Eq. (1.13)], in which the angle brackets denote the time average. If the brackets are removed, i.e., the temperature is calculated at a single timestep at time t instead of being averaged over a sufficiently long time interval, the obtained result is known as the instantaneous temperature $T(t)$. The instantaneous temperature is rarely equal to the target temperature T but always fluctuates around it. In the velocity rescaling method, to keep the instantaneous temperature $T(t)$ constant, the atomic velocities $v_{i,\alpha}$ are rescaled after each update by the time-integrator according to the following equation:

$$v'_{i,\alpha} = \sqrt{\frac{T}{T(t)}}v_{i,\alpha}. \tag{3.1}$$

Although the velocity rescaling method restricts the system temperature to a preset value, it leads to severe discontinuities in the atomic momenta, and thus the dynamics of the system are critically interrupted. It can be proven that the simple velocity rescaling method cannot reproduce the canonical ensemble but only an isokinetic ensemble in which the kinetic energy of the system is conserved. Hence, this method is usually adopted for a quick system equilibration at the desired temperature. Once the target temperature is obtained, the velocity rescaling ceases, and more sophisticated thermostats capable of reproducing the canonical ensemble are applied to investigate the dynamic behaviors of the system.

3.3.2 Gaussian thermostat

The Gaussian thermostat can be regarded as an extension of the simple velocity rescaling method by implementing a constraint in the Newtonian equations of motion of the atoms/molecules to keep the system temperature fixed. It is named the Gaussian thermostat because it applies the Gaussian principle of least constraint, which states that a force applied to restrict a particle's motion on a constraint hypersurface should be normal to the surface in a realistic dynamic setting. The objective of applying the Gaussian principle is to obtain the smallest perturbative constraint force required to maintain the instantaneous temperature (Hünenberger, 2005). According to Newton's laws of motion, the force imposed on each atom is given by $\boldsymbol{F}_i(\boldsymbol{R}^N) = -\partial U(\boldsymbol{R}^N)/\partial \boldsymbol{r}_i$ [see Eq. (1.2)]. To apply the Gaussian thermostat, a constraint force term $-\alpha m_i\dot{\boldsymbol{r}}_i$ is added to $\boldsymbol{F}_i(\boldsymbol{R}^N)$, and its Newtonian equation of motion becomes

$$m_i\ddot{\boldsymbol{r}}_i = -\frac{\partial U(\boldsymbol{R}^N)}{\partial \boldsymbol{r}_i} - \alpha m_i\dot{\boldsymbol{r}}_i, \tag{3.2}$$

where m_i denotes the mass of the ith atom; $\dot{\boldsymbol{r}}_i$ and $\ddot{\boldsymbol{r}}_i$ denote the first- and second-order derivatives of $\boldsymbol{r}_i$ with respect to time, respectively. The parameter α is determined by fulfilling the requirement that the temperature or the total kinetic energy of the system be kept constant, i.e.,

$$\sum m_i\boldsymbol{v}_i \cdot \frac{d\boldsymbol{v}_i}{dt} = 0. \tag{3.3}$$

Thus, α is obtained as

$$\alpha = -\sum \boldsymbol{v}_i \cdot \frac{\partial U(\boldsymbol{R}^N)/\partial \boldsymbol{r}_i}{\sum m_i v_i^2}, \tag{3.4}$$

where $\boldsymbol{v}_i$ denotes the velocity vector of the ith atom. The Gaussian thermostat does not reproduce the canonical ensemble either; it only reproduces the isokinetic ensemble. However, the isokinetic ensemble samples the same configuration phase space as the canonical ensemble. Hence, the position-dependent equilibrium properties of a system can be reliably investigated with the Gaussian thermostat. However, it should be avoided for the study of velocity-related properties.

3.3.3 Berendsen thermostat

The main issue of the velocity rescaling thermostat is that it does not allow for temperature fluctuations that are present in the canonical ensemble. To overcome this problem, Berendsen et al. (1984) proposed a temperature control method by introducing a weak coupling of a simulated system with a "thermal bath." Hence, the Berendsen thermostat is also known as the weak coupling thermostat. In this thermostat algorithm, deviations of the instantaneous system temperature $T(t)$ from the target temperature T are corrected by multiplying the atomic velocities at each timestep by a factor ζ, so that the system dynamics are steered toward one corresponding to T. The choice of the factor ζ should ensure that the rate of change of the instantaneous temperature $T(t)$ at time t is proportional to the difference between $T(t)$ and the target temperature T, which is given by

$$\frac{\mathrm{d}T(t)}{\mathrm{d}t} = \frac{1}{\tau_T}[T - T(t)]. \tag{3.5}$$

Here, τ_T is the coupling parameter that determines how strongly the system is coupled with the "thermal bath." The relationship can be understood by assuming that Fourier's Law, which governs the heat transfer between two systems with different temperatures at the macroscopic scale, also holds at the microscopic scale. The solution of Eq. (3.5) leads to the instantaneous temperature approaching the target temperature in an exponential manner, which is given by

$$T(t) = T - C\exp\left(-\frac{t}{\tau_T}\right), \tag{3.6}$$

where C is a constant. The finite-difference form of Eq. (3.5) is written as

$$\Delta T(t) = \frac{\Delta t}{\tau_T}[T - T(t)], \tag{3.7}$$

where Δt is the timestep. Eq. (3.7) implies that energy will be added to the system if $T(t) < T$ and subtracted from the system if $T(t) > T$. In addition, for a given timestep Δt, a smaller value of τ_T leads to a more rapid decay of the temperature difference. In other words, the system temperature is more tightly regulated, and smaller temperature fluctuations are allowed. Hence, τ_T is also known as the temperature relaxation time that determines the relaxation period between two successive energy exchanges between the system and the "thermal bath."

Provided that the velocity rescaling is expressed by $v'_{i,\alpha} = \zeta v_{i,\alpha}$, the energy added to or subtracted from the system can be calculated as

$$\Delta E = \frac{1}{2}\sum_{i,\alpha}\left(m_i v'^2_{i,\alpha} - m_i v^2_{i,\alpha}\right) = \frac{1}{2}(\zeta^2 - 1)\sum_{i,\alpha} m_i v^2_{i,\alpha} = (\zeta^2 - 1)\frac{3Nk_B T(t)}{2}, \tag{3.8}$$

where $v_{i,\alpha}$ is the component of the velocity $\boldsymbol{v}_i$ in the α (x, y, or z) direction, N is the number of atoms in the system, and k_B is the Boltzmann constant (1.380649×10^{-23} J/K). The combination of Eqs. (3.7) and (3.8) leads to the expression of ζ, which is given by

$$\zeta^2 = 1 + \frac{\Delta t}{\tau_T}\left(\frac{T}{T(t)} - 1\right). \tag{3.9}$$

It is apparent that the parameter ζ varies with time and depends on the timestep Δt and the coupling parameter τ_T. The incorporation of velocity rescaling into the Newtonian equations of motion of the atoms/molecules leads to

$$m_i\ddot{\boldsymbol{r}}_i = -\frac{\partial U(\boldsymbol{R}^N)}{\partial \boldsymbol{r}_i} - \frac{1}{2\tau_T}\left(\frac{T}{T(t)} - 1\right)m_i\dot{\boldsymbol{r}}_i. \tag{3.10}$$

For macroscopic problems, the coupling parameter τ_T between the system and the "thermal bath" is defined by the relation $\tau_T = \xi^{-1}V^{-1/3}C_V\lambda^{-1}$, with ξ denoting a dimensionless constant that depends on the shape of the system and the temperature inhomogeneity within it, V is the system volume, C_V is the isochoric heat capacity, and λ is the thermal conductivity. However, in MD simulations, τ_T is not rigorously defined and is adopted as an empirical parameter; its value is carefully selected to tune the coupling strength. If τ_T is too large, the system will be so weakly coupled with the thermostat that a systematic temperature drift might occur. If τ_T approaches an arbitrarily large value, the rescaling factor ζ will approach unity and the velocity rescaling is not performed. As a result, the atomic dynamics will correspond to that of the microcanonical ensemble. If τ_T is too small, the system temperature will be severely constrained and unrealistically small temperature fluctuations will ensue. For the special case in which $\tau_T = \Delta t$, no temperature fluctuation is allowed, and the Berendsen thermostat is reduced to the simple velocity rescaling thermostat. Values of τ_T in the order of 0.1 ps are typically employed in MD simulations of condensed-phase systems.

The Berendsen thermostat does not strictly fix the temperature of the system but implements an exponential relaxation of its instantaneous temperature. The atomic dynamics produced by the Berendsen thermostat corresponds to neither the canonical ensemble nor the microcanonical ensemble. Hence, the Berendsen thermostat is generally not recommended for modern MD simulations. However, it has high efficiency in controlling the temperature and can be adopted to produce relatively accurate estimations of statistical properties of an MD simulation system, particularly when the system contains a large number of atoms.

3.3.4 Bussi–Donadio–Parrinello thermostat

To solve the problem associated with both the simple velocity rescaling method and the Berendsen thermostat in reproducing the canonical ensemble, Bussi et al. (2007) modified the simple velocity rescaling method to enforce a canonical distribution of the kinetic energy of the system. Instead of restricting the kinetic energy at each timestep to being exactly equal to a single value that corresponds to the target temperature, the kinetic energy is chosen stochastically from the Maxwell–Boltzmann distribution [Eq. (1.14)] of the kinetic energy at the canonical equilibrium. To this end, the velocity rescaling factor ζ is taken as

$$\zeta = \sqrt{\frac{K_{\mathrm{st}}}{K(t)}}, \tag{3.11}$$

where $K(t)$ denotes the instantaneous kinetic energy, and K_{st} denotes the stochastic kinetic energy. As a result, the Bussi velocity rescaling thermostat incorporates a canonical probability distribution. Between two successive velocity rescaling operations, the atomic dynamics evolve according to the classical Newton's laws of motion for the microcanonical ensemble, and the time interval for this evolution can be fixed at a constant number of timesteps or randomly varied.

The main issue with the Bussi thermostat is that it considerably disturbs the atomic velocities. A smoother approach for implementing the Bussi thermostat has thus been proposed, in which the choice of K_{st} is based on the value of the instantaneous kinetic energy $K(t)$ (Bussi et al., 2007). In this approach, the velocity rescaling is carried out over a number of timesteps, and the general procedure is described as follows. First, the system is allowed to evolve for a single timestep using a time-reversible integrator such as the velocity Verlet approach (see Chapter 1: Fundamentals of Classical Molecular Dynamics Simulation). Second, the instantaneous kinetic energy $K(t)$ is calculated. Third, $K(t)$ evolves for a period corresponding to a single timestep using auxiliary continuous stochastic dynamics. Finally, the atomic velocities are rescaled according to Eq. (3.11) by assigning the updated $K(t)$ to K_{st} at the end of the third step. For more details of this smoothed Bussi thermostat, the reader may refer to the work of Bussi et al. (2007).

3.3.5 Andersen thermostat

The Andersen thermostat is a stochastic coupling method proposed by Andersen (1980). In this approach, the simulation system is explicitly coupled with a "thermal bath": the atoms or molecules undergo random collisions with fictitious particles in the "thermal bath" at low probabilities, thereby effectively gaining momenta extracted from a Maxwell–Boltzmann distribution that corresponds to the canonical ensemble at the target temperature T. Specifically, if atom i is selected to collide with the particles in the "thermal bath," its velocity will be updated by assigning random values to each Cartesian component $v_{i,\alpha}$ following the Maxwell–Boltzmann distribution. For the other atoms that do not undergo such a collision, their dynamics are left unaltered and are determined only by their Newtonian equations of motion. Hence, in contrast to the thermostat based on the simple velocity rescaling method and the Berendsen thermostat, both of which are global thermostats coupled with the entire system to enforce a uniform velocity rescaling procedure on all the constituent atoms at each timestep, the Andersen thermostat is local and functions by imposing a "massive" thermostat to randomly selected atoms and individually reassigning their velocities.

The time interval $\Delta\tau$ between two successive collisions for a given atom is determined based on the probability distribution $p(\Delta\tau) = \upsilon\exp(-\upsilon\Delta\tau)$. Here, υ is a constant that denotes the collision frequency and determines the coupling strength between the system and the "thermal bath." In principle, a set of random values of $\Delta\tau$ can be generated for each atom based on this probability distribution before the start of the simulation. This set of values is then applied to determine when the corresponding atom collides with the particles in the "thermal bath." In practice, this process can be simplified when the simulation timestep $\Delta t \ll \upsilon^{-1}$, specifically, when the collisions occur rarely. For each timestep, a random number between 0 and 1 is generated for each atom. If this random number is larger than $\upsilon\Delta t$ for a given atom, it will undergo a collision and subsequently a velocity reassignment.

Similarly to the coupling parameter for the Berendsen thermostat, the collision frequency υ should also be selected from an appropriate range. On the one hand, too small a value of υ indicates weak coupling and thus results in poor temperature control. In the limiting case where υ approaches zero, collisions between the MD system and the "thermal bath" do not occur, and the system is reduced to the microcanonical ensemble without any temperature control, which ultimately defeats the purpose of the thermostat. In principle, arbitrarily small collision frequencies are sufficient to ensure the reproduction of the canonical ensemble. However, a long simulation time is required in such cases, and systematic energy drifts caused by the accumulation of numerical errors are expected to occur. On the other hand, a large value of υ may severely disturb the atomic dynamics, although the magnitude of temperature fluctuation does not depend on the collision frequency.

It can be rigorously proven that the Andersen thermostat with a nonzero collision frequency υ reproduces the canonical ensemble, based on the fact that mixing deterministic Newtonian dynamics with the stochastic collisions generates a Markov chain of microstates in the phase space. Here, a Markov chain refers to a stochastic model describing a sequence of possible events in which the probability of each event depends only on the state attained in the previous event. However, it is worth noting that a system coupled with the Andersen thermostat is nondeterministic and time-irreversible because it randomly decorrelates atomic velocities. Furthermore, the stochastic characteristic of the Andersen thermostat introduces discontinuities in the atomic trajectories, and it also leads to the nonconservation of the total kinetic energy and momentum of the system. More precisely, the Andersen thermostat results in a series of microcanonical simulations of the system that resets whenever a collision between the system and the "thermal bath" occurs. After each reset, the total kinetic energy of the system is shifted to a new level. Hence, the Andersen thermostat should be avoided to study the dynamic properties of a system and only be adopted to investigate its structural properties.

3.3.6 Langevin thermostat

The Langevin thermostat was developed based on the Langevin equation, which is the predecessor of most isothermal equations of motion (Schneider and Stoll, 1978). The general form of the Langevin equation reads as

$$m_i \ddot{\boldsymbol{r}}_i = -\frac{\partial U(\boldsymbol{R}^N)}{\partial \boldsymbol{r}_i} - \gamma_i m_i \dot{\boldsymbol{r}}_i + \eta_i \boldsymbol{\xi}_i. \tag{3.12}$$

Here, γ_i is defined to be a positive atomic friction coefficient that dissipates energy from the system; ξ_i is a noise term that imposes a stochastic force on atom i, resulting in energy input to the system; η_i is a coefficient that determines the magnitude of the stochastic force imposed on atom i. For the Andersen thermostat, a stochastic collision is artificially imposed to reproduce the canonical ensemble and does not amount to any actual physical process. In contrast, the Langevin thermostat incorporates a stochastic force with a physical meaning into the Newtonian equations of motion of the atoms/molecules. In fact, the Langevin equation can be applied to describe Brownian dynamics or the dynamics of solutes. In both cases, the system can be modeled as a group of large particles moving within a sea of much smaller particles. When the dynamic properties or behaviors of the smaller particles are not of interest, they are usually modeled implicitly; this practice is common in MD simulations of many solute–solvent systems, such as proteins, deoxyribonucleic acids (DNAs), and nanoparticles in a solution. In these cases, the smaller particles impose a damping force [represented by the term $\gamma_i m_i \dot{\boldsymbol{r}}_i$ in Eq. (3.12)] on the larger particles as they push the smaller particles out of their way. At the same time, the interactions between the smaller particles and the larger particles also result in the transfer or removal of kinetic energy from the latter, which is reflected by the stochastic term $\eta_i \boldsymbol{\xi}_i$.

The stochastic term $\boldsymbol{\xi}_i$ has the following properties. First, it is time-independent and uncorrelated with the particle velocities and the forces acting on them at previous instances of time. Second, the time average of $\boldsymbol{\xi}_i$ is zero. Third, the component $\xi_{i,\alpha}$ of $\boldsymbol{\xi}_i$ along the Cartesian axis α is uncorrelated with any component $\xi_{j,\beta}$ of $\boldsymbol{\xi}_j$ for $i \neq j$. Mathematically, these properties can be expressed as

$$< \boldsymbol{\xi}_i > = 0 \text{ and } < \xi_{i,\alpha}(t)\xi_{j,\beta}(t') > = \delta_{ij}\delta_{\alpha\beta}\delta(t - t'). \tag{3.13}$$

Here, δ_{ij} and $\delta_{\alpha\beta}$ denote the Kronecker delta with α and β being $x, y,$ or z and $\delta(t - t')$ denotes the Dirac delta. To reproduce the canonical ensemble, the parameters η and γ_i are correlated based on the fluctuation–dissipation theorem, which is given by

$$\eta_i^2 = 2\gamma_i m_i k_B T. \tag{3.14}$$

The change in momentum caused by the stochastic force follows a Gaussian distribution, with its standard deviation equal to η_i.

One of the main advantages of the Langevin thermostat is that it improves simulation efficiency. Fewer computations are performed in each timestep because of the implicit treatment of smaller particles. Since the faster motions of the smaller particles are now represented by stochastic terms, and only the slower motions of the larger particles have to be explicitly resolved, a much larger timestep can be adopted. However, the Langevin thermostat also introduces a problem for which the volume effects of the smaller particles, such as solvent molecules, are excluded from the MD simulation. For solute–solvent systems, the solvent molecules must be much smaller than the solute molecules for the Langevin thermostat to be valid.

3.3.7 Nosé–Hoover thermostat

The Nosé–Hoover thermostat was first proposed by Nosé (1984a) based on the extended-system method. A simpler formulation of this thermostat was later proposed by Nosé (1984b) and Hoover (1985) simultaneously and is more widely adopted, and thus it is generally referred to as the Nosé–Hoover thermostat. The basic idea behind this thermostat is that it replaces the "thermal bath" in the previous thermostats with a fictitious degree of freedom s that is added to the energy function. This additional degree of freedom can be regarded as the frictional term in the Langevin equation, and it functions as a "thermal bath" to slow down or accelerate particles until the system temperature reaches the target value. The potential energy associated with s is expressed as $(3N+1)k_B T \ln s$, with $3N+1$ being the total degrees of freedom. Note that there is one more degree of freedom than the usual $3N$ degrees of freedom because the fictitious degree of freedom s is included. The kinetic energy related to s is given by $Q(\mathrm{d}s/\mathrm{d}t)^2/2$, with Q denoting the fictitious mass of s. The equations of motion of the atoms/molecules of an MD system coupled with the Nosé–Hoover thermostat can be written as

$$m_i \ddot{\boldsymbol{r}}_i = -\frac{\partial U(\boldsymbol{R}^N)}{\partial \boldsymbol{r}_i} - \gamma_i m_i \dot{\boldsymbol{r}}_i, \tag{3.15}$$

$$\frac{\mathrm{d}\gamma_i(t)}{\mathrm{d}t} = \frac{1}{Q}\left[\sum_{i=1} m_i \frac{v_i \cdot v_i}{2} - \frac{3N+1}{2} k_B T\right]. \tag{3.16}$$

It can be rigorously proven that the Nosé–Hoover thermostat reproduces the canonical ensemble in the initial system (the system without the fictitious degree of freedom s). The strength of coupling between the system and the thermostat, which depends on the temperature relaxation time τ_T of the system [defined in Eq. (3.7)], is determined by the fictitious mass Q.

As with the choice of the coupling parameter τ_T in the Berendsen thermostat, the value of Q should be selected from an appropriate range. A small value of Q will lead to rapid fluctuations of the kinetic energy of the fictitious degree of freedom s and frequent energy exchange between s and the initial system. On the one hand, too small a value of Q should be avoided, as an extraordinarily high frequency of the fictitious degree of freedom s will cause it to exhibit off-resonance with the characteristic frequencies of the initial system, leading to the decoupling of the system with the thermostat. On the other hand, too large a value of Q will greatly inhibit the energy exchange between the initial system and the thermostat, thereby leading to poor temperature control. A proper choice of Q can be selected by comparing the oscillating frequency of the kinetic energy of s with the inverse of the temperature relaxation time τ_T. For simplicity, Q can be theoretically approximated by the equation $Q \approx 1.4(3N + 1)k_B T \tau_T^2$. In a practical MD simulation, a nominal thermal relaxation time is usually defined instead of Q. For instance, in the LAMMPS (large-scale atomic/molecular massively parallel simulator) package, a damping parameter in the time unit is specified to determine how rapidly the system temperature should relax, and a value about 100 times the timestep is generally recommended.

3.3.8 Thermostat effects in equilibrium molecular dynamics simulations

Generally, all the above-listed thermostats operate either by altering the Newtonian equations of motion of the atoms/molecules or by directly modifying their velocities during the simulation so that the kinetic energy of the system is steered toward a target value. In either approach, disturbing the particle dynamics is inevitable, although the extent of such disturbance depends on the specific thermostat applied.

To characterize the effects of typical thermostats on the atomic dynamic behaviors in MD simulations, Basconi and Shirts (2013) have conducted a systematic comparative study on the dynamic behaviors of three molecular systems in which different thermostats are applied for each of them. In this study, the thermostats tested are classified into two categories: thermostats that work by *velocity randomizing*, such as the Andersen thermostat and the Langevin thermostat, and thermostats that work by *velocity rescaling*, such as the Berendsen thermostat, the Nosé–Hoover thermostat, and the stochastic velocity rescaling approach of Bussi et al. (2007). The atomic dynamics in the NVE ensemble are selected as a benchmark for comparison because the molecular system evolves according to Newton's laws of motion without any artificial disturbance. The three molecular systems chosen are pure water, a solvated methane molecule in water, and a solvated uncharged polymer chain (containing 100 monomers) in water. The thermostat effects have been investigated by comparing their dynamic properties measured on various timescales, including their shear viscosities (with a timescale of 0–1 ps), translational diffusion rates, and rotational diffusion rates (with a timescale of 0–10 ps). The dynamic structure factors of the polymer chain (with a timescale of 0–1000 ps) have also been studied.

It is discovered that the time-correlations of atomic velocities in MD simulations are significantly disrupted by velocity-randomizing thermostats, including the Andersen thermostat that stochastically reassigns velocities to the atoms at regular intervals, and the Langevin thermostat that includes additional terms representing stochastic noise and friction in the atomic equations of motion. In particular, when the coupling strength between the simulated system and the thermostat is increased, the atomic dynamics will be severely dampened as compared with that of the same system described by the NVE ensemble. For velocity-rescaling thermostats, including the Berendsen thermostat and the Nosé–Hoover thermostat, the dynamic properties of the system can be reproduced to be in good agreement with those obtained in the NVE ensemble, and their natural time-correlations can be preserved.

For aqueous solutions, aside from dampening the dynamics of the solvent (water), velocity-randomizing thermostats slow the dynamic behaviors of the water molecules. In contrast, velocity-rescaling thermostats exhibit no such effect regardless of the coupling strength. Moreover, a close relationship is determined to exist between the size of the solute molecules and the dampening effect imposed by the thermostat. A more significant dampening effect is observed in the self-diffusion process of solutes with larger molecules. Hence, if a velocity-randomizing thermostat is adopted to model the dynamic behavior of a solution with large solute molecules, a weak coupling between the solute molecules and the thermostat might be necessary to obtain accurate diffusion properties of the system.

To assess the effects of a thermostat on the dynamic behaviors of a solute with large molecules, one should take into account the specific modes of motion of the molecules, which are observed to exhibit different levels of sensitivity to the thermostat. It is suggested that, in principle, if a thermostat that could potentially disrupt the velocity-time correlation of the atoms is employed, the coupling strength between the system and the thermostat should be carefully modulated.

3.3.9 Temperature control in nonequilibrium molecular dynamics simulations

All the thermostats mentioned above are employed to reproduce the canonical ensemble and describe systems in thermodynamic equilibrium. However, these thermostats are also widely adopted for temperature control in a selection of nonequilibrium problems, for example, a crystal surface irradiated with energetic particles, a crack propagating in a solid under applied stress, two surfaces that slide against each other, and a nonequilibrium heat transfer process. A common characteristic shared by these problems is that they exhibit localized processes, and a certain amount of energy is given or taken from the system. Therefore, an increasingly widely adopted approach for applying a thermostat to model such nonequilibrium problems is to only couple the thermostat with the outer region or boundaries of the simulated system (regions where localized processes do not occur) instead of the entire system so that the atomic dynamics of the localized processes are not disturbed. The temperature of the inner region is controlled by exchanging energy with the outer region that is coupled with the thermostat. Since the stochastic and frictional forces imposed by the thermostat now only apply to the outer region or boundaries of the simulated system, this technique is generally referred to as the stochastic boundary conditions (SBCs) (Toton et al., 2010).

Fig. 3.1 shows the schematics for the application of SBCs for various cases: (1) a three-dimensional (3D) bulk material, (2) the collision between a diatomic molecule and a planar surface of bulk Si, and (3) the friction between two surfaces with adsorbed molecules. To apply SBCs, each system is divided into three regions. The atoms in the outmost layer in

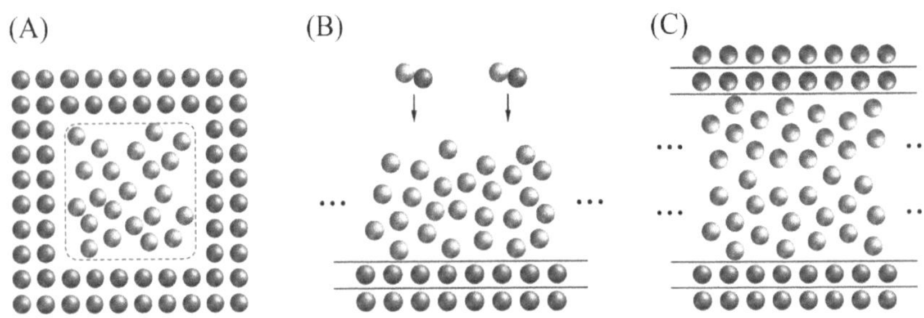

FIGURE 3.1 Schematics of simulation systems treated with SBCs: (A) a bulk system; (B) a diatomic molecule incident normally on a Si (001) surface; (C) two surfaces with adsorbed molecules sliding against each other. The atoms in the outmost layer in part A, the bottom layers in parts B, and the topmost and bottommost layers in part C (colored in blue) are fixed in place or moving with a constant velocity. While the atoms in gold are allowed to move freely according to their Newtonian equations of motion. The atoms in a layer next to the fixed layer (i.e., the atoms colored in red). *SBCs*, Stochastic boundary conditions.

Fig. 3.1A, the bottommost layer in Fig. 3.1B, and the topmost and bottommost layers in Fig. 3.1C (i.e., the atoms colored in blue) are either fixed in place (Fig. 3.1A and B) or moving at a constant velocity (Fig. 3.1C). The atoms in the layer next to the outmost fixed region (i.e., the atoms colored in red) are coupled with a thermostat for constant-temperature control. The inner atoms colored in gold are allowed to move freely according to their Newtonian equations of motion. The SBCs are proposed based on an intuitive notion that only atoms at the boundaries of a system should be affected by the thermostat because only these atoms can "feel" the presence of an isothermal environment. Meanwhile, the inner atoms only experience physical interactions from their surrounding atoms.

By comparing the simulation results obtained from applying SBCs with those obtained by coupling a thermostat with the entire system, Toton et al. (2010) have discovered that the SBCs create a buffer layer between the fixed boundary atoms and the freely moving atoms, thereby reproducing a more realistic temperature relaxation and more accurate dynamic behaviors of these atoms. For example, when a thermostat is coupled with the entire simulated system of a diatomic molecule striking a Si (001) surface, the molecule cools down too quickly upon impact because of the additional frictional force imposed on it by the thermostat. In addition, the surface atoms also lose kinetic energy at a higher rate than the surface atoms in the same system coupled with the SBCs.

However, there is a problem associated with SBCs. Since the outermost atoms are fixed in place, the inner free atoms near the boundary will be partially slowed down if their distances from these fixed atoms fall below the interaction cutoff distance specified by the potential energy function adopted. As a result, the dynamic behavior of the inner free atoms near the fixed outer atoms at the boundary will be dampened to a certain degree even if they are in equilibrium. To ameliorate this problem, the thickness of the thermostat-coupled region can be increased so that its buffering effect can be enhanced, and the inner free atoms can stay beyond the cutoff distance of the fixed outer atoms.

3.4 Pressure control

In many experiments, the thermodynamic properties of interest are measured under open-air conditions, which means that a constant pressure is also required for their corresponding MD simulations besides a constant temperature. In other situations, a nonatmospheric pressure is set up in experiments to study the behaviors of molecular systems under different ambient pressures. To realize such a nonatmospheric pressure, certain devices, such as a container with a piston filled with an inert gas, can be used in experiments. In the corresponding MD simulations, such devices are replaced by a barostat. Barostat algorithms only control the pressure of a system, and thus if the target ensemble is an NPT ensemble, a thermostat should be applied in tandem with a barostat. If a barostat is applied alone, only the number of atoms, pressure, and enthalpy of the system will be kept constant, resulting in the NPH ensemble.

A well-controlled pressure in MD simulation is useful in several situations. For instance, compared with the NVE ensemble, the NPT ensemble with a constant pressure allows energy and density fluctuations to occur within a system, thus enabling the study of processes that involve large energy fluctuations, such as crystal nucleation and phase separation. A controlled pressure also facilitates the investigation of system responses regarding finite rates of heating, cooling, compression, and expansion, such as the glass transition that occurs when a liquid is cooled and the formation of bubbles during the heating of a liquid. Nonequilibrium processes that release large amounts of energy and lead to significant volume changes, including chemical reactions and phase changes, can also be simulated.

The pressure of an MD system is commonly measured using the virial equation as given by Eq. (1.53). Since the virial term is a function of the atomic positions and their interaction forces, which are both time-dependent, the system pressure is also a time-dependent variable. As with the system temperature, if the system pressure is calculated at any single timestep, an instantaneous pressure will be obtained. In other words, the instantaneous pressure is rarely equal to the target pressure but fluctuates around it. Several barostat algorithms have been developed, and a few popular and historically significant barostats are introduced in the following sections.

3.4.1 Berendsen barostat

The most straightforward method of tuning the pressure of an MD system is to directly rescale the system volume by remapping the atomic coordinates. However, this approach does not result in a proper thermodynamic ensemble, and instead it introduces severe unphysical artifacts to the dynamic behaviors of the system. In addition, the simple volume rescaling method does not allow for pressure fluctuations, and there could be abrupt jumps in the total potential energy of the system. The Berendsen barostat (Berendsen et al., 1984) invokes the weak coupling method to tackle the problem encountered by the simple volume rescaling method, which is analogous to the algorithm adopted by the Berendsen thermostat. For the Berendsen barostat, deviations of the instantaneous system pressure $P(t)$ from the target pressure P are corrected by multiplying the atomic coordinates at each timestep by a factor of ζ to steer the system state toward the one corresponding to P. The choice of the factor ζ ensures that the rate of change of the instantaneous pressure $P(t)$ is proportional to the difference between $P(t)$ and P according to the following formula:

$$\frac{\mathrm{d}P(t)}{\mathrm{d}t} = \frac{1}{\tau_P}[P - P(t)], \tag{3.17}$$

which is similar to Eq. (3.5). Here, τ_P is the coupling parameter that determines how closely the system is coupled with the barostat. To determine ζ, an additional term proportional to $r_{i,\alpha}$ is added to the right-hand side of the equation $\dot{r}_{i,\alpha} = v_{i,\alpha}$, which leads to

$$\dot{r}_{i,\alpha} = v_{i,\alpha} + \chi r_{i,\alpha}. \tag{3.18}$$

The system volume V then changes accordingly based on the relationship $\dot{V} = 3\chi V$. The pressure change is then related to the isothermal compressibility β, and the relationship is described by

$$\frac{\mathrm{d}P(t)}{\mathrm{d}t} = -\frac{1}{\beta V}\frac{\mathrm{d}V}{\mathrm{d}t} = -\frac{3\chi}{\beta}. \tag{3.19}$$

Consequently, the parameter χ can be determined as

$$\chi = -\frac{\beta[P - P(t)]}{3\tau_P}. \tag{3.20}$$

Thus, the Newtonian equations of motion are modified to be

$$\dot{r}_{i,\alpha} = v_{i,\alpha} - \frac{\beta[P - P(t)]}{3\tau_P} r_{i,\alpha}. \tag{3.21}$$

The above equation indicates that at each timestep, the atomic coordinates $r_{i,\alpha}$ and the box length l (assuming an isotropic system in a cubic box) are multiplied by a scaling coefficient ζ that is written as

$$\zeta = 1 - \frac{\beta \Delta t}{3\tau_P}[P - P(t)]. \tag{3.22}$$

Based on the relationship $\Delta V = (\zeta^3 - 1)V = 3\chi V \Delta t$, an equivalent expression can be written as

$$\zeta = \left[1 - \frac{\beta \Delta t}{\tau_P}[P - P(t)]\right]^{1/3}. \tag{3.23}$$

In principle, the choice of τ_P can be arbitrary. However, the value of τ_P determines the efficiency of the Berendsen barostat in pressure control, which is analogous to the choice of τ_T for the Berendsen thermostat.

The expressions of Eqs. (3.22) and (3.23) apply to isotropic systems. After certain modifications, they can also be applied to anisotropic systems. The pressure of an anisotropic system is expressed as a tensor instead of a scalar:

$$P(t) = \frac{1}{V}\left\{\sum_i m_i v_i^{\mathrm{T}} v_i + \sum_{i<j} F_{ij}^{\mathrm{T}} r_{ij}\right\}. \tag{3.24}$$

The volume V is the determinant of the matrix $h = (a, b, c)^{\mathrm{T}}$, which is formed by three vectors a, b, and c that represent the edges of the unit cell in a Cartesian coordinate system:

$$V = \det h = a \cdot (b \times c). \tag{3.25}$$

If the isothermal compressibility β is a scalar, Eq. (3.22) for an anisotropic system becomes a tensorial equation:

$$\zeta = I_3 - \frac{\beta \Delta t}{3\tau_P}[P - P(t)], \tag{3.26}$$

where I_3 is a 3×3 unit matrix. The atomic coordinate vectors are now scaled to $r_i \zeta$.

Compared with the simple volume scaling method, coupling a system with the Berendsen barostat allows for pressure fluctuations, and the system pressure approaches the target value more realistically. However, the resulting ensemble is not well defined, and neither the NPT ensemble nor the NPH ensemble can be reproduced. Hence, the Berendsen barostat may be a good choice for performing an initial pressure equilibration of the system. However, it will lead to inaccuracies in the investigation of atomic dynamics under a constant pressure.

3.4.2 Andersen barostat

In the Andersen barostat, an additional degree of freedom is added to the Newtonian equations of motion of the atoms/molecules (Andersen, 1980), which is similar to the strategy used for developing the Nosé–Hoover thermostat. Specifically, in the Andersen barostat, the coordinate vector r_i is replaced by the scaled coordinate vector ς_i that is defined as $\varsigma_i = r_i / V^{1/3}$. By adding the extra degree of freedom Q into the Lagrangian of the MD system, it becomes

$$L\left(\varsigma^N, \dot{\varsigma}^N, Q, \dot{Q}\right) = \frac{1}{2} Q^{2/3} \sum_{i=1}^{N} m_i \dot{\varsigma} \cdot \dot{\varsigma} - \sum_{i<j=1}^{N} U\left(Q^{1/3} \varsigma_{ij}\right) + \frac{1}{2} M \dot{Q}^2 - \eta Q, \tag{3.27}$$

where $\varsigma^N = \{\varsigma_1, \varsigma_2, \varsigma_3 \cdots \varsigma_N\}$, $\dot{\varsigma}^N = \{\dot{\varsigma}_1, \dot{\varsigma}_2, \dot{\varsigma}_3 \cdots \dot{\varsigma}_N\}$, and $\varsigma_{ij} = \varsigma_i - \varsigma_j$; $\dot{\varsigma}_i$ denotes the derivative of ς_i with respect to time; M is the fictitious mass having the extra degree of freedom Q; η is a constant. The first two terms on the right-hand side are simply the Lagrangian of the unscaled system. The third and fourth terms are the kinetic energy and potential energy associated with Q, respectively. Physically, this Lagrangian can be understood by considering a fluid constrained in a container with a variable volume. Hence, Q, whose value is given by the volume of the container, can be interpreted as the position of a piston that exerts a pressure on the fluid. Meanwhile, ηQ, a potential similar to PV, is derived from an external pressure η that acts on the piston with a mass denoted by M. Such a piston is not a conventional piston in a cylindrical container that compresses or expands the system along one direction; instead, it causes an isotropic expansion or contraction along all directions upon a change in Q.

Based on the Lagrangian presented in Eq. (3.27), the momentum conjugate to ς_i is denoted by π_i, with the expression:

$$\pi_i = \frac{\partial L}{\partial \dot{\varsigma}_i} = m_i Q^{2/3}\dot{\varsigma}_i, \tag{3.28}$$

where m_i is the mass of the ith atom. The momentum conjugate to Q is denoted by Π, which is given by

$$\Pi = \frac{\partial L}{\partial \dot{Q}} = M\dot{Q}. \tag{3.29}$$

The Hamiltonian is then written as

$$\begin{aligned} H(\varsigma^N, \pi^N, Q, \Pi) &= \sum_{i=1}^{N} \dot{\varsigma}_i \cdot \pi_i + \dot{Q}\Pi - L\left(\varsigma^N, \dot{\varsigma}^N, Q, \dot{Q}\right) \\ &= \left(2Q^{2/3}\right)^{-1} \sum_{i=1}^{N} \pi_i \cdot \pi_i / m_i + \sum_{i<j=1}^{N} U\left(Q^{1/3}\varsigma_{ij}\right) + (2M)^{-1}\Pi^2 + \eta Q. \end{aligned} \tag{3.30}$$

The Hamiltonian equations of motion are

$$\frac{\mathrm{d}\boldsymbol{p}_i}{\mathrm{d}t} = \frac{\partial H}{\partial \pi_i} = \frac{\pi_i}{m_i Q^{2/3}}, \tag{3.31a}$$

$$\frac{\mathrm{d}\pi_i}{\mathrm{d}t} = -\frac{\partial H}{\partial \varsigma_i} = -Q^{1/3} \sum_{j(\neq i)=1}^{N} \frac{\varsigma_{ij}}{|\varsigma_{ij}|} \frac{\partial U\left(Q^{1/3}|\varsigma_{ij}|\right)}{\partial |\varsigma_{ij}|}, \tag{3.31b}$$

$$\frac{\mathrm{d}Q}{\mathrm{d}t} = \frac{\partial H}{\partial \Pi} = \frac{\Pi}{M}, \tag{3.31c}$$

$$\frac{\mathrm{d}\Pi}{\mathrm{d}t} = -\frac{\partial H}{\partial Q} = -(3Q)^{-1}\left[-2\left(2Q^{2/3}\right)^{-1} \sum_{i=1}^{N} \pi_i \cdot \pi_i / m_i + Q^{1/3} \sum_{i<j} \varsigma_{ij} \frac{\partial U\left(Q^{1/3}\varsigma_{ij}\right)}{\partial\left(Q^{1/3}\right)} + 3\eta Q\right]. \tag{3.31d}$$

These equations of motion for the scaled system can be solved numerically to obtain the atomic coordinates and momenta as functions of time. To obtain a direct description of the motion of the atoms in the initial system without the barostat, the following correspondence can be applied:

$$V(t) = Q(t), \tag{3.32a}$$

$$\boldsymbol{r}_i(t) = [Q(t)]^{1/3}\varsigma_i(t), \tag{3.32b}$$

$$\boldsymbol{p}_i(t) = \pi_i(t)/[Q(t)]^{1/3}. \tag{3.32c}$$

The combination of Eqs. (3.31) and (3.32) leads to the expression for the Newtonian equations of motion of the atoms in the initial system, which reads as

$$\frac{\mathrm{d}\boldsymbol{r}_i}{\mathrm{d}t} = \frac{\boldsymbol{p}_i}{m_i} + \frac{1}{3}\boldsymbol{r}_i\frac{\mathrm{d}\ln V}{dt}, \tag{3.33a}$$

$$\frac{\mathrm{d}\boldsymbol{p}_i}{\mathrm{d}t} = -\sum_{j(\neq i)=1}^{N} \hat{\boldsymbol{r}}_{ij}U'(\boldsymbol{r}_{ij}) - \frac{1}{3}\boldsymbol{p}_i\frac{\mathrm{d}\ln V}{dt}, \tag{3.33b}$$

$$\frac{M\mathrm{d}^2V}{\mathrm{d}t^2} = -\eta + \left(\frac{2}{3}\sum_{i=1}^{N}\frac{\boldsymbol{p}_i\cdot\boldsymbol{p}_i}{2m_i} - \frac{1}{3}\sum_{i<j=1}^{N}\boldsymbol{r}_{ij}\cdot U'(\boldsymbol{r}_{ij})\right)/V. \tag{3.33c}$$

Here, $\hat{\boldsymbol{r}}_{ij}$ is a unit vector parallel to the vector $\boldsymbol{r}_{ij}$, and $U'(\boldsymbol{r}_{ij})$ denotes the derivative of $U(\boldsymbol{r}_{ij})$ with respect to $\boldsymbol{r}_{ij}$.

It can be proven that the Andersen barostat can accurately reproduce the NPH and NPT ensembles. In practical applications, the constants M and η must be specified. The parameter η can be adopted as the target pressure, while caution must be taken in selecting the value of M. In principle, any finite and positive value of M can be chosen if the objective of the simulation is to calculate the thermodynamic properties of the system at equilibrium. However, if the objective is to simulate the dynamics of atoms within a small volume under constant pressure, selecting an appropriate value of M will be imperative. On the right-hand side of Eq. (3.33c), the first term η represents the external pressure exerted by the fictitious piston, and the second term represents the internal pressure due to the simulated particles, which fluctuates as the number of particles is not sufficiently large. An imbalance between these two pressures will result in an acceleration of the piston, whose position will fluctuate with the internal pressure. The timescale for the volume fluctuation is determined by the mass M of the piston. If a sample of fluid (assuming a cubic shape) is surrounded by a much larger fluid sample, the volume of the smaller sample will fluctuate in response to an imbalance between its internal pressure and external pressure. The timescale for the fluctuation of the volume of the small fluid sample is approximately equal to the side length of the cubic sample divided by the speed of sound propagating through it. As a result, it is desirable to choose such a value of M that the timescale for the fluctuation of Q in the scaled system is approximately equal to $Q^{1/3}$ divided by the speed of sound in the fluid. A value of M that significantly deviates from $Q^{1/3}$ will lead to failure in the coupling of the small fluid sample with its surroundings and hence its pressure control.

3.4.3 Parrinello–Rahman barostat

Although the Andersen barostat can accurately reproduce the NPH and NPT ensembles, it possesses a disadvantage; its isotropic nature does not permit the modeling of anisotropic systems or the application of nonuniform pressure to the system. To resolve this problem, Parrinello and Rahman (1981) extended the Andersen barostat algorithm to allow the simulated system to change both its shape and volume. Similarly to the treatment of an anisotropic system by the Berendsen barostat, the volume of an arbitrary entity is described by three vectors, $\boldsymbol{a}$, $\boldsymbol{b}$, and $\boldsymbol{c}$, with different lengths and arbitrary mutual orientations, as described by Eq. (3.25). The position $\boldsymbol{r}_i$ of atom i can be written in terms of $\boldsymbol{h}$ and a row vector $\boldsymbol{s}_i$ with components l_i, m_i, and n_i and it is given by

$$\boldsymbol{r}_i = \boldsymbol{s}_i\boldsymbol{h} = l_i\boldsymbol{a} + m_i\boldsymbol{b} + n_i\boldsymbol{c}, \tag{3.34}$$

where $0 \le l_i, m_i, n_i \le 1$. The square of the distance between atoms i and j is given by

$$r_{ij}^2 = (\boldsymbol{s}_i - \boldsymbol{s}_j)\boldsymbol{G}(\boldsymbol{s}_i - \boldsymbol{s}_j)^{\mathrm{T}}, \tag{3.35}$$

where the metric tensor $\boldsymbol{G}$ is given by

$$\boldsymbol{G} = \boldsymbol{h}\boldsymbol{h}^{\mathrm{T}}. \tag{3.36}$$

Based on this notation, the Lagrangian of the system coupled with the Andersen barostat is now written as

$$L = \frac{1}{2}\sum_i m_i\dot{\boldsymbol{s}}_i\boldsymbol{G}\dot{\boldsymbol{s}}_i^{\mathrm{T}} - \sum_i\sum_j U(r_{ij}) + \frac{1}{2}M\mathrm{Tr}\left(\dot{\boldsymbol{h}}_i\dot{\boldsymbol{h}}_i^{\mathrm{T}}\right) - pV, \tag{3.37}$$

where Tr(.) denotes the trace of a matrix; M, p, and V have the same physical meanings as those for the Andersen barostat. Compared with the Andersen barostat, the Parrinello–Rahman barostat is useful in solid simulations in which phase transformation occurs and leads to dramatic changes in the shape of the system. One issue of the Parrinello–Rahman barostat is that it utilizes PBCs because it was designed for large systems from the outset, which may lead to inaccuracies when it is coupled with smaller systems.

3.4.4 Martyna–Tuckerman–Tobias–Klein barostat

The MTTK barostat was developed by modifying Hoover's equations of motion and is substantially similar to the Parrinello–Rahman and Andersen barostats. The MTTK barostat combines both isotropic volume fluctuations and the full flexibility of the simulation cell, and it is usually regarded as an improvement over the Parrinello–Rahman barostat. The MTTK barostat is now widely adopted in MD simulations to reproduce the NPT and NPH ensembles. In the LAMMPS package, this barostat is also adopted as the standard pressure control algorithm; a damping parameter in units of time is specified to determine how rapidly the system pressure should relax, and a value about 1000 times the timestep is generally recommended. Readers who are interested in further details of the MTTK barostat can refer to the works of Martyna et al. (1994, 1996).

3.5 Boundary conditions

In MD simulations, there are circumstances in which PBCs are not applicable. For instance, if one studies an interface formed by two materials, periodicity may be present in directions parallel to the interface but not in the direction normal to it. In MD systems containing randomly distributed nonpoint defects (e.g., dislocations and grain boundaries), the application of PBCs will also lead to an inaccurate description of the system. In other cases in which the investigation focuses on low-dimensional nanomaterials (e.g., nanodots, nanotubes, and nanofilms) instead of macroscopic bulk materials, the effects of a free surface and the interactions of the nanomaterials with their surrounding environment have to be considered. In such circumstances, other non-PBCs should be applied.

A widely applied non-PBC is the *repulsive boundary* that serves as a reflecting wall to confine particles within a certain domain. Upon colliding with the repulsive boundary, the velocities of particles are reversed, and thus their kinetic energy is conserved. The repulsive boundary is usually realized by imposing a continuous barrier potential at the boundary of the system. The simplest case is the flat repulsive boundary illustrated in Fig. 3.2. In this case, the potential energy E_{rep} of an atom

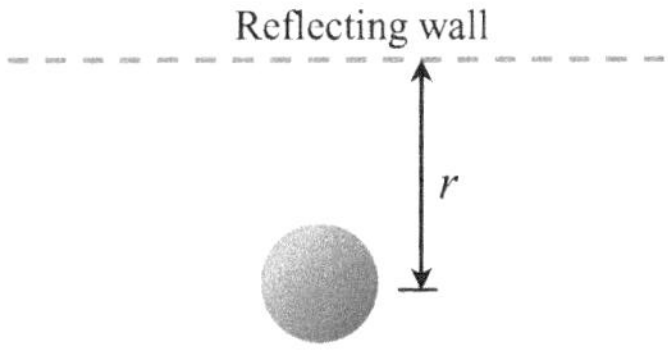

FIGURE 3.2 Schematic of a flat repulsive boundary.

interacting with the repulsive wall can be described by a simple harmonic spring potential that reads as

$$E_{\text{rep}} = \varepsilon(r - r_0)^2. \tag{3.38}$$

Here, ε denotes the effective spring constant that determines the interaction strength, and r_0 is the equilibrium distance.

A virtual barrier texture can be introduced through more complex expressions of the potential function, such as the Lennard–Jones (LJ) potential function. For example, the 10–4–3 LJ potential can be adopted to impose a repulsive wall with the curvature representing the (100) plane of a face-centered cubic (FCC) crystal lattice, and its formula is given by

$$E_{\text{rep}} = 2\pi\varepsilon\left[\frac{2}{5}\left(\frac{\sigma}{r}\right)^{10} - \left(\frac{\sigma}{r}\right)^{4} - \frac{\sqrt{2}\sigma^3}{3\left(r + \left(0.61/\sqrt{2}\right)\sigma\right)^3}\right], \tag{3.39}$$

where r denotes the distance between the repulsive wall and a particle interacting with it, ε denotes the interaction strength, and σ denotes the size of the particle. The three numbers in the 10–4–3 LJ potential refer to the exponents involved in the three terms of its formula. Note that the physical meaning of σ in Eq. (3.39) is different from that in the 12–6 LJ potential for the van der Waals interactions among particles. In the 12–6 LJ potential, σ denotes the zero-crossing distance [see Eq. (2.12)].

Up to six walls can be introduced to fully wrap up a simulation box. The positions of the walls can be specified in various ways. The most straightforward way would be to set repulsive walls at the simulation box boundaries. In the case where a vacuum region outside the boundaries of the simulation box is required to enable system expansion, a numerical constant is specified, and a wall is placed at the position defined by this constant along the corresponding axis. In the above cases, the walls are fixed in place. To allow the simulation of a change in the volume or shape of the system, the wall position can be set as a variable. This variable is evaluated for each timestep, and the result becomes the new position of the repulsive wall; i.e., the position of the wall is updated via an iterative process.

Another frequently used non-PBC is the *fixed boundary* condition in which a small portion of the atoms, particularly those in the outer shell of the simulation system, are fixed in place. The other atoms can move freely according to the Newtonian equations of motion, as shown in Fig. 3.3. The fixed boundary condition is useful when the total momentum of a modeled system is set at zero, for example, in the case of a suspended thin film.

The fixed boundary condition will inevitably introduce an unphysical disturbance to the atomic dynamics because it is unscientific to completely freeze the movement of an actual particle. To mitigate the effects of such an artificial disturbance, the system size is usually required to be sufficiently large so that particles beyond a certain distance from the boundary will hardly "feel" the boundary. For MD systems in which the atomic interactions only involve short-range forces and/or long-range forces that have been cut off beyond certain distances, the unphysical disturbance to the atomic dynamics might not be a major issue. Under such circumstances, the disturbance can usually be ignored for atoms with a distance from the boundary that is several times the cutoff distance, particularly for a condensed system.

For both repulsive and fixed boundaries, the total energy of the system is conserved, and no energy exchange with the environment is allowed, provided that no thermostats or other external constraints are applied. However, it is desirable to model the energy exchange between the simulated system and its environment in certain circumstances. To achieve this purpose, the *semirigid atomistic boundary condition* can be adopted. The implementation of this boundary condition can be realized through two approaches. In the first approach, also known as the massive boundary approach, the outmost atoms in Fig. 3.3 are allowed to move, but their masses are set at values much larger than those of the inner atoms. As a result, the boundary atoms move much slower than the free atoms, but energy exchange between the system and the external environment is permitted. One issue with the massive boundary approach is that the boundary atoms might drift away from the system during the MD simulation when they occasionally gain sufficient kinetic energy. In this case, the

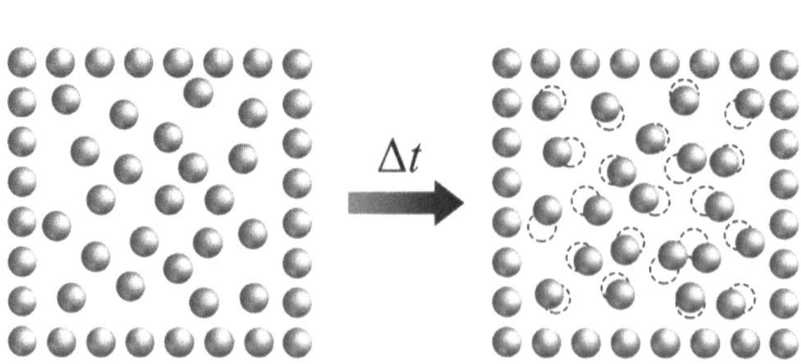

FIGURE 3.3 Schematic of the fixed boundary condition. The solid circles in the outmost region (colored in cyan) are fixed atoms and those in the inner region (colored in gold) can move freely. The empty circles denote the initial positions of the free atoms.

boundary atoms could be lost as there are no sufficiently large interaction forces to pull them back into the system because of their large mass.

To solve this problem, instead of modifying the mass of the boundary atoms, an alternative approach imposes a spring potential on each boundary atom, which reproduces the situation in which they are bounded to their equilibrium sites by an imaginary spring, as illustrated in Fig. 3.4. This spring potential is normally a harmonic function $E_{st} = k(r - r_0)^2/2$ with k denoting the force constant. By tuning the force constant, the flexibility of the boundary can be modified.

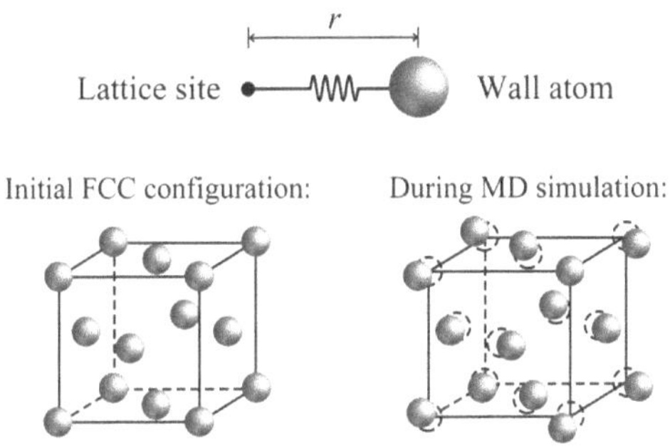

FIGURE 3.4 Illustration of the semirigid atomistic boundary condition in which the boundary atoms are tethered to their equilibrium sites by an imaginary spring. Taking a unit cell with the face-centered cubic (FCC) lattice structure as an example, the atoms are constrained to vibrate around their equilibrium lattice sites as indicated by the dotted circles.

3.6 Rigid bond constraints

In MD simulations of organic molecules containing light atoms, such as H atoms, vibrations of the bonds formed with these light atoms typically possess the highest frequencies, which can be used to set the upper limit of the timestep that is employed for integrating the Newtonian equations of motion. A timestep larger than this upper limit can lead to the loss of information on the atomic dynamic behaviors. As a result, the total simulation time is significantly shortened for the same computational cost in terms of the number of central processing unit (CPU) cores and running hours. One possible way to increase the simulation timestep and thus the total time duration modeled is to remove the degrees of freedom with the highest frequency by rigidly constraining the bond lengths associated with these light atoms. This is a common practice for many MD models, particularly for those containing water molecules.

Upon constraining the bond vibrations associated with light atoms, undesirable effects will be introduced to the energy and relevant dynamic properties of the simulated system. One reason for this phenomenon is that a vibrating bond, harmonic or nonharmonic, always contributes to the total potential energy of a system, while a rigidly constrained bond does not. A popular method of reducing such effects is to add additional Lagrange multipliers to the Lagrangian of the system. The basic ideas behind this method are illustrated as follows.

Consider a simple case in which the bond between atoms 1 and 2 is constrained to a length of d in an MD simulation. When the two atoms are held apart at a fixed distance, a corresponding bond-conserving force acts along the bond length direction, counteracting any interatomic forces altering the separation. To consider this bond-conserving force, the potential energy of the simulated system can be rewritten as

$$U_{\text{Rigid}}\left(\boldsymbol{R}^N\right) = U\left(\boldsymbol{R}^N\right) + \lambda\sigma(\boldsymbol{r}_1, \boldsymbol{r}_2). \tag{3.40}$$

Here, $\sigma(\boldsymbol{r}_1, \boldsymbol{r}_2)$ is a constraint equation defined as

$$\sigma(\boldsymbol{r}_1, \boldsymbol{r}_2) = r_{12}^2 - d^2, \tag{3.41}$$

and the variable λ is a time-dependent Lagrange multiplier that ensures $\lambda\sigma(\boldsymbol{r}_1, \boldsymbol{r}_2)$ equals zero. Provided that the additional term $\lambda\sigma(\boldsymbol{r}_1, \boldsymbol{r}_2)$ is kept at zero, the original potential energy $U\left(\boldsymbol{R}^N\right)$ of the system can be preserved. With the additional bond length constraint, the force acting on atom 1 can be obtained by

$$f_1' = -\frac{\partial U_{\text{Rigid}}\left(\boldsymbol{R}^N\right)}{\partial \boldsymbol{r}_1} = -f_1 + 2\lambda(\boldsymbol{r}_2 - \boldsymbol{r}_1). \tag{3.42}$$

Similarly, the force acting on atom 2 can be written as

$$f_2' = -\frac{\partial U_{\text{Rigid}}\left(\boldsymbol{R}^N\right)}{\partial \boldsymbol{r}_2} = -f_2 + 2\lambda(\boldsymbol{r}_1 - \boldsymbol{r}_2). \tag{3.43}$$

The integration of the Newtonian equations of motion of atom 1 with the simple Verlet algorithm gives the position of atom 1 at the time Δt as

$$\begin{aligned} \boldsymbol{r}_1(\Delta t) &= 2\boldsymbol{r}_1(0) - \boldsymbol{r}_1(-\Delta t) + \frac{f_1'}{m_1}\Delta t^2 \\ &= \tilde{\boldsymbol{r}}_1(\Delta t) + \frac{2\lambda}{m_1}[\boldsymbol{r}_1(0) - \boldsymbol{r}_2(0)]\Delta t^2. \end{aligned} \tag{3.44}$$

Here, $\tilde{\boldsymbol{r}}_1(\Delta t)$ gives the position of atom 1 without considering the bond constraint between atoms 1 and 2. In the same way, the position of atom 2 under the bond constraint at the time Δt can also be obtained. To keep the bond length constant, the following requirement should be fulfilled:

$$\left|\boldsymbol{r}_1(\Delta t)-\boldsymbol{r}_2(\Delta t)\right|^2=d^2. \tag{3.45}$$

The substitution of $\boldsymbol{r}_1(\Delta t)$ and $\boldsymbol{r}_2(\Delta t)$ in Eq. (3.45) with their respective expressions leads to

$$\left|\tilde{\boldsymbol{r}}_1(\Delta t)-\tilde{\boldsymbol{r}}_2(\Delta t)+\frac{4\lambda}{m_1}[\boldsymbol{r}_1(0)-\boldsymbol{r}_2(0)]\Delta t^2\right|^2=d^2. \tag{3.46}$$

The solution of this equation provides the value of λ, based on which the atomic positions can be predicted from their known positions before the application of the rigid bond constraint.

For multiple bond constraints, the above approach can be generalized by introducing a Lagrange multiplier λ_k for each constraint. For MD systems involving more than one constraint for each atom, the solutions to all the multipliers λ_k require solving a system of various coupled nonlinear equations. Similarly to the bond length, bond angles and torsion angles can also be constrained. For instance, a water molecule is usually constrained to exhibit a constant O–H bond length and H–O–H bond angle.

The SHAKE algorithm, proposed by Ryckaert et al. (1977), is a numerical procedure designed to facilitate the solution of the Lagrange multipliers for bond constraints and is now frequently adopted in MD simulations in which bond constraints are applied. The SHAKE algorithm is an iterative approach and considers all rigid bonds in succession. The value of λ_k is computed by keeping the other Lagrange multipliers constant based on the assumption of the algorithm that two atoms involved in a bond constraint are not involved in any other constraint. This process is repeated iteratively until all the multipliers no longer change beyond a certain tolerance.

References

Andersen, H.C., 1980. Molecular dynamics simulations at constant pressure and/or temperature. J. Chem. Phys. 72, 2384–2393.

Basconi, J.E., Shirts, M.R., 2013. Effects of temperature control algorithms on transport properties and kinetics in molecular dynamics simulations. J. Chem. Theory Comput. 9, 2887–2899.

Berendsen, H.J.C., Postma, J.P.M., van Gunsteren, W.F., DiNola, A., Haak, J.R., 1984. Molecular dynamics with coupling to an external bath. J. Chem. Phys. 81, 3684–3690.

Bussi, G., Donadio, D., Parrinello, M., 2007. Canonical sampling through velocity rescaling. J. Chem. Phys. 126, 014101-1–014101-7.

Hoover, W.G., 1985. Canonical dynamics: equilibrium phase-space distributions. Phys. Rev. A 31, 1695–1697.
Hünenberger, P.H., 2005. Thermostat algorithms for molecular dynamics simulations. Advanced Computer Simulation Approaches. Adv. Polym. Sci. 173, 105–149.
Martyna, G.J., Tobias, D.J., Klein, M.L., 1994. Constant pressure molecular dynamics algorithms. J. Chem. Phys. 101, 4177–4189.
Martyna, G.J., Tuckerman, M.E., Tobias, D.J., Klein, M.L., 1996. Explicit reversible integrators for extended systems dynamics. Mol. Phys. 87, 1117–1157.
Nosé, S., 1984a. A molecular dynamics method for simulations in the canonical ensemble. Mol. Phys. 52, 255–268.
Nosé, S., 1984b. A unified formulation of the constant temperature molecular dynamics methods. J. Chem. Phys. 81, 511–519.
Parrinello, M., Rahman, A., 1981. Polymorphic transitions in single crystals: a new molecular dynamics method. J. Appl. Phys. 52, 7182–7190.
Ryckaert, J.P., Ciccotti, G., Berendsen, H.J.C., 1977. Numerical integration of the Cartesian equations of motion of a system with constraints: molecular dynamics of n-alkanes. J. Comput. Phys. 23, 327–341.
Schneider, T., Stoll, E., 1978. Molecular dynamics study of a three-dimensional one-component model for distortive phase transitions. Phys. Rev. B 17, 1302–1322.
Toton, D., Lorenz, C.D., Rompotis, N., Martsinovich, N., Kantorovich, L., 2010. Temperature control in molecular dynamic simulations of non-equilibrium processes. J. Physics: Condens. Matter. 22, 074205-1–074205-12.

CHAPTER

4

Advanced ab initio molecular dynamics and coarse-grained molecular dynamics

This chapter introduces two classes of advanced molecular dynamics (MD) methods: ab initio MD (AIMD) methods and coarse-grained MD (CGMD) methods. It begins with the derivation of classical MD from the time-dependent Schrödinger equations of an atomic system. Two typical AIMD methods, namely, the Born–Oppenheimer and Car–Parrinello methods, are then presented with their respective fundamentals highlighted. Afterward, three widely adopted bottom-up approaches for the CGMD methods, namely, the iterative Boltzmann inversion (IBI) approach, multiscale CG approach, and relative entropy optimization approach, are discussed.

4.1 Motivations for the development of advanced molecular dynamics simulation methods

Aided by the rapid development of powerful computing techniques, MD simulation has advanced to a level where the microscopic details of many physical and chemical processes can now be well modeled to provide a direct observation of the atomic dynamics. A core step in conducting an MD simulation is to compute the interaction forces within a simulated system based on a carefully selected potential energy function. In most MD practices, such a potential energy function is constructed based on an empirical model where simple mathematical forms are employed to describe various types of interactions within an atomic or molecular system, such as bond stretching, bond angle bending, bond torsion, van der Waals interactions, and electrostatic attraction/repulsion. The relevant parameters are obtained by fitting experimental

Molecular Dynamics Simulation
DOI: https://doi.org/10.1016/B978-0-12-816419-8.00009-X

data or high-level ab initio calculations. This approach for constructing potential energy functions has attained great success in simulating systems ranging from simple solids and liquids to complex polymers and biological systems. Despite their success, MD simulations suffer from several limitations, which presents a demand for the development of advanced MD simulation techniques.

One of these limitations is that the reliability of an MD simulation highly depends on the potential energy function adopted. When a system involving many different materials is modeled, a myriad of different interatomic interactions needs to be parameterized. This requirement is because the potential energy functions currently available have been collectively developed for specific materials or classes of materials that possess the same elements or similar structures. When different types of materials are involved in one single system, a reparameterization of their respective potential functions is needed to consider the intermaterial interactions within the system. One may argue that some generalized force fields, such as CHARMM (Chemistry at Harvard Macromolecular Mechanics), AMBER (assisted model building with energy refinement), CVFF (consistent-valence force field), and universal force field, can be applied for these complex systems. Unfortunately, in exchange for their wide applicability, the accuracy of these force fields is compromised when they are used to describe the behavior of complex systems on the atomic scale.

In addition, the potential energy functions assume predefined bonding among atoms within a system. Therefore bond breaking and forming events cannot be captured by MD simulations. Although this problem can be tackled using the empirical valance bond method and other semiempirical approaches, potential energy functions developed based on these approaches are not transferable and are still required to undergo reparameterization for each type of bond breaking or forming process. Similarly, charges in MD simulations are preassigned and are typically untransferable during the entire simulation, and thus charge transfer and electronic polarization effects are not considered. Attempts have been made to rectify this problem using polarized models in which charges and/or induced dipoles are allowed to fluctuate in response to a changing environment. However, these models also suffer from a lack of transferability and standardization.

One of the most important recent developments to improve the accuracy, transferability, and applicability of the classical MD simulation is to combine it with ab initio calculations, which led to the advent of the AIMD technique. The basic idea underlying every AIMD simulation is to compute the forces acting on the nuclei in a system through ab initio calculations of its electronic structure, which are performed on the fly with an MD simulation. In this manner, the degrees of freedom of electrons are treated explicitly. Thus, the accuracy of the description of many-body forces, electronic polarization, and bond breaking and forming events is now solely dependent on the representation of the electronic structure. However, this dependence implies that the arduous task of selecting or constructing an appropriate potential energy function is now replaced by the search for a suitable approximation for the solution to the Schrödinger equation of the electronic structure because it would be virtually impossible to obtain an exact solution.

Another noteworthy limitation of MD simulation pertains to its temporal and spatial scales discussed in Chapter 1, Fundamentals of Classical Molecular Dynamics Simulation. A common approach for addressing this limitation is the CGMD method. In contrast to the AIMD method that increases the degrees of freedom of a system by taking into account the dynamics of the electrons, the CGMD method reduces the system's degrees of freedom by clustering a group of atoms to form a pseudoatom or a CG site. With the aid of coarse-graining, MD simulations can be performed on larger systems and for much longer durations. The accuracy of the CGMD method is preserved because the driving forces behind many phenomena of interest often manifest on larger spatial and temporal scales, thereby eliminating the need for a fine-grained model to describe the underlying mechanisms of these phenomena. In this chapter, both AIMD and CGMD methods are briefly introduced with their main principles highlighted.

4.2 Ab initio molecular dynamics

4.2.1 Quantum mechanics foundation of classical molecular dynamics

To understand how the AIMD method integrates both ab initio calculations and MD simulations, one begins by deriving classical MD from the nonrelativistic time-dependent Schrödinger equation. Consider a system of N_n nuclei described by their position vectors $\boldsymbol{R}^{N_n} = \{\boldsymbol{R}_1, \ldots, \boldsymbol{R}_{N_n}\}$, momentum vectors $\boldsymbol{P}^{N_n} = \{\boldsymbol{P}_1, \ldots, \boldsymbol{P}_{N_n}\}$ and mass $\{M_1, \ldots, M_{N_n}\}$, and N_e electrons described by their position vectors $\boldsymbol{r}^{N_e} = \{\boldsymbol{r}_1, \ldots, \boldsymbol{r}_{N_e}\}$ and momentum vectors $\boldsymbol{p}^{N_e} = \{\boldsymbol{p}_1, \ldots, \boldsymbol{p}_{N_e}\}$. Nuclear spin is ignored here. The time-dependent Schrödinger equation of this system can be written as

$$\mathrm{i}\hbar \frac{\partial}{\partial t} \Phi(\boldsymbol{r}^{N_e}, \boldsymbol{R}^{N_n}, t) = H\Phi(\boldsymbol{r}^{N_e}, \boldsymbol{R}^{N_n}, t). \tag{4.1}$$

Here, $\hbar$ denotes the reduced Planck constant (1.054572×10^{-34} J s); $\mathrm{i}^2 = -1$; $\Phi(\boldsymbol{r}^{N_e}, \boldsymbol{R}^{N_n}, t)$ is the complete wave function of the system contributed by both its nuclei and electrons, and H is the standard Hamiltonian that can be expressed as

$$\begin{aligned} H &= -\sum_I \frac{\hbar^2}{2M_I}\nabla_I^2 - \sum_i \frac{\hbar^2}{2m_e}\nabla_i^2 + \sum_{i<j}\frac{e^2}{|\boldsymbol{r}_i - \boldsymbol{r}_j|} - \sum_{I,i}\frac{e^2 Z_I}{|\boldsymbol{R}_I - \boldsymbol{r}_i|} + \sum_{I,J}\frac{e^2 Z_I Z_J}{|\boldsymbol{R}_I - \boldsymbol{R}_J|} \\ &= -\sum_I \frac{\hbar^2}{2M_I}\nabla_I^2 - \sum_i \frac{\hbar^2}{2m_e}\nabla_i^2 + V_{ne}(\boldsymbol{r}^{N_e}, \boldsymbol{R}^{N_n}) \\ &= -\sum_I \frac{\hbar^2}{2M_I}\nabla_I^2 + H_e(\boldsymbol{r}^{N_e}, \boldsymbol{R}^{N_n}), \end{aligned} \tag{4.2}$$

where $H_e(\boldsymbol{r}^{N_e}, \boldsymbol{R}^{N_n}) = -\sum_i \left[\left(\hbar^2/2m_e\right)\nabla_i^2\right] + V_{ne}(\boldsymbol{r}^{N_e}, \boldsymbol{R}^{N_n})$ is a part of the standard Hamiltonian; M_I and Z_I denote the mass and the atomic number of nucleus I, respectively; m_e denotes the mass of an electron (9.109383×10^{-31} kg); e denotes the elementary electron charge (1.602176×10^{-19} C). The first and second terms on the right-hand side of the first line denote the nuclear and electronic kinetic energy operators, respectively, and the last three terms represent the potential energy operators attributed to the electron − electron, electron − nucleus, and nucleus − nucleus interactions, respectively. The third term $V_{ne}(\boldsymbol{r}^{N_e}, \boldsymbol{R}^{N_n})$ in the second line denotes the operator for the total potential energy of the nuclei and electrons. To solve the Schrödinger equation in Eq. (4.1), the complete wave function $\Phi(\boldsymbol{r}^{N_e}, \boldsymbol{R}^{N_n}, t)$ that depends on both the nuclear and electronic degrees of freedom is separated by invoking the Born–Oppenheimer assumption (Tully, 1998), given by

$$\Phi(\boldsymbol{r}^{N_e}, \boldsymbol{R}^{N_n}, t) \approx \Psi(\boldsymbol{r}^{N_e}, t)\chi(\boldsymbol{R}^{N_n}, t)\exp\left[\frac{i}{\hbar}\int_{t_0}^{t} dt' \tilde{E}_e(t')\right], \tag{4.3}$$

where $\Psi(\boldsymbol{r}^{N_e}, t)$ and $\chi(\boldsymbol{R}^{N_n}, t)$ are the electronic and nuclear wave functions, respectively, and are normalized to unity at any given time, that is, $\langle \chi; t | \chi; t\rangle = 1$ and $\langle \Psi; t | \Psi; t\rangle = 1$. Here, the operator "$\langle|\rangle$" indicates the inner product of two functions. For example, $\langle f(x) | g(x)\rangle$ denotes the inner product of two functions $f(x)$ and $g(x)$, and it can be calculated as $\langle f(x) | g(x)\rangle \equiv \int f(x)^* g(x) dx$. Here, $f(x)^*$ is the complex conjugate of $f(x)$. The term $\tilde{E}_e(t')$ is a phase factor introduced in Eq. (4.3) to simplify its expression and reads as

$$\tilde{E}_e(t') = \int\int d\boldsymbol{r}^{N_e} d\boldsymbol{R}^{N_n} \Psi^*(\boldsymbol{r}^{N_e}, t)\chi^*(\boldsymbol{R}^{N_n}, t) H_e \Psi(\boldsymbol{r}^{N_e}, t)\chi(\boldsymbol{R}^{N_n}, t). \tag{4.4}$$

The integral $\int\int d\boldsymbol{r}^{N_e} d\boldsymbol{R}^{N_n}$ refers to integration over all the electronic and nuclear positions.

The quantum-mechanical average (or expectation value) of the standard Hamiltonian H is $\langle H\rangle = \int\int d\boldsymbol{r}^{N_e} d\boldsymbol{R}^{N_n} \Phi^*(\boldsymbol{r}^{N_e}, \boldsymbol{R}^{N_n}, t) H \Phi(\boldsymbol{r}^{N_e}, \boldsymbol{R}^{N_n}, t)$. Substituting the expressions of H and Φ in Eqs. (4.2) and (4.3), respectively, into Eq. (4.1) and imposing the condition of energy conservation $d\langle H\rangle/dt \equiv 0$ yield the following relations:

$$i\hbar\frac{\partial\Psi}{\partial t} = -\sum_i \frac{\hbar^2}{2m_e}\nabla_i^2\Psi + \left\{\int d\boldsymbol{R}^{N_n}\chi^*(\boldsymbol{R}^{N_n},t)V_{ne}(\boldsymbol{r}^{N_e},\boldsymbol{R}^{N_n})\chi(\boldsymbol{R}^{N_n},t)\right\}\Psi, \quad (4.5)$$

$$i\hbar\frac{\partial\chi}{\partial t} = -\sum_i \frac{\hbar^2}{2M_I}\nabla_i^2\chi + \left\{\int d\boldsymbol{r}^{N_e}\Psi^*(\boldsymbol{r}^{N_e},t)H_e\Psi(\boldsymbol{r}^{N_e},t)\right\}\chi. \quad (4.6)$$

This set of coupled equations defines the basis of the time-dependent self-consistent field method (Deumens et al., 1994). Hence, the nuclei (electrons) move according to the time-dependent effective potential determined by the electronic (nuclear) wave functions.

If the nuclear quantum effect is ignored, the dynamic quantities of the nuclei can be extracted from the above quantum-mechanical expression by assuming $\chi(\boldsymbol{R}^{N_n},t)$ to be of the form:

$$\chi(\boldsymbol{R}^{N_n},t) = A(\boldsymbol{R}^{N_n},t)\exp\left[iS(\boldsymbol{R}^{N_n},t)/\hbar\right], \quad (4.7)$$

where the amplitude factor $A(\boldsymbol{R}^{N_n},t)$ and phase factor $S(\boldsymbol{R}^{N_n},t)$ are both considered to be real. The substitution of $\chi(\boldsymbol{R}^{N_n},t)$ into Eq. (4.6) with its expression in Eq. (4.7) and the separation of its real and imaginary parts yield the following equations for the nuclei:

$$\frac{\partial S}{\partial t} + \sum_I \frac{1}{2M_I}(\nabla_I S)^2 + \int d\boldsymbol{r}^{N_e}\Psi^* H_e\Psi = \hbar^2\sum_I \frac{1}{2M_I}\frac{\nabla_I^2 A}{A}, \quad (4.8)$$

$$\frac{\partial A}{\partial t} + \sum_I \frac{1}{M_I}(\nabla_I A)(\nabla_I S) + \sum_I \frac{1}{2M_I}A(\nabla_I^2 S) = 0. \quad (4.9)$$

In the classical limit, the right-hand side of Eq. (4.8) approaches zero, and Eq. (4.8) is reduced to

$$\frac{\partial S}{\partial t} + \sum_I \frac{1}{2M_I}(\nabla_I S)^2 + \int d\boldsymbol{r}^{N_e}\Psi^* H_e\Psi = 0, \quad (4.10)$$

which can be recast into the equations of motion in the classical Hamiltonian–Jacobi formulation:

$$\frac{\partial S}{\partial t} + H(\boldsymbol{R}^{N_n},\boldsymbol{P}^{N_n}) = 0. \quad (4.11)$$

Here, the nuclear Hamiltonian is expressed as

$$H(\boldsymbol{R}^{N_n},\boldsymbol{P}^{N_n}) = \sum_I \frac{1}{2M_I}(\nabla_I S)^2 + \int d\boldsymbol{r}^{N_e}\Psi^* H_e\Psi, \quad (4.12)$$

with the momentum $P_I = \nabla_I S$. Thus, the Newtonian equations of motion for the nuclei corresponding to Eq. (4.10) can be written as

$$M_I R_I(t) = -\nabla_I \int \mathrm{d}r^{N_e} \Psi^* H_e \Psi = -\nabla_I V_e^{\mathrm{E}}(R^{N_n}(t)). \tag{4.13}$$

As a result, the nuclei move according to classical mechanics in an effective potential $V_e^E(R^{N_n})$ caused by the electrons. This effective potential is a function of only the nuclear coordinates R^{N_n} at time t and is calculated based on a quantum-mechanical average of H_e by computing its expectation value over the degrees of freedom of the electrons when the nuclei are fixed at their instantaneous positions. However, the nuclear wave functions still appear in Eq. (4.5) and have to be replaced by the positions of the nuclei for consistency. In the classical limit, the position operator $\int \mathrm{d}R^{N_n} \chi^*(R^{N_n}, t) R_I \chi(R^{N_n}, t)$ can be approximated by the instantaneous positions of the nuclei $R_I(t)$ given by Eq. (4.13). Therefore the time-dependent wave equation for the electrons in Eq. (4.5) transforms into

$$\begin{aligned} i\hbar \frac{\partial \Psi}{\partial t} &= -\sum_i \frac{\hbar^2}{2m_e} \nabla_i^2 \Psi + V_{ne}(r^{N_e}, R^{N_n}(t))\Psi \\ &= H_e(r^{N_e}, R^{N_n}(t))\Psi(r^{N_e}, R^{N_n}, t), \end{aligned} \tag{4.14}$$

and evolves as the nuclei move classically according to Eq. (4.13). It is noted that the partial Hamiltonian H_e and the electronic wave function $\Psi(r^{N_e}, R^{N_n}, t)$ are parametrically dependent on the instantaneous classical nuclear positions $R^{N_n}(t)$ at time t through $V_{ne}(r^{N_e}, R^{N_n}(t))$. This relationship indicates that the classical nuclear degrees of freedom and the quantum electronic degrees of freedom are coupled in both directions. The approach for describing the dynamics of an atomic system via Eqs. (4.13) and (4.14) is sometimes known as the Ehrenfest MD approach in honor of Paul Ehrenfest who first proposed the problem of deriving Newtonian classical dynamics from the Schrödinger equation. Therefore the Ehrenfest MD approach is also regarded as the oldest AIMD method.

In the case where the energy difference between the ground state Ψ_0 and the first excited state Ψ_1 is large compared with the amount of thermal energy $k_B T$, the total electronic wave function Ψ can be safely assumed to be restricted to the ground state at every instance of time. As a result, the nuclei can be approximated to move on a single potential energy surface as

$$V_e^{\mathrm{E}}(R^{N_n}(t)) \approx \int \mathrm{d}r^{N_e} \Psi_0^* H_e \Psi_0 = \langle \Psi_0 | H_e | \Psi_0 \rangle \equiv E_0(R^{N_n}(t)), \tag{4.15}$$

where E_0 can be computed by solving the time-independent Schrödinger equation of the electrons, where

$$H_e\Psi_0 = E_0\Psi_0. \tag{4.16}$$

The Newtonian equations of motion for the nuclei described by Eq. (4.13) can be transformed into

$$M_I\ddot{\boldsymbol{R}}_I(t) = -\nabla_I\langle\Psi_0|H_e|\Psi_0\rangle. \tag{4.17}$$

Therefore the task of generating the nuclear dynamics is decoupled from calculating the potential energy surface. First, E_0 is computed for many configurations of nuclear positions by solving Eq. (4.16). Then, these computed values of E_0 are fitted to an analytical form to yield a global potential energy surface from which the gradients can be calculated analytically to obtain the forces acting on the nuclei. Afterward, the Newtonian equations of motion in Eq. (4.13) are solved for this surface under many different initial conditions to produce the trajectories of the nuclei according to classical dynamics.

However, calculations of the global potential energy surface are often met with the issue of a "dimensionality bottleneck" as the number of active nuclear degrees of freedom increases. This problem arises because the number of electronic structure calculations according to Eq. (4.16) increases exponentially at a rate of $\sim 10^N$ for an unconstrained N-body system. A way of circumventing the "dimensionality bottleneck" is to approximate the global potential energy surface by the sum of several truncated many-body contributions as

$$V_e^E(\boldsymbol{R}^{N_n}(t)) = \sum_I^N V_1(\boldsymbol{R}_I) + \sum_{I<J}^N V_2(\boldsymbol{R}_I, \boldsymbol{R}_J) + \sum_{I<J<K}^N V_3(\boldsymbol{R}_I, \boldsymbol{R}_J, \boldsymbol{R}_K) + \cdots. \tag{4.18}$$

By applying this approximation, the electronic degrees of freedom are replaced by the interaction potentials instead of explicit degrees of freedom in the equations of motion, which forms the basis for the classical MD simulation approach.

4.2.2 Born–Oppenheimer molecular dynamics

Besides the Ehrenfest MD approach, another approach for including the electronic structure in an MD simulation is the Born–Oppenheimer MD approach, which directly solves the issue of a static electronic structure in each MD step by taking into account the set of fixed nuclear positions at that instance of time. As a result, the time-dependent quantum problem of the electronic wave function is reduced to a time-independent problem by solving the static Schrödinger equation. The time-dependence of the electronic structure is thus a consequence of the nuclear motion and is not intrinsic as in the Ehrenfest MD approach. The equations of motion for the Born–Oppenheimer MD for the electronic ground state are expressed as

$$M_I \ddot{\boldsymbol{R}}_I(t) = -\nabla_I \min_{\Psi_0} \{\langle \Psi_0 | H_e | \Psi_0 \rangle\}, \tag{4.19}$$

where Ψ_0 satisfies the time-independent electronic Schrödinger equation in Eq. (4.16), and $\min_{\Psi_0}\{\langle \Psi_0 | H_e | \Psi_0 \rangle\}$ denotes the minimum of $\langle \Psi_0 | H_e | \Psi_0 \rangle$. The above equation indicates that the minimum of $\langle H_e \rangle$ must be reached at each timestep for the Born–Oppenheimer MD, which is much different from the Ehrenfest MD approach where a wave function that minimizes $\langle H_e \rangle$ initially will stay invariable as the nuclei move according to Eq. (4.17).

It is noted that Ψ_0 in Eq. (4.19) does not correspond to the exact electronic structure, the solution of which generally requires certain approximations to be made. The Hartree–Fock approximation is a good example that presents the total electronic wave function as the determinant of a single Slater matrix $\Psi_0 = \det\{\psi_i\}$ with ψ_i being the one-particle orbital subjected to the orthonormal constraint $\langle \psi_i | \psi_j \rangle = \delta_{ij}$, where δ_{ij} denotes the Kronecker delta function. The corresponding one-particle Schrödinger equation can be written as

$$\left[-\frac{\hbar^2}{2m_e} \nabla^2 + V(\boldsymbol{r}) + V_H(\boldsymbol{r}) \right] \psi_i(\boldsymbol{r}) = E_i \psi_i, \tag{4.20}$$

where $V(\boldsymbol{r})$ denotes the electron–nuclei interaction potential energy, and $V_H(\boldsymbol{r})$ is the Hartree potential energy defined as

$$V_H(r) = e^2 \int \frac{n(r')}{|r - r'|} dr'. \tag{4.21}$$

Here, $n(\boldsymbol{r}')$ is the electron density distribution. It is obvious that the Hartree–Fock approach is a mean-field method. For a system containing N electrons, the N orbitals with the lowest energies are used to construct the Slater matrix whose determinant produces the total wave function.

The largest timescale for the Born–Oppenheimer MD approach is significantly larger than that of the Ehrenfest MD approach. This difference is because the timestep used in the latter approach to simultaneously integrate Eqs. (4.13) and (4.14) is dictated by the intrinsic dynamics of electrons that evolve much faster than those of nuclei. In contrast to the Ehrenfest MD approach, the Born–Oppenheimer MD approach does not involve electronic dynamics in solving Eqs. (4.16) and (4.19), and thus these equations can be integrated on a timescale determined by the nuclear motion. However, the computational cost for the Born–Oppenheimer MD approach is much higher than that of the Ehrenfest MD approach because a minimization process must be performed for the electronic structure at each MD step for the former. This process is unnecessary for the latter because of the possibility of propagating the electronic wave function by applying the Hamiltonian operator to an initial wave function.

It is desirable to develop an ideal algorithm that combines the advantages of both the Born–Oppenheimer and Ehrenfest MD approaches so that the integration of the equations of motion could be performed on a large timescale determined by the nuclear motion while the smooth time evolution of the dynamically evolving electronic degrees of freedom is preserved. The Car–Parrinello MD (CPMD) approach (Car and Parrinello, 1985) is an efficient method developed to achieve this goal.

4.2.3 Car–Parrinello molecular dynamics

The basic idea of the CPMD approach is to transform the quantum-mechanical adiabatic timescale separation of the fast electronic and slow nuclear motion into a classical mechanical adiabatic energy-scale separation through the framework of dynamic system theory. To understand how the CPMD approach works, consider a simple one-dimensional (1D) chain model consisting of two particles 1 and 2 with their positions denoted by x_1 and x_2, respectively, and their mass m_1 and m_2, respectively. The motions of these two particles are coupled through a potential $V(x_1, x_2)$, and the Hamiltonian of the system is taken to be in the form (Rosso et al., 2002):

$$H = \frac{p_1^2}{2m_1} + \frac{p_2^2}{2m_2} + V(x_1, x_2). \tag{4.22}$$

Given that ω_1 and ω_2 are the characteristic frequencies of the motions of particle 1 and particle 2, respectively, an adiabatic separation between these two particles can be achieved by imposing the condition $\omega_1 \gg \omega_2$, which can be realized dynamically by setting the associated mass in a manner in which $m_1 \ll m_2$. In addition, the temperatures T_1 and T_2 associated with these two particles are controlled by coupling them to two separate thermostats so that the condition $T_1 \ll T_2$ is satisfied. Since $m_1 \ll m_2$, the motion of particle 1 will instantaneously follow the motion of particle 2, which is in turn driven by a time-averaged force over particle 1. Hence, given that Δt is a characteristic time interval of the motion of particle 2, a numerical evolution step for x_2 can be written as

$$x_2(\Delta t) = x_2(0) + \Delta t v_2(0) + \frac{1}{2}a(\Delta t)^2, a = \frac{2}{m_2 \Delta t}\int_0^{\Delta t/2} dt F_2\left[x_{1,\text{adi}}(x_1(0), v_1(0), x_2(0), t)\right], \tag{4.23}$$

where $x_{1,\mathrm{adi}}(x_1(0), v_1(0), x_2(0), t)$ denotes the value of x_1 at t when x_1 evolves adiabatically over a period of $\Delta t/2$ under the initial conditions of $x_1(0)$ and $v_1(0)$ with x_2 being fixed at its initial value $x_2(0)$; F_2 is the force imposed on particle 2 due to its interaction with particle 1. Provided that the motion of particle 1 is ergodic over the period of $\Delta t/2$, the time average in Eq. (4.23) can be replaced by a phase space average as

$$\begin{aligned}\frac{2}{\Delta t}\int_0^{\Delta t/2} \mathrm{d}tF_2\left[x_{1,\mathrm{adi}}(x_1(0), v_1(0), x_2(0), t)\right] &= \frac{\int \mathrm{d}x_1 F_2(x_1, x_2)\exp\left[-\beta_1 V(x_1, x_2)\right]}{\int \mathrm{d}x_1 \exp\left[-\beta_1 V(x_1, x_2)\right]} \\ &= \frac{\partial}{\partial x_2}\frac{1}{\beta_1}\ln Z_1(x_2;\beta_1).\end{aligned} \tag{4.24}$$

Here, β_1 equals $1/(k_B T_1)$, and $Z_1(x_2, \beta_1)$ is an effective configuration partition function of x_2 and is expressed as $Z_1(x_2, \beta_1) = \int \mathrm{d}x_1 \exp\left[-\beta_1 V(x_1, x_2)\right]$. Therefore, in the adiabatic limit, an effective Hamiltonian for particle 2 can be written as

$$H_{2,\mathrm{adi}}(p_2, x_2) = K_2 - \frac{1}{\beta_1}\ln Z_1(x_2, \beta_1), \tag{4.25}$$

where $K_2 = p_2^2/(2m_2)$ is the kinetic energy of particle 2. Following statistical mechanics, the partition function of x_2, corrected in the adiabatic limit, can be written based on the expression of $H_{2,\mathrm{adi}}(p_2, x_2)$ as

$$\begin{aligned}Q(\beta_1, \beta_2) &= \iint \mathrm{d}p_2 \mathrm{d}x_2 \exp\left(-\beta_2 H_{2,\mathrm{adi}}(p_2, x_2)\right) \\ &= \int \mathrm{d}p_2 \exp\left(-\beta_2 K_2\right)\int \mathrm{d}x_2 \exp\left(-\beta_2\left(-\frac{1}{\beta_1}\ln Z_1(x_2, \beta_1)\right)\right) \\ &= \int \mathrm{d}p_2 \exp\left(-\beta_2 K_2\right)\int \mathrm{d}x_2 \left[Z_1(x_2, \beta_1)\right]^{\beta_2/\beta_1},\end{aligned} \tag{4.26}$$

where $\beta_2 = 1/(k_B T_2)$. Therefore the probability distribution function of x_2 in the adiabatic limit becomes

$$P_{\mathrm{adi}}(x_2) = \frac{\left[Z_1(x_2, \beta_1)\right]^{\beta_2/\beta_1}}{Q(\beta_1, \beta_2)}\int \mathrm{d}p_2 \exp\left(-\beta_2 K_2\right), \tag{4.27}$$

and it is apparent that $P_{\mathrm{adi}}(x_2)$ is proportional to $\left[\int \mathrm{d}x_1 \exp\left[-\beta_1 V(x_1, x_2)\right]\right]^{\beta_2/\beta_1}$. In the limit where $T_1 \to 0$ and $\beta_1 \to \infty$, the integral $\int \mathrm{d}x_1 \exp\left[-\beta_1 V(x_1, x_2)\right]$ can be simply replaced by $\exp\left[-\beta_1 \min_{x_1} V(x_1, x_2)\right]$ because the system would most probably stay at its lowest energy state of $\min_{x_1} V(x_1, x_2)$ when the kinetic energy of particle 1 is nearly zero. Here, $\min_{x_1} V(x_1, x_2)$ denotes the minimized potential energy $V(x_1, x_2)$ with respect to x_1 for a fixed value of x_2. Thus, the distribution $P_{\mathrm{adi}}(x_2)$ can be reduced to be proportional to $\exp\left[-\beta_2 \min_{x_1} V(x_1, x_2)\right]$. This relation indicates that the motion of particle 2 can be described by an effective Hamiltonian $H_2 = p_2^2/(2m_2) + \min_{x_1} V(x_1, x_2)$, which is the desired form for AIMD. Hence, under the condition $T_1 \ll T_2$, only small fluctuations around $\min_{x_1} V(x_1, x_2)$ will contribute to the probability distribution function of x_2, and particle 2 will move on a well-approximated potential energy surface that is close to the minimized one.

This idea can be realized in AIMD by introducing fictitious dynamics to the electrons via a set of orbital "velocities" $\{\dot{\psi}_i(\boldsymbol{r})\}$ and a corresponding fictitious electronic "kinetic energy" that is denoted by

$$K_{fict} = \mu \sum_i \langle \dot{\psi}_i | \dot{\psi}_i \rangle, \tag{4.28}$$

where μ is a fictitious mass parameter (with the unit of eV s^2) that controls the timescale on which the electrons move. The set of electronic orbitals $\{\psi_i(\boldsymbol{r})\}$ (corresponding to an initially minimized total electronic wave function Ψ_0) is incorporated into an extended dynamic system described by a Lagrangian of the form:

$$L_{CP} = \mu \sum_i \langle \dot{\psi}_i | \dot{\psi}_i \rangle + \frac{1}{2} \sum_I^N M_I \dot{\boldsymbol{R}}_I \cdot \dot{\boldsymbol{R}}_I - \langle \Psi_0 | H_e | \Psi_0 \rangle + \sum_{i,j} \left[\Lambda_{ij} \left(\langle \psi_i | \psi_j \rangle - \delta_{ij} \right) \right]. \tag{4.29}$$

The last term containing the matrix Λ_{ij} is a set of Lagrange multipliers introduced to ensure that the condition $\langle \psi_i | \psi_j \rangle = \delta_{ij}$ is satisfied dynamically. This Lagrangian specifies the true dynamics of the nuclei and the fictitious adiabatic dynamics of the electrons that instantaneously follow the nuclear motion, and it generates the instantaneous forces acting on the nuclei from the approximately minimized electronic configuration. The corresponding Newtonian equations of motion are obtained from the associated Euler–Lagrange equations:

$$\frac{\mathrm{d}}{\mathrm{d}t} \left(\frac{\delta L}{\delta \dot{\psi}_i^*(\boldsymbol{r})} \right) - \frac{\delta L}{\delta \psi_i^*(\boldsymbol{r})} = 0, \tag{4.30}$$

$$\frac{\mathrm{d}}{\mathrm{d}t} \left(\frac{\partial L}{\partial \dot{\boldsymbol{R}}_I} \right) - \frac{\partial L}{\partial \boldsymbol{R}_I} = 0. \tag{4.31}$$

The generic Car–Parrinello equations of motion can be written in the following form:

$$M_I \ddot{\boldsymbol{R}}_I(t) = -\frac{\partial}{\partial \boldsymbol{R}_I} \langle \Psi_0 | H_e | \Psi_0 \rangle + \sum_{ij} \Lambda_{ij} \frac{\partial}{\partial \boldsymbol{R}_I} \left\langle \psi_i | \psi_j \right\rangle, \tag{4.32}$$

$$\mu \ddot{\psi}_i(r) = -\frac{\delta}{\delta \dot{\psi}_i^*(\boldsymbol{r})} \langle \Psi_0 | H_e | \Psi_0 \rangle + \sum_j \Lambda_{ij} \psi_j(\boldsymbol{r}). \tag{4.33}$$

The detailed routine for updating the velocities and positions of both the nuclei and electrons is similar to that in classical MD simulation. However, because of the orthogonal constraint $\langle\psi_i|\psi_j\rangle = \delta_{ij}$, the updating process becomes more complex. For more technical details of the updating process, readers may refer to the work of Tuckerman (2002).

According to Eqs. (4.32) and (4.33), the nuclei evolve in time at a certain (instantaneous) physical temperature T_{nucl} proportional to $\sum_I M_I \dot{\boldsymbol{R}}_I \cdot \dot{\boldsymbol{R}}_I$, whereas a "fictitious" temperature T_{elec} proportional to $\sum_i \mu\langle\dot{\psi}_i|\dot{\psi}_i\rangle$ is associated with the electronic degrees of freedom. Under the condition $T_{\text{elec}} << T_{\text{nucl}}$, the electrons stay "cold," and the electronic subsystem is close to its instantaneous minimum energy $\min_{\{\psi_i\}}\langle\Psi_0|H_e|\Psi_0\rangle$, i.e., it is close to the exact Born–Oppenheimer surface. As a result, the ground-state wave function that is initially minimized for a given nuclear configuration will remain close to its ground state during the subsequent time evolution if it is subjected to a sufficiently low temperature.

The CPMD approach highly relies on the assumption that an adiabatic separation between the fictitious electron dynamics and the nuclear dynamics can be maintained. Under this assumption, the fast electronic subsystem can stay at a low temperature for a sufficiently long time while following the slow nuclear motion adiabatically. Simultaneously, the nuclei can be kept at a much higher temperature. This condition is achievable when the power spectra stemming from both the nuclear and electronic dynamics do not substantially overlap in the frequency domain, so that energy transfer from the "hot" nuclei to the "cold" electrons is practically inhibited. By invoking the harmonic approximation, the frequency spectrum of the electronic dynamics described by Eq. (4.33) can be written as (Pastore et al., 1991)

$$\omega_{ij} = \left(\frac{2(\varepsilon_i - \varepsilon_j)}{\mu}\right)^{1/2}, \tag{4.34}$$

where ε_i and ε_j are the eigenvalues of the unoccupied and occupied orbitals, respectively. Hence, an analytic estimation of the lowest possible electronic frequency $\omega_e^{\min}$ has the following relationship:

$$\omega_e^{\min} \propto \left(\frac{E_{\text{gap}}}{\mu}\right)^{1/2}, \tag{4.35}$$

where E_{gap} is the energy difference between the lowest unoccupied and the highest occupied orbital. Eq. (4.35) indicates that $\omega_e^{\min}$ increases with a decreasing fictitious mass parameter μ.

To guarantee the adiabatic energy separation, the difference between the lowest frequency $\omega_e^{\min}$ for the electronic dynamics and the highest frequency

ω_n^{max} for the nuclear dynamics should be sufficiently large. Since both E_{gap} and ω_n^{max} are invariable quantities that are intrinsically determined by the physical properties of the system, only the fictitious mass parameter μ is tunable for achieving a proper separation of the frequency spectra, and it is known as the "adiabaticity parameter." However, μ cannot be reduced arbitrarily. Such a reduction not only shifts the electronic spectrum to a high-frequency range but also stretches the entire frequency spectrum according to Eq. (4.34) and thus increases the highest frequency ω_e^{max} based on $\omega_e^{max} \propto (E_{cut}/\mu)^{1/2}$, where E_{cut} is the highest kinetic energy in an expansion of the wave function in terms of an electronic orbital set. The maximum value of the AIMD timestep Δt^{max} is thus reduced accordingly because it is inversely proportional to ω_e^{max}. Hence, a compromise must be made between efficiency and accuracy when it comes to choosing the parameter μ.

A problem of CPMD arises when the electronic energy gap E_{gap} is small or even nonexistent, which commonly occurs in metallic systems. In this case, all the arguments presented above break down because of the occurrence of zero-frequency electronic modes in the power spectrum according to Eq. (4.35), which necessarily leads to an overlap of the nuclear and electronic power spectra. To overcome this problem, the coupling of separate Nosé–Hoover thermostats (refer to more discussions on the Nosé–Hoover thermostat in Section 3.3.7) to the nuclear and electronic subsystems has been proposed to maintain the adiabatic energy separation by counter-balancing the energy flow from the nuclei to the electrons so that they can remain "cold" (Blöchl and Parrinello, 1992).

4.3 Coarse-grained molecular dynamics

In contrast to AIMD that aims to provide a more accurate description of interatomic interactions by including the electronic degrees of freedom, CGMD is generally adopted to expand the temporal and spatial scales of an MD simulation by reducing the number of degrees of freedom. To this end, two or more atoms in an all-atom MD simulation model are grouped to form effective interacting mass points, which are usually referred to as CG sites, CG beads, or pseudoatoms.

There are both practical and philosophical reasons for performing CGMD simulations. Practically, the size and duration of an all-atom MD simulation model are limited not only by its characteristic time and length scales but also by the speed and memory storage capacities of available computing resources. State-of-the-art MD simulations that are executed on supercomputer operating systems can handle about 10^9 atoms (which amount to less than a cubic micron of a solid) for a maximum of several nanoseconds when highly efficient simulation packages such as LAMMPS (large-scale atomic/molecular massively parallel simulator), GROMACS, and NAMD (nanoscale MD) are utilized. However, many practical physical and chemical processes have temporal and spatial scales that are much larger than what an all-atom MD simulation model can handle. For instance, protein folding, which is crucial for most biological processes, spans over 10 orders of magnitude on the timescale ranging from several nanoseconds to tens of seconds; the design of advanced materials with fine microstructures requires an understanding of material behaviors over six orders of magnitude on the spatial scale from angstroms to millimeters. Therefore it is unlikely for MD simulation to be capable of modeling, anytime soon, complex molecular systems or processes in engineering practices that involve physical processes occurring across large time and length scales. These limitations form the practical motivation of developing CGMD methods simply because the fewer degrees of freedom of CGMD models usually permit a high computational efficiency.

Philosophically, the development of the CGMD method is desirable simply because the driving forces behind many phenomena of interest often manifest on temporal and spatial scales that are larger than those on which all-atom MD simulations are normally conducted. Consequently, all-atom MD simulations produce more detailed information than is needed, and a CGMD model becomes a more preferable tool that can offer considerable advantages. Ideally, the CGMD model can provide significantly more insights into the behavior of the system because highly complex atomic-scale information is reduced to a simpler representation but with reasonable accuracy. Thus, the CGMD model can be viewed as the compression of atomistic information into lower-resolution models that seek to retain the key physical features of the system of interest. In this sense, the CGMD method is established similarly to classical phenomenological theories, such as classical thermodynamics and classical mechanics, because in these theories, a majority of degrees of freedom are eliminated, but the most essential physical laws are retained. Hence, the need for CGMD modeling will remain strong in the future, although the capability of all-atom MD simulation will advance significantly in the near future to cover much wider temporal and spatial scales than what it currently does.

There are two approaches for constructing a CGMD model: bottom-up and top-down. The first approach relies on a reference atomistic model, such as an MD or Monte Carlo model, to parameterize the CGMD model. The second approach relies on fitting relevant parameters in the CGMD model based on experimental data or ab initio calculation results. In addition, the dissipative particle dynamics (DPD) method developed by Hoogerbrugge and Koelman (1992) can be regarded as a CGMD method because it is a mesoscale particle method that bridges the gap between microscopic and macroscopic simulations. It has been successfully applied to different areas of interest, especially in modeling the hydrodynamic behavior of complex fluids in mesoscale. However, DPD is established upon a different algorithm from that of the classical MD simulation, and there is also a lack of close connection between a DPD simulation model and its corresponding all-atom model. Hence, this chapter does not discuss the DPD method. Instead, it mainly focuses on discussing typical bottom-up methods, including the IBI method, the multiscale CG (MSCG) method, and the relative entropy optimization method.

4.3.1 Theoretical formulation

The construction of a CGMD model of a system via either the bottom-up or top-down way can be divided into four major steps (Saunders and Voth, 2013): (1) choosing the resolution of the model, (2) assigning a mapping function for the CG sites, (3) determining the types of active interactions, and (4) parameterizing these interactions to optimally describe the properties of the system.

For a given system, there is no optimal value for the number of CG sites, and the appropriate resolution of the model depends largely on the types of behaviors and properties that it intends to simulate, the size of the system, and the computational resources available. For instance, for the polyleucine molecule shown in Fig. 4.1, there are three possible CG resolution levels. With the highest resolution, only the degrees of freedom of the H atoms are subtracted from the all-atom model. To

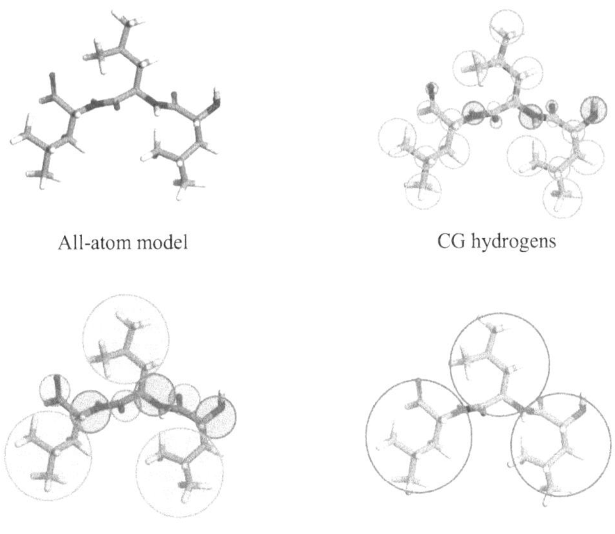

FIGURE 4.1 Schematic illustrations showing the three possible CG resolution levels for a polyleucine molecule.

further lower the resolution, atoms in the same functional group or amino acid residue can be grouped to form a single CG site.

Once a resolution is selected, a mapping function between the CG sites and the corresponding atoms in the reference all-atom model will be selected. In alignment with the denotation rule mentioned in the discussion of the AIMD method, the spatial coordinates of the n atoms of the detailed atomistic system are denoted by $\boldsymbol{r}^n = \{\boldsymbol{r}_1, \ldots, \boldsymbol{r}_n\}$, and those of the N CG sites are denoted by $\boldsymbol{R}^N = \{\boldsymbol{R}_1, \ldots, \boldsymbol{R}_N\}$ with $N < n$. The mapping function of the coordinates can be expressed as $\boldsymbol{M}_{\boldsymbol{R}}^N(\boldsymbol{r}^n) = \{\boldsymbol{M}_{R_1}(\boldsymbol{r}^n), \cdots, \boldsymbol{M}_{R_N}(\boldsymbol{r}^n)\}$ with $\boldsymbol{M}_{R_I}(\boldsymbol{r}^n)$ being a linear combination of $\boldsymbol{r}_i$ as

$$\boldsymbol{M}_{R_I}(\boldsymbol{r}^n) = \sum_{i=1}^{n} c_{Ii}\boldsymbol{r}_i, \tag{4.36}$$

where c_{Ii} is the linear coefficient. Typically, the position $\boldsymbol{R}_I$ of the Ith CG site is defined as the center of mass of the corresponding group of atoms in the reference all-atom system, so that $\boldsymbol{M}_{R_I}(\boldsymbol{r}^n)$ is defined as

$$\boldsymbol{M}_{R_I}(\boldsymbol{r}^n) = \frac{\sum_{i=1}^{n} \wp_{Ii} m_i \boldsymbol{r}_i}{\sum_{i=1}^{n} \wp_{Ii} m_i}, \tag{4.37}$$

where $\wp_{Ii} = 1$ if atom i is part of the CG site I, and $\wp_{Ii} = 0$ otherwise; m_i is the mass of atom i.

Upon determining the CG site mapping, the types of interactions among the CG sites need to be identified. Generally, the bonded interactions between two neighboring CG sites within the same subunit (e.g., a molecular chain for a polymeric system) and the nonbonded interactions between two CG sites from different subunits are considered. In some cases, the nonbonded interactions between nonneighboring CG sites in the same subunit are also considered. Bonded interactions are typically assumed to be two-body, which could lead to a low resolution and low accuracy in the description of the dynamic behaviors of the system. Hence, higher-order interactions, such as three-body and four-body interactions, should also be considered to address the effects of bond angle bending and torsion to achieve a better sampling of the conformation space and thus higher physical accuracy. Nonbonded interactions can generally be divided into two classes: short-range repulsive interactions and long-range electrostatic interactions. The former can be considered to account for the space-filling properties of a molecular system, such as its density and radial distribution function (RDF), while the latter accounts for the interactions between charged species.

After determining the types of interactions, all that remains is to construct effective potentials U_{CG} to optimally reproduce the underlying physics of the system. As discussed previously, typical methods for constructing such potential functions include the IBI, MSCG, and relative entropy optimization methods. For all these methods, a reference all-atom MD system is initiated beforehand so that potentials U_{CG} can be constructed to reproduce certain properties of the reference system. These properties can be structure-based (e.g., the distributions of bond distances and angles), force-based (e.g., the effective forces between the CG sites), or thermodynamic property-based (e.g., the density and entropy). One should not expect any one of these methods to be capable of reproducing all the properties of the reference system accurately. The choice of a specific method thus depends on the properties of interest pertaining to the CGMD model.

At this stage, one remaining problem is the criteria of assessing whether a CG potential is "optimal." This issue can be addressed from a thermodynamic perspective by examining whether consistency is obtained between the CGMD model and the reference all-atom MD model in their configuration phase space (Noid et al., 2008a). For a canonical ensemble, the equilibrium probability density of the dynamical states of an all-atom system described by the position vectors $\boldsymbol{r}^n = \{\boldsymbol{r}_1, \ldots, \boldsymbol{r}_n\}$ and momentum vectors $\boldsymbol{p}^n = \{\boldsymbol{p}_1, \ldots, \boldsymbol{p}_n\}$ is given as

$$p_{rp}(\boldsymbol{r}^n, \boldsymbol{p}^n) = p_r(\boldsymbol{r}^n) p_p(\boldsymbol{p}^n), \tag{4.38}$$

$$p_r(\boldsymbol{r}^n) \propto \exp\left(-\frac{u(\boldsymbol{r}^n)}{k_B T}\right), \tag{4.39}$$

$$p_p(\boldsymbol{p}^n) \propto \exp\left(-\sum_{i=1}^{n} \frac{\boldsymbol{p}_i \cdot \boldsymbol{p}_i}{2m_i k_B T}\right), \tag{4.40}$$

where $u(\boldsymbol{r}^n)$ is the atomistic potential, k_B is the Boltzmann constant (1.380649×10^{-23} J/K), and T is the temperature. For the CGMD model, the equilibrium probability density of its dynamical states has the same functional form in terms of the site position vectors $\boldsymbol{R}^N = \{\boldsymbol{R}_1, \ldots, \boldsymbol{R}_N\}$ and momentum vectors $\boldsymbol{P}^N = \{\boldsymbol{P}_1, \ldots, \boldsymbol{P}_N\}$ and is given as

$$P_{RP}(\boldsymbol{R}^N, \boldsymbol{P}^N) = P_R(\boldsymbol{R}^N) P_P(\boldsymbol{P}^N), \tag{4.41}$$

$$P_R(\boldsymbol{R}^N) \propto \exp\left(-\frac{U(\boldsymbol{R}^N)}{k_B T}\right), \tag{4.42}$$

$$P_P(\boldsymbol{P}^N) \propto \exp\left(-\sum_{I=1}^{N} \frac{\boldsymbol{P}_I \cdot \boldsymbol{P}_I}{2M_I k_B T}\right), \tag{4.43}$$

where $U(\boldsymbol{R}^N)$ is the CG potential energy function. Here, the momenta of the CG sites can also be expressed as a linear combination of the atomistic momenta through a mapping function $\boldsymbol{M}_{P_I}(\boldsymbol{p}^n)$ as

$$\boldsymbol{M}_{P_I}(\boldsymbol{p}^n) = M_I \sum_{i=1}^{n} \frac{c_{Ii} \boldsymbol{p}_i}{m_i}. \tag{4.44}$$

Based on Eqs. (4.38)–(4.43), the CGMD model can thus be regarded as consistent with the all-atom MD model in the phase space if the equilibrium joint probability density of the CG coordinates and momenta, as given by Eq. (4.41), is equal to that implied by the atomistic probability density in Eq. (4.38) and the mapping functions in Eqs. (4.36) and (4.44). In addition, if only the equilibrium probability density of the CG coordinates in Eq. (4.42) is equal to the atomistic coordinate distribution in Eq. (4.39), consistency in the configuration space can still be obtained. Here, the configuration space represents the collection of all the possible configurations of the system, i.e., positions of the constituent particles. A phase space represents the collection of all the possible states of the system, i.e., the positions and velocities of the constituent particles. It is intuitively clear that to achieve consistency between the CGMD model and the all-atom reference system, there must be a relationship between the CG potential function $U(\boldsymbol{R}^N)$ and the atomistic potential function $u(\boldsymbol{r}^n)$. As shown below, $u(\boldsymbol{r}^n)$ and the mapping functions would determine a unique $U(\boldsymbol{R}^N)$ for a consistent CGMD model.

The atomistic equilibrium probability density, which is expressed in terms of the atomistic variables $\boldsymbol{r}^n$ and $\boldsymbol{p}^n$ shown by Eq. (4.38), can be recast in terms of the CG variables $\boldsymbol{R}^N$ and $\boldsymbol{P}^N$ by implementing the mapping functions $\boldsymbol{M}_{\boldsymbol{R}}^N(\boldsymbol{r}^n)$ and $\boldsymbol{M}_{\boldsymbol{P}}^N(\boldsymbol{p}^n)$ as

$$p_{RP}(\boldsymbol{R}^N,\boldsymbol{P}^N) = \int d\boldsymbol{r}^n \int d\boldsymbol{p}^n p_{rp}(\boldsymbol{r}^n,\boldsymbol{p}^n)\delta\left(\boldsymbol{M}_{\boldsymbol{R}}^N(\boldsymbol{r}^n) - \boldsymbol{R}^N\right)\delta\left(\boldsymbol{M}_{\boldsymbol{P}}^N(\boldsymbol{p}^n) - \boldsymbol{P}^N\right)$$
$$= p_R(\boldsymbol{R}^N)p_P(\boldsymbol{P}^N), \tag{4.45}$$

where

$$p_R(\boldsymbol{R}^N) = \int d\boldsymbol{r}^n p_r(\boldsymbol{r}^n)\delta\left(\boldsymbol{M}_{\boldsymbol{R}}^N(\boldsymbol{r}^n) - \boldsymbol{R}^N\right), \tag{4.46}$$

$$p_P(\boldsymbol{P}^N) = \int d\boldsymbol{p}^n p_p(\boldsymbol{p}^n)\delta\left(\boldsymbol{M}_{\boldsymbol{P}}^N(\boldsymbol{p}^n) - \boldsymbol{P}^N\right). \tag{4.47}$$

Here, the delta function serves to select all the configurations in the atomistic space $\boldsymbol{r}^n$ (or $\boldsymbol{p}^n$) that are mapped to the same CG configuration $\boldsymbol{R}^N$ (or $\boldsymbol{P}^N$). It is noted that the probability is denoted by a lowercase "p" to indicate that it is calculated based on the trajectory produced by an all-atom simulation. By definition, the following relationships are satisfied:

$$\delta\left(\boldsymbol{M}_{\boldsymbol{R}}^N(\boldsymbol{r}^n) - \boldsymbol{R}^N\right) \equiv \prod_{I=1}^{N} \delta\left(\boldsymbol{M}_{\boldsymbol{R}_I}(\boldsymbol{r}^n) - \boldsymbol{R}_I\right), \tag{4.48}$$

$$\delta\left(\boldsymbol{M}_{\boldsymbol{P}}^N(\boldsymbol{p}^n) - \boldsymbol{P}^N\right) \equiv \prod_{I=1}^{N} \delta\left(\boldsymbol{M}_{\boldsymbol{P}_I}(\boldsymbol{p}^n) - \boldsymbol{P}_I\right). \tag{4.49}$$

The CGMD model will be consistent with the reference atomistic system in their phase spaces, if and only if $P_{RP}(\boldsymbol{R}^N, \boldsymbol{P}^N)$ in Eq. (4.41) equals $p_{RP}(\boldsymbol{R}^N, \boldsymbol{P}^N)$ in Eq. (4.45), i.e.,

$$P_{RP}(\boldsymbol{R}^N, \boldsymbol{P}^N) = p_{RP}(\boldsymbol{R}^N, \boldsymbol{P}^N). \tag{4.50}$$

Since both expressions can be factorized into position- and momentum-dependent parts, the CG model will be consistent with the reference atomistic system in their phase spaces, if and only if the following two relationships hold true:

$$P_R(\boldsymbol{R}^N) = p_R(\boldsymbol{R}^N), \tag{4.51}$$

$$P_P(\boldsymbol{P}^N) = p_P(\boldsymbol{P}^N). \tag{4.52}$$

These relationships are equivalent to the following formulae:

$$\exp\left(-\frac{U(\boldsymbol{R}^N)}{k_B T}\right) \propto \int \mathrm{d}\boldsymbol{r}^n \exp\left(-\frac{u(\boldsymbol{r}^n)}{k_B T}\right) \delta\left(\boldsymbol{M}_R^N(\boldsymbol{r}^n) - \boldsymbol{R}^N\right), \tag{4.53}$$

$$\exp\left(-\sum_{I=1}^{N} \frac{\boldsymbol{P}_I \cdot \boldsymbol{P}_I}{2M_I k_B T}\right) \propto \int \mathrm{d}\boldsymbol{p}^n \exp\left(-\sum_{i=1}^{n} \frac{\boldsymbol{p}_i \cdot \boldsymbol{p}_i}{2m_i k_B T}\right) \delta\left(\boldsymbol{M}_P^N(\boldsymbol{p}^n) - \boldsymbol{P}^N\right). \tag{4.54}$$

Similarly, the CGMD model will be consistent with the reference atomistic system in their configuration spaces, if and only if Eq. (4.53) holds true. By taking the logarithm of both sides of Eq. (4.53), a direct connection between $u(\boldsymbol{r}^n)$ and $U(\boldsymbol{R}^N)$ is revealed as

$$U(\boldsymbol{R}^N) = -k_B T \ln\, z(\boldsymbol{R}^N) + C, \tag{4.55}$$

with

$$z(\boldsymbol{R}^N) \equiv \int \mathrm{d}\boldsymbol{r}^n \exp\left(-\frac{u(\boldsymbol{r}^n)}{k_B T}\right) \delta\left(\boldsymbol{M}_R^N(\boldsymbol{r}^n) - \boldsymbol{R}^N\right), \tag{4.56}$$

where C is an integration constant. Notice that the first term on the right-hand side of Eq. (4.55) simply gives the free energy corresponding to a given set of CG site coordinates $\boldsymbol{R}^N$. In other words, the potential $U(\boldsymbol{R}^N)$ for a CGMD model that is consistent with the corresponding all-atom MD model well represents a many-dimensional free energy surface or a many-body potential of mean force (PMF) along the degrees of freedom $\boldsymbol{R}^N$. As a result, all averaged thermodynamic properties in the CG system should equal their values in the corresponding reference atomistic system because the configuration distributions of the CGMD model and the reference all-atom system are the same. Note that there is no guarantee that $U(\boldsymbol{R}^N)$ exhibits a particular functional form. In particular, there is no prior reason that $U(\boldsymbol{R}^N)$ should be treated as a two-body additive sum of interactions between the CG sites, although this treatment is usually adopted for developing a CG potential function to achieve high computational efficiency.

4.3.2 Iterative Boltzmann inversion method

The IBI method is one of the iterative inverse CG methods used to derive effective CGMD potentials for reproducing certain target properties of an all-atom reference system (Reith et al., 2003). A general method to develop such a CGMD potential $U(\boldsymbol{R}^N)$ is to approximate it with a sum of two-body potentials as

$$U(\boldsymbol{R}^N) = \sum_{M<N} \sum_{I<J} U_{MN}(R_{IJ}). \tag{4.57}$$

Here, the indices M and N refer to the types of CG sites, while I and J refer to the indices of a specific CG site type, and R_{IJ} denotes their distance. The RDF is commonly chosen as the target property because the Henderson uniqueness theorem indicates that for a given pair RDF $g(R)$, there exists a unique underlying two-body potential $U(R)$ that produces this RDF. To increase the physical accuracy of the CGMD model, the bonded and nonbonded interactions can be treated separately. In other words, although the nonbonded interactions are kept two-body, the bonded potentials are not necessarily required to be approximated as two-body interactions, and many-body interactions can be included. For these bonded interactions, the normalized probability distribution $P(q)$ of a particular degree of freedom q that is related to the bonded interactions can be utilized as the target property instead of the RDF. For instance, q can be adopted as the bond length, bond angle, or dihedral angle.

To illustrate the IBI process, the development of a nonbonded two-body potential $U_{MN}(R)$ is taken as an example. An iterative scheme is applied, in which an initial guess for the effective two-body potential is made. This potential is then continuously updated until a certain consistency between the RDFs of the CGMD model and the reference all-atom MD model is achieved. The initial guess of $U_{MN}^0(R)$ can be set as the direct Boltzmann inversion of the target RDF $g_{MN}^{\text{ref}}(R)$ determined by calculations based on the reference atomistic model as

$$U_{MN}^0(R) = -k_B T \ln g_{MN}^{\text{ref}}(R). \tag{4.58}$$

It is noted that to calculate $g_{MN}^{\text{ref}}(R)$, the instantaneous configurations in the trajectory frames of the reference atomistic simulation are required to be converted to those corresponding to the CGMD model. The initial guess is usually a poor estimation and is unable to reproduce the target RDF. Hence, $U_{MN}^0(R)$ is updated iteratively until the RDF generated from the effective potential converges to the target RDF. At the $k+1$ step of the iterative process, a CGMD simulation is performed in the canonical ensemble that utilizes the effective potential $U_{MN}^{k}(R)$ obtained from the k step, and the pair RDF $g_{MN}^{k}(R)$ is then calculated. The potential $U_{MN}(R)$ is updated accordingly as

$$U_{MN}^{k+1}(R) = U_{MN}^{k}(R) + k_{B}T\ln\left[\frac{g_{MN}^{k}(R)}{g_{MN}^{\text{ref}}(R)}\right] = U_{MN}^{k}(R) + \Delta U_{MN}(R). \quad (4.59)$$

This equation indicates that when $g_{MN}^{k}(R) < g_{MN}^{\text{ref}}(R)$, the current CGMD potential function underestimates the interaction of two CG sites separated by a distance of R, and hence, the two-body potential energy $U_{MN}^{k}(R)$ is lowered (because ln $\left[g_{MN}^{k}(R)/g_{MN}^{\text{ref}}(R)\right] < 0$ in this case) at this distance during the k+1 iteration step, making the two-body interaction more favorable there.

RDFs are predominantly determined by short-range repulsive interactions. Hence, a match in the RDF can only guarantee a match in the repulsive part of the potential between the CG and reference atomistic models. However, thermodynamic properties such as pressure are mainly determined by long-range attractive interactions, which are not represented in the potential obtained through the IBI method. As a result, IBI-based potentials tend to overestimate the virial pressure of the system. To account for the long-range interactions, thermodynamic constraints are incorporated into the IBI method.

In general, incorporating thermodynamic constraints into the IBI method interferes with the Henderson theorem because there is only one effective two-body potential that reproduces the RDF. Therefore these constraints can only be incorporated by reducing the accuracy with which the RDF is being matched. However, it is difficult to achieve such relaxation for the IBI method because the whole iteration process is targeted at reproducing an RDF that well matches the one obtained from the reference atomistic model. Accordingly, the thermodynamic constraints can be enforced by incorporating certain artificial correction steps into the IBI iteration. For example, after a certain number of IBI steps, a linear pressure correction term $\Delta U_{P}(R)$ can be applied to all the two-body potentials $U_{MN}(R)$ simultaneously as

$$\Delta U_{P}(R) = A\left(1 - \frac{R}{R_{\text{cut}}}\right), \quad (4.60)$$

with

$$A = \text{sgn }(\Delta P)0.1k_{B}T\text{min }(1, |f\Delta P|). \quad (4.61)$$

Here, R_{cut} is the cutoff distance for the two-body interaction, ΔP is the difference in pressure between the reference all-atom MD and CGMD systems, and f is a scaling factor that is selected to prevent the pressure from oscillating around its desired value. By including the correction of the pressure, Eq. (4.59) takes on the new form:

$$U_{MN}^{k+1}(R) = U_{MN}^{k}(R) + \Delta U_{MN}(R) + \Delta U_{P}(R). \tag{4.62}$$

Since the magnitude of the pressure correction term $\Delta U_P(R)$ depends on ΔP, the correction is synergically iterated with the IBI process until the pressure of the CG system reaches the target value.

When strong correlations between the various bonded and nonbonded interactions exist, an iterative process beyond the IBI process is still required. This need is because each update of a particular bonded potential will inevitably alter the pair RDFs produced by the previously obtained nonbonded potentials that might have already been considered to meet the requirement of convergence, and vice versa. Hence, iterative readjustments are necessary until all the pair RDFs and normalized probability distributions $P(q)$ simultaneously reach their respective targets determined by the reference all-atom MD model. Generally, such readjustments are performed by conducting another IBI process to update one particular potential (bonded or nonbonded) while keeping the other constant. Hence, the task of developing an effective potential for a CG system is often arduous and time-consuming, particularly when there are a large number of interactions to be modeled.

Besides the RDF, the Kirkwood–Buff integrals (KBI) can be introduced as a target property (similar to pressure) to derive the nonbonded potential. These integrals can be related to various other thermodynamic properties of multicomponent systems such as activity coefficients, solvation free energies, partial molar volumes, and isothermal compressibility. A KBI is defined as the integral of the RDF over a finite volume as

$$G_{MN}^{V} = \frac{1}{V}\int_{V}\int_{V}\left[g_{MN}^{\mu VT}(R_{IJ}) - 1\right]\mathrm{d}\boldsymbol{R}_I\mathrm{d}\boldsymbol{R}_J. \tag{4.63}$$

Here, $g_{MN}^{\mu VT}(R_{IJ})$ is the RDF in the μVT ensemble (constant chemical potential, volume, and temperature). In the limit where $V \rightarrow \infty$, the integral is transformed into the following manner:

$$G_{MN}^{\infty} = 4\pi\int_{0}^{\infty}\left[g_{MN}(R) - 1\right]R^2\mathrm{d}R, \tag{4.64}$$

with G_{MN}^{∞} being the KBI in the thermodynamic limit. The KBI in Eq. (4.64) should be formally calculated in the μVT ensemble, but it has been shown that calculations of KBIs are also possible in NPT (constant number of species, pressure, and temperature) or NVT (constant number of species, volume, and temperature) ensembles (Weerasinghe and

Smith, 2003a,b). To this end, the running KBI (RKBI) is defined by setting a finite integration upper limit R for the KBI as

$$G_{MN}^{V} = 4\pi \int_0^R \left[g_{MN}(S) - 1\right] S^2 \mathrm{d}S. \tag{4.65}$$

The limit G_{MN}^{∞} can be approximated by averaging G_{MN}^{V} over a suitable range of R (typically between 1.0 nm and 2.0 nm in aqueous solution mixtures) where the RKBI reaches a limiting plateau value. The validity of this approximation can be justified by the fact that density fluctuations are normally localized, and the RDF generally approaches unity over a certain distance, and thus it does not contribute to the KBI. Physically, if the number density of CG particles of type N is denoted as ρ_N, $\rho_N G_{MN}^{V}$ can be interpreted as the change in the number of CG particles of type N in a spherical region of radius R after placing a CG particle of type M at the center of that region. Hence, G_{MN}^{V} can be regarded as a local quantity that can serve as a measure of the affinity between species M and N.

By incorporating the RKBI into the IBI framework, Ganguly et al. (2012) have proposed the Kirkwood–Buff IBI (KB-IBI) method that aims to reproduce not only the structural properties but also the thermodynamic properties of a CGMD model involving multiple species. To obtain a consistent RKBI, a correction term is added to the CGMD potential:

$$\Delta U^{\mathrm{KBI}}(R) = \alpha' \left[G_{MN}^{k}(R) - G_{MN}^{\mathrm{ref}}(R)\right]\left(1 - \frac{R}{R_{\mathrm{cut}}}\right), \tag{4.66}$$

where $G_{MN}^{\mathrm{ref}}(R)$ is the plateau value of the RKBI calculated from the reference all-atom system, and $G_{MN}^{k}(R)$ is the plateau value of the RKBI after the nth IBI iteration. The parameter α' is a system-specific prefactor that is generally estimated to lie in the range of 0.01–0.1 kJ nm^{-3} mol^{-1}. Similar to the pressure ramp correction in Eq. (4.62), the KB-IBI ramp correction term $\Delta U^{\mathrm{KBI}}(R)$ can directly be combined with the IBI update term $\Delta U_{MN}(R)$ as

$$U_{MN}^{k+1}(R) = U_{MN}^{k}(R) + \Delta U_{MN}(R) + \Delta U^{\mathrm{KBI}}(R). \tag{4.67}$$

One concern regarding the potentials derived from the IBI method is that they are discrete and lack a functional form, and thus tabulated data sets are generally used in CGMD simulations to calculate CG site interactions. Although a tabulated potential can reproduce accurate target distribution functions, it inevitably loses the computational efficiency, ability to physically interpret adjustable parameters, and mathematical transferability of the potential form. In all-atom MD simulations, structural properties, such as the probability distribution of the distance between two adjacent atoms, can be well reproduced by a simple analytical potential. However, in CGMD models, the reduction in the degrees of freedom and the intrinsic complex nature of modeled molecules (e.g., large polymeric or biologic molecules) lead to great difficulties in constructing their analytical potential. Li et al. (2017) have proposed a method to address this issue by developing the analytical potential for CGMD based on numerically fitting the distributions of structural properties with multiple Gaussian functions. When the peaks of these Gaussian distributions are well separated, the final CG potential obtained by this approach can be further approximated by a combination of several simple harmonic oscillator potentials, which facilitate their implementation in practical MD simulations.

4.3.3 Multiscale coarse-grained method

The MSCG method is also referred to as the force-matching method and was initially developed by Izvekov and Voth (2005). This method constructs CGMD potentials by matching the forces that CG sites exert on one another with those from a reference all-atom system based on the variational principle. The MSCG method has been developed based on the fact that in a consistent model, the CGMD potential energy function $U(\boldsymbol{R}^N)$ is a many-body PMF determined by a reference atomistic interaction potential $u(\boldsymbol{r}^n)$. The CGMD force field determined by the gradients of the PMF is thus related to the atomistic force field accordingly. The forces acting on the CG sites are determined from the gradients of $U(\boldsymbol{R}^N)$ as

$$\boldsymbol{F}_I(\boldsymbol{R}^N) \equiv -\frac{\partial U(\boldsymbol{R}^N)}{\partial \boldsymbol{R}_I}, \tag{4.68}$$

where $\boldsymbol{F}_I$ represents the force acting on the Ith CG site.

On the one hand, $\boldsymbol{F}_I$ can be calculated as the specially selected ensemble-average of the atomistic forces in an equilibrium all-atom MD simulation. For instance, $\boldsymbol{F}_I$ can be represented by the sum of forces acting on the set of atoms that are mapped to the CG site located at $\boldsymbol{R}_I$. In addition, to consider the different atomic configurations formed by this set of atoms, an average of $\boldsymbol{F}_I$ is performed over all the configurations $\boldsymbol{r}^n$ in the reference all-atom MD model with $\delta\left(\mathbf{M}_{\boldsymbol{R}}^N(\boldsymbol{r}^n) - \boldsymbol{R}^N\right)$ possessing a nonzero value. Mathematically, $\boldsymbol{F}_I(\boldsymbol{R}^N)$ is expressed as

$$\boldsymbol{F}_I(\boldsymbol{R}^N) = \left\langle \boldsymbol{f}_I(\boldsymbol{r}^n) \right\rangle_{\boldsymbol{R}^N}, \tag{4.69a}$$

where

$$\boldsymbol{f}_I = \sum_{i=1}^{n} \wp_{Ii} \boldsymbol{f}_i. \tag{4.69b}$$

The subscript $\boldsymbol{R}^N$ on the right-hand side of Eq. (4.69a) denotes that the averaging is only performed on those atomistic configurations mapped onto $\boldsymbol{R}^N$ according to Eq. (4.37). The symbol f_i represents the total atomistic force acting on atom i.

On the other hand, $\boldsymbol{F}_I$ can be approximated by an effective force $\boldsymbol{F}_I^{\mathrm{MS}}$ on the Ith CG site obtained from an effective potential function for the CGMD model. This effective potential function can be approximated as the combination of a set of basis functions, and their negative gradients with respect to the CG site positions constitute the basis functions for the CGMD force field. To construct these basis functions, the potential energy of the CGMD model can be decomposed into several parts, each of which corresponds to an intramolecular or intermolecular interaction in an all-atom MD model. Typically, these interactions include electrostatic interactions between pairs of charged sites, nonbonded and bonded interactions between pairs of sites, bond angle bending, and bond torsion. Each of these interactions can usually be mathematically described by a functional form. Then, for each type of interaction (except for the electrostatic interactions), flexible basis functions that can appropriately represent its functional form are constructed. Thus, the final MSCG MD potential is expressed as a linear combination of these basis functions with a set of variationally determined parameters $l = \{l_1, l_2, l_3 \cdots l_{N_k}\}$.

According to the variational principle, the optimal CGMD potential function $U(\boldsymbol{R}^N)$ can be determined by minimizing an objective residual function χ^2 that is related to the difference between the PMF of the reference all-atom MD model and that of the CGMD model. Specifically, χ^2 is defined as

$$\begin{aligned}\chi^2[l] &= \frac{1}{3n_tN}\sum_{i=1}^{n_t}\sum_{I=1}^{N}\left|\boldsymbol{f}_I(\boldsymbol{r}^{n,i}) - \boldsymbol{F}_I^{MS}(\boldsymbol{M}_{\boldsymbol{R}}^N(\boldsymbol{r}^{n,i}), l)\right|^2 \\ &= \frac{1}{3N}\left\langle \sum_{I=1}^{N}\left|\boldsymbol{f}_I(\boldsymbol{r}^{n,i}) - \boldsymbol{F}_I^{MS}(\boldsymbol{M}_{\boldsymbol{R}}^N(\boldsymbol{r}^{n,i}), l)\right|^2 \right\rangle_t.\end{aligned} \tag{4.70}$$

Here, n_t is the total number of configurations sampled during the entire course of the all-atom MD simulation, and $\boldsymbol{r}^{n,i}$ corresponds to the ith atomically detailed configuration. The best solution corresponds to the minimum of the residual, i.e., where $\chi^2 = 0$. The fact that the residual is quadratic indicates that it possesses only one minimum. More details on the rigorous theoretical foundation, numerical implementations, and applications of the MSCG method can be found in the works of Ayton et al. (2007) and Noid et al. (2008a, b).

4.3.4 Relative entropy optimization method

One disadvantage of the IBI and MSCG methods is that they cannot provide a universal metric to assess the quality of the CGMD model; i.e., there is no practical way to compare the relative quality of CGMD models with different levels of resolution. Shell (2008) identified a new quantity, the relative entropy, to be of fundamental physical significance to the development of CGMD models, and this quantity can be adopted as a universal metric for quantifying CGMD models on different scales. The relative entropy has stemmed from information theory and been specifically introduced to handle a class of tasks that aim to identify the thermodynamic ensemble (i.e., a particular system described by an appropriate interaction potential under a set of state conditions) that is the best at reproducing features of an existing ensemble. Provided that the reproductive and existing ensembles are the modeled and targeted ensembles, respectively, it can be shown that the modeled ensemble represents the target ensemble most optimally when their relative entropy is minimized, as given by

$$S_{\mathrm{rel}} = \sum_i p_T(i) \ln \frac{p_T(i)}{p_M(i)}. \tag{4.71}$$

Here, the summation includes all possible configurations, $p(i)$ is the probability of configuration i occurring in an ensemble, and the subscripts "T" and "M" denote the target and model, respectively. To find the CGMD potential that best reproduces the thermodynamic properties of a reference all-atom system, Eq. (4.71) can be rewritten in terms of the configuration probabilities $p_r(\boldsymbol{r}^n)$ and $P_R(\boldsymbol{R}^N)$ defined previously as

$$S_{\mathrm{rel}} = \sum_i p_r(\boldsymbol{r}^{n,i}) \ln \frac{p_r(\boldsymbol{r}^{n,i})}{P'_R(\boldsymbol{r}^{n,i})}, \tag{4.72}$$

with

$$P'_R(\boldsymbol{r}^{n,i}) = \frac{P_R(\boldsymbol{R}^{N,i})}{\Omega_{\mathrm{map}}(\boldsymbol{R}^{N,i})} = \frac{P_R(\boldsymbol{M}_{\boldsymbol{R}}^N(\boldsymbol{r}^{n,i}))}{\Omega_{\mathrm{map}}(\boldsymbol{M}_{\boldsymbol{R}}^N(\boldsymbol{r}^{n,i}))}. \tag{4.73}$$

Here, $\boldsymbol{r}^{n,i}$ denotes the ith all-atom configuration as stated previously, $\boldsymbol{R}^{N,i}$ is the corresponding CG site configuration obtained by applying the configuration mapping function $\boldsymbol{M}_{\boldsymbol{R}}^N(\boldsymbol{r}^{n,i})$, $\Omega_{\mathrm{map}}(\boldsymbol{M}_{\boldsymbol{R}}^N(\boldsymbol{r}^{n,i}))$ is the degree of degeneracy of $\boldsymbol{R}^{N,i}$ as multiple atomic configurations $\boldsymbol{r}^{n,i}$ of the all-atom MD model can be mapped to a single CG configuration $\boldsymbol{R}^{N,i}$ of the CGMD model. Mathematically, $\Omega_{\mathrm{map}}(\boldsymbol{M}_{\boldsymbol{R}}^N(\boldsymbol{r}^{n,i}))$ can be expressed as

$$\Omega_{\mathrm{map}}(\boldsymbol{M}_{\boldsymbol{R}}^N(\boldsymbol{r}^{n,i})) = \sum_j \delta\left(\boldsymbol{M}_{\boldsymbol{R}}^N(\boldsymbol{r}^{n,j}) - \boldsymbol{R}^{N,i}\right), \tag{4.74}$$

where j indexes all the all-atom configurations $\boldsymbol{r}^{n,j}$ that can be mapped to $\boldsymbol{R}^{N,i}$ after applying the mapping function. The substitution of Eq. (4.73) for $P'_{\mathrm{R}}(\boldsymbol{r}^{n,i})$ into Eq. (4.72) leads to

$$\begin{aligned} S_{rel} &= \sum_i p_r(\boldsymbol{r}^{n,i}) \ln \frac{p_r(\boldsymbol{r}^{n,i})}{P_R(\boldsymbol{M}_{\boldsymbol{R}}^N(\boldsymbol{r}^{n,i}))} + \sum_i p_r(\boldsymbol{r}^{n,i}) \ln \Omega_{\mathrm{map}}(\boldsymbol{M}_{\boldsymbol{R}}^N(\boldsymbol{r}^{n,i})) \\ &= \sum_i p_r(\boldsymbol{r}^{n,i}) \ln \frac{p_r(\boldsymbol{r}^{n,i})}{P_R(\boldsymbol{M}_{\boldsymbol{R}}^N(\boldsymbol{r}^{n,i}))} + \langle S_{\mathrm{map}} \rangle_{\boldsymbol{r}^n}, \end{aligned} \tag{4.75}$$

where $\langle S_{\mathrm{map}} \rangle_{\boldsymbol{r}^n}$ is defined as the average entropy that results from the degeneracies caused by the configuration mapping. It is noted that this mapping entropy does not depend on the properties of the CGMD model but only on the mapping function $\boldsymbol{M}_{\boldsymbol{R}}^N(\boldsymbol{r}^n)$.

In essence, the relative entropy S_{rel} is a measurement of the overlap or similarity between a model system and the target or reference system and thus possesses several basic properties. First, the value of S_{rel} is always positive or zero because the degree of overlap is a probability-related quantity. Second, for an optimal model, S_{rel} is at a minimum, and the minimum value equals zero in the case where the modeled ensemble perfectly reproduces the reference ensemble.

In the NVT ensemble, the configuration probabilities $p_r(\boldsymbol{r}^{n,i})$ and $P_R(\boldsymbol{R}^{N,i})$ are given by

$$p_r(\boldsymbol{r}^{n,i}) = \exp\left(-\frac{u(\boldsymbol{r}^{n,i})}{k_{\mathrm{B}}T}\right) / z(\boldsymbol{r}^{n,i}), \tag{4.76}$$

$$P_R(\boldsymbol{R}^{N,i}) = \exp\left(-\frac{U(\boldsymbol{R}^{N,i})}{k_{\mathrm{B}}T}\right) / z(\boldsymbol{R}^{N,i}), \tag{4.77}$$

with $z(\boldsymbol{r}^{n,i})$ and $z(\boldsymbol{R}^{N,i})$ denoting their corresponding partition functions. The substitution of $p_r(\boldsymbol{r}^{n,i})$ and $P_R(\boldsymbol{R}^{N,i})$ into Eq. (4.75) with their respective expressions in Eqs. (4.76) and (4.77) leads to

$$S_{\mathrm{rel}} = \beta \langle U(\boldsymbol{R}^{N,i}) - u(\boldsymbol{r}^{n,i}) \rangle_{\boldsymbol{r}^n} - \beta (A(\boldsymbol{R}^{N,i}) - A(\boldsymbol{r}^{n,i})) + \langle S_{\mathrm{map}} \rangle_{\boldsymbol{r}^n}, \tag{4.78}$$

where $\beta = 1/(k_{\mathrm{B}}T)$, A is the configurational component of the Helmholtz free energy, and the average over the potential energy is performed in the reference all-atom MD system, i.e., using a set of all-atom trajectories.

It can be proven that the minimization of the relative entropy S_{rel} in Eq. (4.78) always corresponds to a CGMD potential $U(\boldsymbol{R}^N)$ with a given functional form that optimally reproduces the reference probability $p_R(\boldsymbol{R}^N)$ determined by the corresponding all-atom trajectory. In other words, upon the minimization of S_{rel}, configuration consistency between the CGMD model and reference all-atom system can be expected, as indicated by Eq. (4.53). Consider the case in which the effective potential function $U(\boldsymbol{R}^N)$ can take any form and is not constrained to be a two-body additive or have any other functional dependence. The optimal $U(\boldsymbol{R}^N)$ would then be the one that functionally minimizes the relative entropy for any given set of CG coordinates $\boldsymbol{R}^N$

$$\frac{\delta S_{\mathrm{rel}}\left[U(\boldsymbol{R}^N)\right]}{\delta\left[U(\boldsymbol{R}^N)\right]} = 0. \tag{4.79}$$

To obtain the functional derivative, it is advantageous to express S_{rel} in terms of configuration integrals as

$$\begin{aligned} S_{\mathrm{rel}} &= \beta \int \mathrm{d}\boldsymbol{r}^n p_r(\boldsymbol{r}^n)\left[U(\boldsymbol{M}_R^N(\boldsymbol{r}^n)) - u(\boldsymbol{r}^n)\right] \\ &+ \ln \int \mathrm{d}\boldsymbol{R}^N \exp\left(-\beta U(\boldsymbol{R}^N)\right) - \ln \int \mathrm{d}\boldsymbol{r}^n \exp\left(-\beta u(\boldsymbol{r}^n)\right) + \left\langle S_{\mathrm{map}}\right\rangle_{\boldsymbol{r}^n}. \end{aligned} \tag{4.80}$$

The only integrals with terms of $U(\boldsymbol{R}^N)$ in Eq. (4.80) are the first and second terms on its right-hand side. It is noted that the mapping entropy does not depend on $U(\boldsymbol{R}^N)$ but on the mapping strategy. It can then be shown that the functional derivative of the relative entropy gives (Chaimovich and Shell, 2011)

$$\frac{\delta S_{\mathrm{rel}}\left[U(\boldsymbol{R}^N)\right]}{\delta\left[U(\boldsymbol{R}^N)\right]} = \beta \int \mathrm{d}\boldsymbol{r}^n p_r(\boldsymbol{r}^n)\delta\left[\boldsymbol{M}_R^N(\boldsymbol{r}^n) - \boldsymbol{R}^N\right] - P_R(\boldsymbol{R}^N). \tag{4.81}$$

The first term on the right-hand side of Eq. (4.81) is simply the configuration probability $p_r(\boldsymbol{R}^N)$. When S_{rel} is at the minimum, it follows that

$$p_r(\boldsymbol{R}^N) = P_R(\boldsymbol{R}^N). \tag{4.82}$$

This relation implies that the relative entropy minimization for $U(\boldsymbol{R}^N)$ rigorously reproduces configuration probabilities in the CGMD model that are identical to those in the reference all-atom MD model. Hence, the relative entropy minimization provides an ideal framework for the optimization of $U(\boldsymbol{R}^N)$.

When the CGMD potential $U(\boldsymbol{R}^N)$ is assumed to be represented by a specific functional form with certain variational parameters to be optimized, the problem of minimizing S_{rel} with respect to $U(\boldsymbol{R}^N)$ is now reduced to a problem of numerically minimizing S_{rel} with respect to these parameters. Two approaches can be used to compute S_{rel} for this purpose.

In the first approach, the minima of S_{rel} can be located directly by its calculation over a broad range of the parameter space. For instance, Eq. (4.78) only requires calculations of the potential energy $U(\boldsymbol{R}^N)$ and free energy $A(\boldsymbol{R}^N)$ for all possible values of the parameters to be optimized, and the remaining terms are constant for a fixed mapping function. There are various methods available to calculate the change of $A(\boldsymbol{R}^N)$ with respect to varying parameters, for example, the thermodynamic integration technique, multiple histogram reweighting technique, and flat histogram technique. When multiple parameters are required to be optimized simultaneously, "line searches" can be performed in the parameter space, where a gradient-based minimization procedure is coupled with an on-the-fly computation of free energy changes along parameter vectors.

In the second approach, the local minima of S_{rel} can be determined by applying standard root-finding methods to its derivative, which can be conveniently implemented as an iterative simulation protocol. For instance, by applying the Newton–Raphson iteration algorithm, a simple updating scheme can be established for optimizing a single parameter λ in the NVT ensemble as

$$\lambda_{i+1} = \lambda_i - (\partial S_{\mathrm{rel}}/\partial\lambda)/(\partial^2 S_{\mathrm{rel}}/\partial\lambda^2)$$
$$= \lambda_i - \left[\left\langle\frac{\partial U}{\partial\lambda}\right\rangle_{\boldsymbol{r}^n} - \left\langle\frac{\partial U}{\partial\lambda}\right\rangle_{\boldsymbol{R}^N}\right]\left[\left\langle\frac{\partial^2 U}{\partial\lambda^2}\right\rangle_{\boldsymbol{r}^n} - \left\langle\frac{\partial^2 U}{\partial\lambda^2}\right\rangle_{\boldsymbol{R}^N} + \beta\left\langle\left(\frac{\partial U}{\partial\lambda}\right)^2\right\rangle_{\boldsymbol{R}^N} - \beta\left\langle\frac{\partial U}{\partial\lambda}\right\rangle^2_{\boldsymbol{R}^N}\right]^{-1}. \tag{4.83}$$

Here, U is the effective potential function $U(\boldsymbol{R}^N)$ for the CGMD model, i is an iteration index, and the averaging process in the CGMD model with the subscript "$\boldsymbol{R}^N$" is performed using the trajectory obtained from a test simulation of the CGMD model with the parameter λ_i.

4.3.5 Challenges

There are two main challenges associated with the CGMD method: representability and transferability. The former is concerned with the ability of a CGMD model to represent physical properties at the thermodynamic state at which it is parameterized, while the latter is concerned with the ability of the same model to predict physical properties at different states, i.e., under conditions where parameterization data are unavailable (Potter et al., 2019). These issues exist in CGMD models developed by both the top-down and bottom-up approaches.

The problem of representability arises from the fact that a CGMD model parameterized to reproduce one property would not necessarily be able to do so for other properties at the same thermodynamic state. For instance, in most structure-based CGMD models, such as those obtained with the IBI approach, thermodynamic properties such as pressure are usually not reproduced well unless additional optimizations are performed. Similarly, the MSCG method based on force matching only guarantees the preservation of forces, but the local conformational properties may not be well represented. A particular issue regarding the representability of a CGMD model involves the CG dynamics that generally differs from that of the underlying all-atom system. Due to their fewer degrees of freedom, CGMD models have faster dynamics than their all-atom counterparts. This is a positive feature of the CGMD model when its primary objective is statistical sampling. It would be ludicrous to employ CGMD models to replicate actual dynamics, as efficient statistical sampling inherent in the models would be reduced if they were developed with the prime objective of reproducing the actual dynamics. Therefore it is an enormous challenge to construct CGMD models that are highly representative of their corresponding all-atom systems, dynamically accurate, and computationally efficient at the same time.

The problem of transferability is caused by the state-dependence of the effective potentials and is directly related to the predictive capability of a CGMD model. For example, a CG nonbonded potential developed with the IBI method relies on the precise reproduction of pair RDFs in the reference all-atom system. With a significant change in a certain thermodynamic property such as pressure, temperature, or chemical concentration, the pair RDFs could change dramatically. As a result, the CGMD potential developed at a particular thermodynamic state could become inapplicable to the same system at other states, and observations made by such a CGMD simulation could have low generalizability. Recently, Xia et al. (2017) have developed an energy-renormalization strategy based on the Adam – Gibbs theory to achieve temperature-transferable CGMD models for the study of polymer dynamics in a wide temperature range. Despite this important progress, highly transferable CGMD models that are applicable for molecular systems under variable pressure and chemical conditions remain lacking. Therefore a key challenge of CGMD is to develop a better understanding of transferability between various physical systems under different thermodynamic conditions and develop a more rigorous framework to address the transferability issue.

References

Ayton, G.S., Noid, W.G., Voth, G.A., 2007. Systematic coarse graining of biomolecular and soft-matter systems. MRS Bull. 32, 929–934.

Blöchl P. E., Parrinello M., 1992. Adiabaticity in first-principles molecular dynamics. Phys. Rev. B Condens. Matter. 45, 9413–9416.

Car, R., Parrinello, M., 1985. Unified approach for molecular dynamics and density-functional theory. Phys. Rev. Lett. 55, 2471–2474.

Chaimovich, A., Shell, M.S., 2011. Coarse-graining errors and numerical optimization using a relative entropy framework. J. Chem. Phys. 134, 094112-1–094112-15.

Deumens, E., Diz, A., Longo, R., Öhrn, Y., 1994. Time-dependent theoretical treatments of the dynamics of electrons and nuclei in molecular systems. Rev. Modern Phys. 66, 917–983.

Ganguly, P., Mukherji, D., Junghans, C., van der Vegt, N.F.A., 2012. Kirkwood-Buff coarse-grained force fields for aqueous solutions. J. Chem. Theory Comput. 8, 1802–1807.

Hoogerbrugge, P.J., Koelman, J.M.V.A., 1992. Simulating microscopic hydrodynamic phenomena with dissipative particle dynamics. Europhys. Lett. 19, 155–160.

Izvekov, S., Voth, G.A., 2005. A multiscale coarse-graining method for biomolecular systems. J. Phys. Chem. B 109, 2469–2473.

Li, B., Chen, Q., Huang, S., Liu, H., 2017. Developing structure and thermodynamic properties-consistent coarse-grained model for random copolymer systems. Polymer (Guildf) 123, 107–120.

Noid, W.G., Chu, J.-W., Ayton, G.S., Krishna, V., Izvekov, S., Voth, G.A., et al., 2008a. The multiscale coarse-graining method. I. A rigorous bridge between atomistic and coarse-grained models. J. Chem. Phys. 128, 244114-1–244114-11.

Noid, W.G., Liu, P., Wang, Y., Chu, J.-W., Ayton, G.S., Izvekov, S., et al., 2008b. The multiscale coarse-graining method. II. Numerical implementation for coarse-grained molecular models. J. Chem. Phys. 128, 244115-1–244115-2.

Pastore, G., Smargiassi, E., and Buda. F., 1991. Theory of ab initio molecular-dynamics calculations. Phys. Rev. A 44, 10, 6334–6347

Potter, T.D., Tasche, J., Wilson, M.R., 2019. Assessing the transferability of common top-down and bottom-up coarse-grained molecular models for molecular mixtures. Phys. Chem. Chem. Phys. 21, 1912–1927.

Reith, D., Pütz, M., Müller-Plathe, F., 2003. Deriving effective mesoscale potentials from atomistic simulations. J. Comput. Chem. 24, 1624–1636.

Rosso, L., Mináry, P., Zhu, Z., Tuckerman, M.E., 2002. On the use of the adiabatic molecular dynamics technique in the calculation of free energy profiles. J. Chem. Phys. 116, 4389–4402.

Saunders, M.G., Voth, G.A., 2013. Coarse-graining methods for computational biology. Ann. Rev. Biophys. 42, 73–93.

Shell, M.S., 2008. The relative entropy is fundamental to multiscale and inverse thermodynamic problems. J. Chem. Phys. 129, 144108-1–144108-7.

Tuckerman, M.E., 2002. Ab initio molecular dynamics: basic concepts, current trends and novel applications. J. Phys.: Conden. Matter 14, R1297–R1355.

Tully, J.C., 1998. Classical and Quantum Dynamics in Condensed Phase Simulations. World Scientific., Singapore.

Weerasinghe, S., Smith, P.E., 2003a. Kirkwood-Buff derived force field for mixtures of acetone and water. J. Chem. Phys. 118, 10663–10670.

Weerasinghe, S., Smith, P.E., 2003b. A Kirkwood-Buff derived force field for mixtures of urea and water. J. Phys. Chem. B 107, 3891–3898.

Xia, W., Song, J., Jeong, C., Hsu, D.D., Phelan, F.R., Douglas, J.F., et al., 2017. Energy-renormalization for achieving temperature transferable coarse-graining of polymer dynamics. Macromolecules 50, 8787–8796.

CHAPTER

5

Application of molecular dynamics simulation in mechanical problems

This chapter is dedicated to illustrating general molecular dynamics (MD) simulation methods to calculate the mechanical properties of materials under various loading conditions. It begins by presenting the models for conducting tensile, compressive, shear, nanoindentation, and nanoscratching tests via MD simulation. Afterward, popular MD simulation models for investigating the tribological behaviors of materials are illustrated, followed by the characterization methods for interfacial effects in polymer- and metal-based nanocomposites. Finally, the simulation models for investigating the effects of defects with various dimensions on material properties are discussed.

5.1 Role of molecular dynamics simulation in modeling the mechanical properties of materials

The rapid development of nanomaterials of different dimensions, such as zero-dimensional (0D) fullerenes, one-dimensional (1D) carbon nanotubes (CNTs), and two-dimensional (2D) graphene, has bestowed mankind with a vast variety of materials with superior physical, chemical, and mechanical properties. These low-dimensional nanomaterials are incorporated into polymeric and metallic matrices to form three-dimensional (3D) bulk nanomaterials or nanocomposites with enhanced mechanical strength and toughness.

The excellent properties of nanomaterials render them suitable for a wide range of applications across different fields, including nanoelectronics, energy storage, catalysis, and environmental treatments. In these applications, nanomaterials are subjected to various types of external

Molecular Dynamics Simulation
DOI: https://doi.org/10.1016/B978-0-12-816419-8.00010-6

loading. Strain engineering is usually adopted to tailor the properties of nanomaterials to fulfill specific requirements. An in-depth understanding of the mechanical properties of these materials is crucial to maximizing their applications.

Because of the low-dimensional features of most nanomaterials, achieving a systematic characterization of their mechanical responses through experimental approaches has proved to be a daunting challenge. In such cases, computational simulations are ideal for exploring the mechanical behaviors of these nanomaterials under different types of loading. Compared with other computational simulation methods, such as ab initio calculation and Monte Carlo simulation, MD simulation employs well-developed potential energy functions to capture the interactions among the atoms of a system under investigation and track their movements in detail, and therefore it is capable of accurately predicting the mechanical properties of nanomaterials.

MD simulations are also frequently adopted to understand the mechanical phenomena that occur in macroscopic bulk materials but are highly localized in small regions with nanoscale sizes or occur on a timescale that cannot be observed by experimental equipment. For instance, the mechanical behaviors of interfaces and various types of microdefects, such as cracks and dislocations, can be well captured by conducting MD simulations.

This chapter discusses MD simulation methods that are frequently adopted to study the mechanical properties of materials under various loading conditions, such as tensile loading, compressive loading, shear loading, nanoindentation, and nanoscratching. Popular MD simulation models for investigating tribological behaviors and interfacial effects in nanocomposites are presented. Some special issues such as the effects of defects on the mechanical properties of a material are also discussed.

5.2 Tensile, compressive, and shear tests

The mechanical properties of a material are commonly characterized by parameters such as the elastic modulus, shear modulus, yield strain and strength, ultimate strain, fracture strength, Poisson's ratio, and hardness. In experiments, these parameters can be measured via different tests, such as the tensile test, shear test, and indentation test. In MD simulations, these tests can be modeled to provide an accurate prediction of the mechanical properties of materials.

5.2.1 Tensile tests

A tensile test in an MD simulation is generally modeled through one of the following two approaches.

In the first approach, two boundary layers are initially defined. To apply a tensile load to the system, one of the boundary layers is fixed, while the other moves in the loading direction, as shown in Fig. 5.1. The magnitude of the velocity of the latter determines the strain rate. This approach is commonly referred to as the displacement-control method. In general, the movable boundary layer can be displaced in two manners: the dynamic manner and the incremental manner. In the dynamic manner, the layer is displaced continuously, whereas in the incremental manner, the layer is displaced incrementally, and each displacement increment is followed by a structure relaxation process during which the boundary layers are kept fixed. These two manners typically produce similar predictions for the elastic modulus and the ultimate strain of metallic materials. However, the dynamic manner tends to provide an overestimate of the fracture strength of covalent systems, since covalent bond networks typically exhibit brittle fracture instead of ductile fracture because of insufficient structure relaxation when they are subjected to tensile loading.

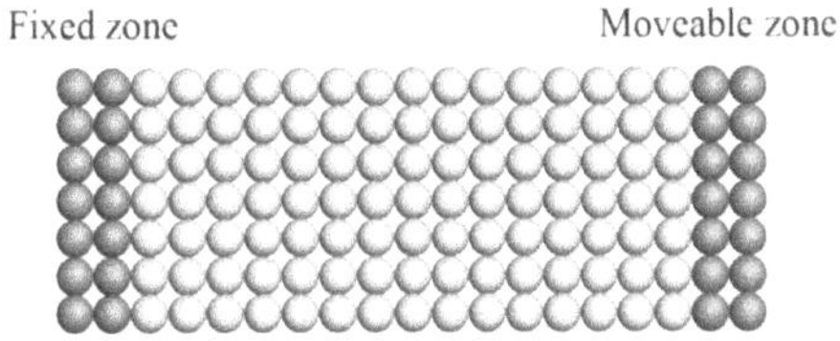

FIGURE 5.1 Schematic of an MD simulation model for tensile tests using the displacement-control method. The boundary layer on the left (blue color) is fixed, while the boundary layer on the right is moveable.

In the second approach, the simulation box is expanded at a constant rate by rescaling the coordinates of atoms along the direction of the tensile load via the relation $x_i = x_i^0[1 + \dot{\varepsilon}_x(t - t_0)]$. Here, x_i^0 denotes the x-coordinate of atom i at the starting time t_0 when the tensile load is applied, and x_i denotes its x-coordinate at time t; $\dot{\varepsilon}_x$ is the strain rate in the x-direction along which the tensile load is applied. Since the coordinates of all the atoms are rescaled by the same factor, their fractional coordinates remain constant during the entire loading process. This approach is known as the rescaling method.

In both approaches, the atomic stress must be calculated to explore the mechanical properties and deformation mechanisms of the modeled system in light of continuum mechanics. There are different methods to calculate the atomic stress in an MD simulation system. A pioneering work was conducted by Born and Huang (1954), who utilized an elastic energy approach to evaluate the stress in crystal lattices using the Cauchy–Born hypothesis for homogeneous deformation. This hypothesis establishes a direct relationship between microscale and macroscale deformations and states that in a crystalline solid subjected to a small strain, the positions of the atoms within the crystal lattice follow the overall strain of the crystal.

Another widely used representation of the atomic stress in MD simulations is the virial stress, which is based on a generalization of the virial theorem. Specifically, the virial stress $\boldsymbol{\sigma}^{\mathrm{virial}}(\boldsymbol{r})$ of a volume element Ω surrounding the position vector $\boldsymbol{r} = (r_x, r_y, r_z)$ consists of two parts and can be expressed as (Shen and Atluri, 2004)

$$\boldsymbol{\sigma}^{\mathrm{virial}}(\boldsymbol{r}) = \frac{1}{\Omega}\sum_{i}\left[-m_i(\dot{\boldsymbol{u}}_i)^{\mathrm{T}} \otimes (\dot{\boldsymbol{u}}_i)^{\mathrm{T}} + \frac{1}{2}\sum_{j\neq i}(\boldsymbol{r}_{ij})^{\mathrm{T}} \otimes \left(\boldsymbol{f}_{ij}\right)^{\mathrm{T}}\right]. \tag{5.1}$$

Here, i and j index the atoms that occupy the volume element Ω; m_i and $\boldsymbol{u}_i$ denote the mass of atom i and its displacement vector with respect to a selected reference point, respectively; $\otimes$ denotes the dyadic product operator of two vectors; $\dot{\boldsymbol{u}}_i$ is the time-derivative of $\boldsymbol{u}_i$; $\boldsymbol{r}_{ij}$ denotes the atomic distance between atoms i and j; $\boldsymbol{f}_{ij}$ denotes the interatomic force imposed on atom i due to atom j, which is calculated by Eq. (1.2). The first term on the right-hand side of Eq. (5.1) is a kinetic energy term, which accounts for mass transport across a surface fixed in space, whereas the second term is a potential energy term. In an MD simulation, the element volume Ω is usually taken to be the volume of the simulation box.

The local virial stress tensor for atom i is given by

$$\boldsymbol{\sigma}_i^{virial} = \frac{1}{\Omega_i}\left[-m_i(\dot{\boldsymbol{u}}_i)^{\mathrm{T}} \otimes (\dot{\boldsymbol{u}}_i)^{\mathrm{T}} + \frac{1}{2}\sum_{j\neq i}(\boldsymbol{r}_{ij})^{\mathrm{T}} \otimes \left(\boldsymbol{f}_{ij}\right)^{\mathrm{T}}\right], \tag{5.2}$$

where Ω_i is the volume surrounding atom i and satisfies the condition $\sum_i \Omega_i = \Omega$. A precise definition of Ω_i is usually difficult to obtain. For a condensed system, Ω_i is simply taken to be the average atomic volume, which is obtained by dividing the total volume of the simulation box by the total number of atoms within it. Note that Eqs. (5.1) and (5.2) are expressions of the instantaneous virial stress, which is useful for studying nonequilibrium mechanical processes in which stress distributions evolve with time. For equilibrium processes, a time average of the virial stresses is usually performed to reduce statistical errors.

In MD simulations, atomic strain is also defined in a manner that is completely different from that in continuum mechanics. According to a study by Liu et al. (2013), the atomic strain of a target atom can be defined based on the displacements of its neighboring atoms. For points in a continuous medium, the relationship $\boldsymbol{v}_{ij} = \boldsymbol{r}_{ij}\boldsymbol{L}$ holds if the velocity vectors $\boldsymbol{v}_{ij}$ of these points form an affine vector space. Here, $\boldsymbol{L}$ denotes a second-order tensor or matrix of deformation rate; $\boldsymbol{v}_{ij}$ and $\boldsymbol{r}_{ij}$ denote the relative velocity and atomic distance between two points i and j, respectively, and are expressed by

$$\boldsymbol{r}_{ij} = \boldsymbol{r}_i - \boldsymbol{r}_j, \tag{5.3}$$

$$\boldsymbol{v}_{ij} = \boldsymbol{v}_i - \boldsymbol{v}_j. \tag{5.4}$$

However, the velocity field in an MD simulation system is nonaffine because of the random motion and local arrangement of the atoms. Therefore, the relationship $\boldsymbol{v}_{ij} = \boldsymbol{r}_{ij}\boldsymbol{L}$ does not hold for the atoms, and the matrix $\boldsymbol{L}$ only serves to minimize the following error function:

$$\begin{aligned}\sum\nolimits_{j\in\Phi_{R_c}^i} \left|\boldsymbol{r}_{ij}\boldsymbol{L}-\boldsymbol{v}_{ij}\right|^2 &= \sum\nolimits_{j\in\Phi_{R_c}^i} \left(\boldsymbol{r}_{ij}\boldsymbol{L}-\boldsymbol{v}_{ij}\right)\left(\boldsymbol{r}_{ij}\boldsymbol{L}-\boldsymbol{v}_{ij}\right)^{\mathrm{T}} \\ &= \mathrm{tr}\left(\sum\nolimits_{j\in\Phi_{R_c}^i} \left(\boldsymbol{r}_{ij}\boldsymbol{L}-\boldsymbol{v}_{ij}\right)^{\mathrm{T}}\left(\boldsymbol{r}_{ij}\boldsymbol{L}-\boldsymbol{v}_{ij}\right)\right),\end{aligned} \tag{5.5}$$

where tr(.) represents the trace of a matrix, and $\Phi_{R_c}^i$ denotes the set of atoms within a sphere of radius R_c with atom i at its center. At the minimum point of the error function, any infinitesimal variation in $\boldsymbol{L}$ should satisfy the condition:

$$\mathrm{tr}\left(\sum_{j\in\Phi_{R_c}^i} \left(\boldsymbol{r}_{ij}\boldsymbol{L}-\boldsymbol{v}_{ij}\right)^{\mathrm{T}}\left(\boldsymbol{r}_{ij}\delta\boldsymbol{L}\right)\right) = \mathrm{tr}\left(\left[\sum_{j\in\Phi_{R_c}^i} \left(\boldsymbol{r}_{ij}\boldsymbol{L}-\boldsymbol{v}_{ij}\right)^{\mathrm{T}}\boldsymbol{r}_{ij}\right]\delta\boldsymbol{L}\right) = 0. \tag{5.6}$$

Since the variant $\delta\boldsymbol{L}$ is arbitrary, Eq. (5.6) is reduced to

$$\sum_{j\in\Phi_{R_c}^i} \left(\boldsymbol{r}_{ij}\boldsymbol{L}-\boldsymbol{v}_{ij}\right)^{\mathrm{T}}\boldsymbol{r}_{ij} = \boldsymbol{0}. \tag{5.7}$$

Therefore, the atomic deformation rate can be obtained as

$$\boldsymbol{L} = \left(\sum_{j\in\Phi_{R_c}^i} \left(\boldsymbol{v}_{ij}\right)^{\mathrm{T}}\boldsymbol{r}_{ij}\right)\left(\sum_{j\in\Phi_{R_c}^i} \left(\boldsymbol{r}_{ij}\right)^{\mathrm{T}}\boldsymbol{r}_{ij}\right)^{-1}. \tag{5.8}$$

The atomic strain rate tensor $\dot{\varepsilon}_i$ is thus the symmetric part of $\boldsymbol{L}$, i.e., $\dot{\varepsilon}_i = \left(\boldsymbol{L}+\boldsymbol{L}^{\mathrm{T}}\right)/2$.

To illustrate the MD simulation procedure for modeling a tensile test, the following two case studies are presented.

The first study is the modeling of the tensile behaviors of a nanofilm. Diamond-like carbon (DLC), which is an allotrope of carbon and possesses an amorphous structure and excellent physical, mechanical, and chemical properties such as high hardness, high wear resistance, and chemical inertness. Hence, DLC is selected for the modeling cases in this section. In this example, tensile strain is applied using the rescaling method. Atomic interactions within the DLC nanofilms are described by the Tersoff potential. The initial structure of DLC is obtained through a melting–quenching procedure with periodic boundary conditions (PBCs) applied along the *x*-, *y*-, and *z*-directions. During the melting–quenching process, a crystalline diamond structure with a lattice parameter of 3.6 Å is first created at a temperature of 300 K and then heated to a temperature beyond the melting point of crystalline diamond in the NVT (constant number of species, volume, and temperature) ensemble before being relaxed at this temperature for 20 ps to achieve thermal equilibrium. Afterward, it is cooled to 300 K at a cooling (or quenching) rate of 1000 K/ps. The system volume is subsequently adjusted at zero pressure and 300 K in the NPT (constant number of species, pressure, and temperature) ensemble to reduce the residual stresses induced during the quenching process. The Nose–Hoover thermostat and the MTTK barostat (see Chapter 3: Control Techniques of Molecular Dynamics Simulation) are adopted for temperature control and pressure control, respectively.

The as-prepared DLC structure in the melting–quenching process is then subjected to tensile loading in the *x*-direction with a strain rate of 0.002 ps^{-1} at 300 K by the rescaling method, as shown in Fig. 5.2. PBCs are applied along the *x*- and *z*-directions, while the free boundary condition is applied in the *y*-direction to model a nanofilm.

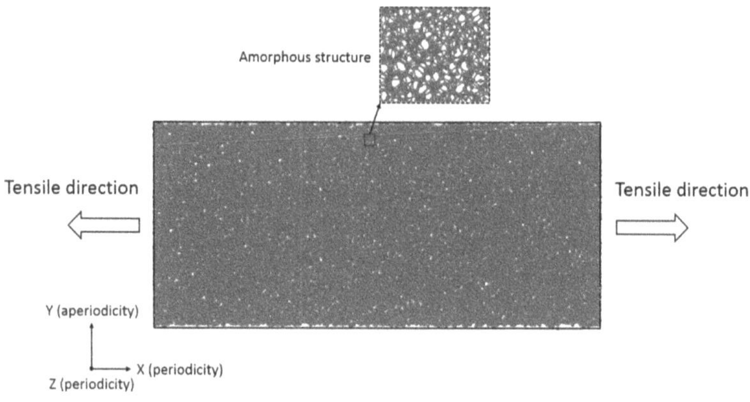

FIGURE 5.2 MD simulation model for the tensile test of a DLC nanofilm in which the simulation box is continuously expanded. Source: *Reprinted with permission from a work by Bai et al. (2016d).*

Fig. 5.3 presents the stress–strain curve obtained from the tensile test simulation. In the initial stage of the tensile test, the stress increases linearly with the strain, which indicates a linear elastic deformation behavior. The deformation is also reversible. The elastic modulus is determined to be ~539 GPa from the gradient of

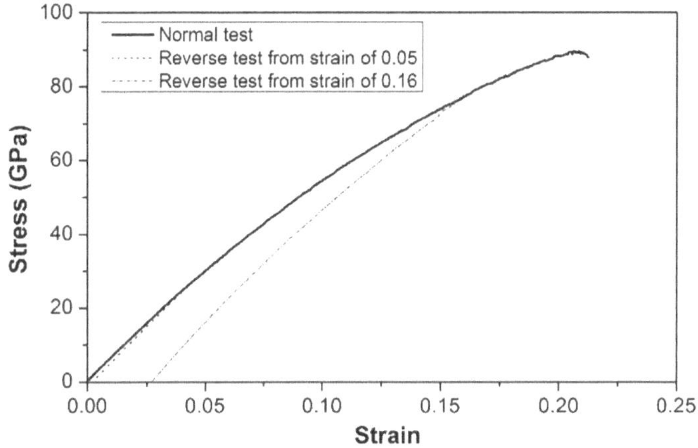

FIGURE 5.3 Stress–strain curves of the normal tensile test and reverse unloading test of a DLC nanofilm. Source: *Reprinted with permission from a work by Bai et al. (2016d).*

the linear relationship. As the strain increases further, the stress–strain curve becomes nonlinear and exhibits a continuously decreasing gradient until fracture of the DLC film occurs at the ultimate strain of 0.21.

In addition to the normal tensile test, an unloading test can be conducted by gradually shrinking the simulation box after it has expanded to a certain tensile strain. The stress–strain curves for unloading tests corresponding to tensile strains of 0.05 and 0.16 are delineated in Fig. 5.3 with a dotted line and a dashed line, respectively. From the unloading stress–strain curves, it can be observed that given the same strain, the stress in the unloading process is lower than that in the loading process, which indicates that microstructural changes occur during the tensile loading process.

To further investigate the microstructural changes of the DLC nanofilm during the tensile loading process, the evolution of its total potential energy, its radial distribution function (RDF), and the distributions of its atomic strain and bond angles are analyzed. Such analyses can be conveniently conducted through freely available packages such as VMD (short for visual MD) and OVITO (short for open visualization tool).

In addition, the coordination numbers of the C atoms are calculated using a cutoff distance of 1.90 Å (a value corresponding to the position of the first minimum in the RDF of C atoms in the DLC nanofilm). In particular, the fourfold C atoms (atoms with a coordination number of 4) and threefold C atoms (atoms with a coordination number of 3) are regarded as sp^3- and sp^2-bonded atoms, respectively.

The simulation results indicate that the DLC nanofilm undergoes graphitization (transition of sp^3-bonded C atoms to sp^2-bonded C atoms) during the tensile test, particularly when a large tensile strain is reached. In addition, graphitization occurs preferentially in regions that contain sp^2-bonded carbon clusters and hence causes atomic strain localization, which in turn further increases the concentration of sp^2-bonded C atoms in these regions and leads to the ultimate fracture of the nanofilm.

The second study pertains to the tensile behaviors of a nanowire. T-carbon, an allotrope of carbon, is modeled in this case study. It exhibits a lattice structure similar to that of diamond, of which each C atom of the diamond lattice is replaced with a tetrahedral carbon nanocluster as shown in Fig. 5.4. In the simulation model, various T-carbon nanowires are tailored to have one of their lattice directions, including [100], [110], and [111], along their axial or longitudinal direction to study the lattice orientation effect on their mechanical properties. The diameter of the nanowires is set at 6 nm, and their length is 40 nm. Since the length of the nanowires is over five times their diameter, the length effect on their microstructural evolution can be neglected (Liu and Shen, 2012).

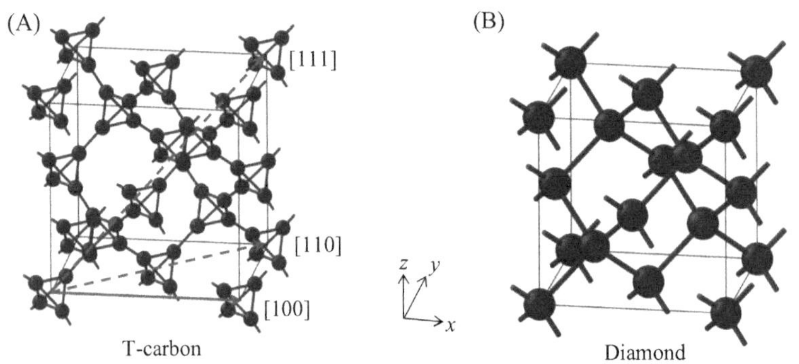

FIGURE 5.4 Lattice structure of (A) T-carbon and (B) diamond. Source: *Reprinted with permission from a work by Bai et al. (2018).*

The interactions among the C atoms are described by the AIREBO (short for adaptive intermolecular reactive empirical bond order) potential, which is capable of accurately predicting the properties of carbon materials, including diamond, graphene, amorphous carbon, and many carbohydrate materials. Since the mechanical properties of bulk carbon materials are highly dependent on their covalent bonds only, the functional term in the AIREBO potential that describes van der Waals forces (see Chapter 2: Potential Energy Functions) is disregarded to reduce computational costs. A time step of 1 fs is set for the simulation.

Initially, a T-carbon nanowire is relaxed at 300 K in the NPT ensemble. Afterward, the simulation is continued in the NVT ensemble, and a tensile load is applied to the nanowire using the displacement-control method in the incremental displacement manner discussed above. Each displacement increment of the movable boundary layer is realized by moving it at a strain rate of 0.01 ps^{-1} for 1 ps and followed by a structure relaxation process for 5 ps.

The simulation results indicate that the T-carbon nanowire exhibits significant anisotropy in its mechanical strength and better ductility than common carbon materials such as diamond and DLC because of the global graphitization induced in T-carbon during the tensile loading. Fig. 5.5 presents the stress–strain ($\sigma - \varepsilon$) curves of T-carbon nanowires with different lattice orientations at 300 K. It can be observed that their tensile response highly depends on the lattice direction. When the longitudinal direction of the nanowire is aligned with the [110] lattice direction, the $\sigma - \varepsilon$ curve can be divided into three stages. The tensile stress σ increases rapidly in Stage I ($\varepsilon < 0.13$) but fluctuates around a constant value in Stage II ($0.13 < \varepsilon < 0.4$). In stage III ($0.4 < \varepsilon$), σ continues to increase until the nanowire fractures.

When the longitudinal direction of the nanowire is oriented along the [111] lattice direction, its $\sigma - \varepsilon$ curve can also be divided into three stages. However, Stage I ($\varepsilon < 0.03$) for the [111] lattice direction is much shorter than that for the [110] lattice direction. In Stage II ($0.03 < \varepsilon < 0.2$), the tensile stress σ keeps increasing but at a lower rate than that in Stage I. For the [100] lattice direction, the $\sigma - \varepsilon$ curve can no longer be divided into different stages, and the tensile stress increases with the tensile strain in a monotonic manner until the nanowire fractures.

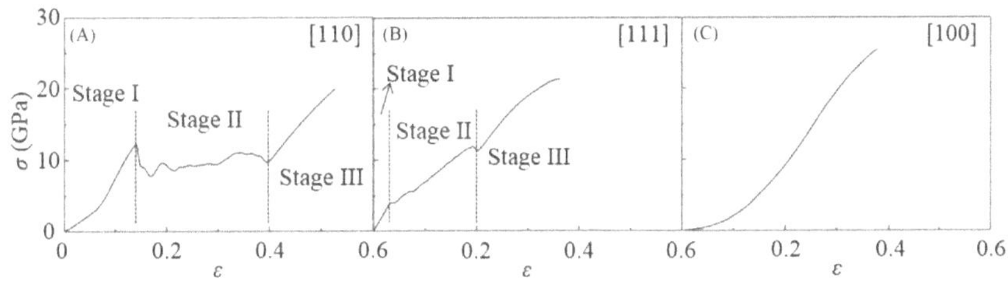

FIGURE 5.5 Stress–strain curves of T-carbon nanowires with the longitudinal direction oriented in different lattice directions: (A) [110], (B) [111], and (C) [100]. Source: *Reprinted with permission from a work by Bai et al. (2018).*

The different tensile behaviors along the different lattice orientations are closely related to the microstructural evolution of the T-carbon nanowire during the tensile loading process. The microstructural evolution of T-carbon mainly involves global graphitization that can be characterized by computing fractions of sp^3- and sp^2-bonded C atoms in the nanowire. The bonding states of the C atoms are evaluated by calculating their atomic coordination number, which is similar to the study of the DLC nanofilm discussed previously. The coordination numbers C_N of 4, 3, and 2 correspond to the sp^3, sp^2, and sp bonding states of a C atom, respectively.

Initially, the fraction of sp^2-bonded C atoms in the nanowire is around 15%. Fig. 5.6 shows the microstructure of the T-carbon nanowire at a tensile strain of 0.3 derived from the bonding states of the C atoms. For the [100] lattice direction, it is evident that the C atoms at the boundaries of the T-carbon nanowire are predominantly sp^2-bonded atoms, while the C atoms within the nanowire are predominantly sp^3-bonded atoms. For the [110] direction, the majority of the internal C atoms (~65%) are sp^2-bonded, indicating that the nanowire experiences a global graphitization. For the [111] direction, the fraction of sp^2-bonded C atoms is even higher (~80%), which indicates a higher graphitization level.

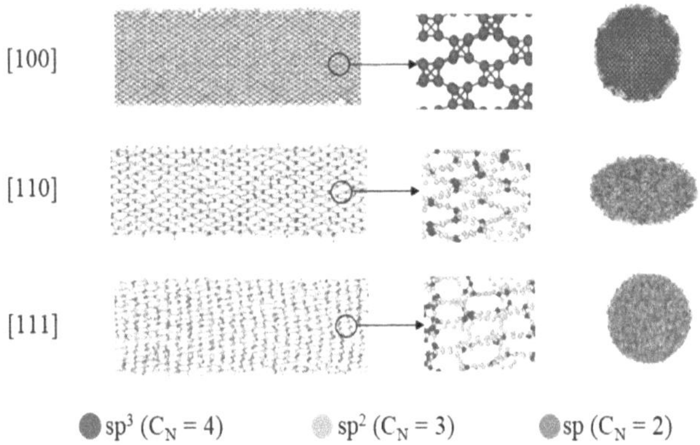

FIGURE 5.6 Microstructure of a T-carbon nanowire with the distributions of sp^3-, sp^2-, and sp-bonded C atoms at a tensile strain of 0.3. Source: *Reprinted with permission from a work by Bai et al. (2018).*

A detailed microstructure analysis indicates that the global graphitization of the T-carbon nanowire occurs through the formation and extension of sp^2-bonded C atom clusters. Specifically, when the tensile strain reaches a critical value, the sp^2-bonded C atom clusters can be induced locally near the boundary of the nanowire. With a further increase in the tensile strain, such sp^2 clusters can subsequently extend into the networks formed by sp^3-bonded C atoms underneath the surface of the nanowire, thereby resulting in the final global graphitization.

Besides T-carbon nanowires, MD simulations have been conducted to investigate the tensile behaviors of many other 1D nanomaterials, and various interesting phenomena have been observed. For instance, it has been demonstrated that the deformation mode of a CNT with a small diameter can be transformed from ductile to brittle by a low temperature, which is a result of defect localization (Tang et al., 2009). The Young's modulus and tensile strength of silicon nanowires have been discovered to be size-dependent, but their failure strain is not (Xu and Kim, 2019). Cast amorphous silica nanowires have been shown to demonstrate a higher stiffness and tensile strength but a lower failure strain than their counterparts obtained from the cutting method, which can be attributed to more surface defects being generated during the cutting process (Yuan and Huang, 2012).

Note that in MD simulations of the tensile tests, the strain rate that is applied to the modeled system is usually in the order of $10^8-10^{10}\ s^{-1}$, which is several orders of magnitude larger than those generally adopted in experimental tensile tests ($0.001-1\ s^{-1}$). The ultrahigh strain rates applied in MD simulations are common and are mainly caused by their time scale limitation. In general, such high strain rates can lead to enhanced brittleness of the investigated materials and changes in their deformation mode. For example, at very high strain rates, tensile strain applied to a covalent material mainly increases its bond lengths, while at low strain rates, the tensile strain induces changes in its bond angles first and subsequently increases its bond lengths. Hence, for practical MD simulations, a low strain rate is preferable for accurately predicting the intrinsic mechanical behaviors of materials. However, a strain rate as low as those adopted in experimental measurements is normally impractical because of the huge computational resources required.

5.2.2 Compressive tests

Compressive tests are conducted to study the mechanical properties of a material under compressive loading. Mechanical parameters including the compressive modulus, yield strength, and compressive strength can be obtained from compressive tests. The mechanical properties of ductile materials obtained based on tensile tests and compressive tests are almost identical. However, brittle materials such as ceramics commonly exhibit different behaviors in compressive tests and tensile tests. Hence, compressive tests are essential for a complete characterization of the mechanical properties of brittle materials.

In MD simulations, compressive tests can be modeled in a manner similar to that of tensile tests, i.e, by shrinking the simulation box via rescaling the atomic coordinates or via moving one boundary layer while fixing the other in place. Take the uniaxial compression of a 1D structure as an example. Kang et al. (2010) have applied MD simulation to investigate the buckling of an intramolecular junction (IMJ) formed by two CNTs with different diameters, as shown in Fig. 5.7. This type of IMJ can serve as crucial functional building blocks for novel nanoelectronics. In such a simulation, the Tersoff potential (see Chapter 2: Potential Energy Functions) is used to describe the interatomic interactions of the C atoms. The entire simulation is divided into two stages. In the first stage, the IMJ is relaxed by conducting energy minimization using the conjugate gradient algorithm under free boundary conditions. During the structure relaxation, the temperature of the IMJ is kept at 0 K by coupling the system with a Nose–Hoover thermostat to minimize the effects of thermal vibrations. In the second stage, a compressive strain is applied to the IMJ by using the displacement-control method in the dynamic displacement manner. To investigate the effect of strain rate, the order of its magnitude is adjusted from 1 ps^{-1} to 10^{-6} ps^{-1}. The temperature of the simulation system is maintained at 300 K by coupling a Nose–Hoover thermostat to a local region that is adjacent to the fixed boundary.

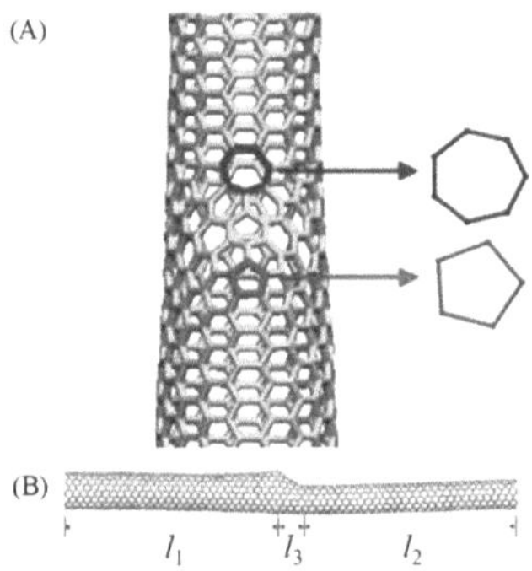

FIGURE 5.7 Schematic of an IMJ formed by two CNTs with different diameters: (A) local views of the connecting region and (B) geometrical structures. Source: *Reprinted with permission from a work by Kang et al. (2010).*

The simulation results indicate that the critical compressive strain that initiates buckling of the IMJ exhibits a weak dependence on a low strain rate but a strong dependence on a high strain rate. The critical strain also depends on the total length and average diameter of the IMJ.

Besides the standard compressive test, MD simulation is also capable of modeling cyclic compressive–unloading or tensile–compressive tests, which are important for characterizing the fatigue behavior of a material. The simulation models for such tests are the same as the model for a monotonic compressive test, except that the displacement direction of the movable boundary in Fig. 5.1 undergoes cyclic changes in the cases of cyclic tests. For example, an MD simulation has been conducted to explore the tension–compression fatigue behavior of $Cu_{50}Zr_{50}$ metallic glass, and the following questions have been answered (Sha et al., 2015). (1) How are the shear bands formed and how do they propagate under cyclic loading? (2) What are the differences between the shear band formation under cyclic loading and that under monotonic loading? (3) What are the basic deformation characteristics and underlying deformation mechanisms of metallic glasses under cyclic loading? (4) What are the effects of the cyclic frequency on the fatigue behaviors of metallic glasses?

5.2.3 Shear tests

Shear testing is commonly conducted to determine the shear modulus and shear strength of a material. It is not only a vital procedure in the design of mechanical components such as bolts and screws but also a useful method for characterizing the mechanical properties of novel materials, which include rigid materials and layered composites.

Experimentally, shear testing is conducted by applying a load to a test specimen to induce shear deformation, during which its two surfaces move in opposite directions parallel to the shear loading movement. Similarly to the tensile and compressive tests, shear tests in an MD simulation can also be modeled by either the displacement-control method or the rescaling method.

Take a rectangular simulation box as an example. As shown in Fig. 5.8, the initially orthogonal simulation box is deformed along the x-direction in the $x-y$ plane, in which the tangent of the tilt angle θ between the $x-z$ plane and $y-z$ plane is denoted by the tilt factor T_f. The shear strain rate is defined as the changing rate of the tilt factor T_f.

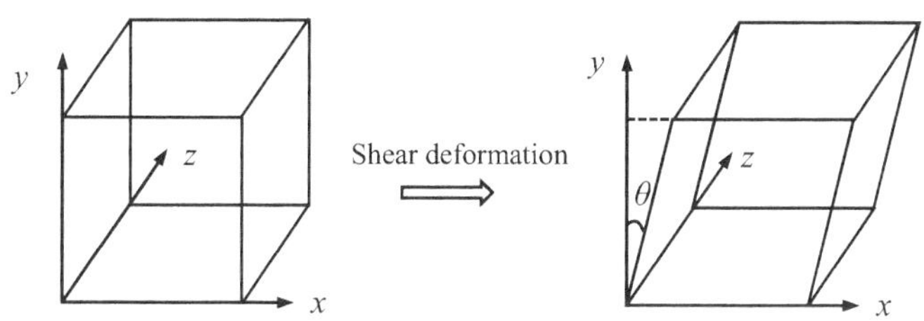

FIGURE 5.8 Schematic of the shape change of an orthogonal simulation box in response to shear loading along the x-direction in the $x-y$ plane.

For a given shear strain rate $\dot{\varepsilon}_{xy}$, the tilt factor at time t can be expressed by the relationship $T_f = T_f^0 + \dot{\varepsilon}_{xy}(t - t_0)$. Here, T_f^0 is the initial tilt factor at $t = t_0$, and its value equals zero for an initially orthogonal simulation box.

For the rescaling method, at each time step, the x-coordinate of atom i is rescaled according to its y-coordinate through the relationship $x_i(t) = x_i(t_0) + (T_f - T_f^0)y_i(t_0)$. Here, $x_i(t_0)$ and $y_i(t_0)$ denote the coordinates of atom i at $t = t_0$. For the displacement-control method, the simulation model is the same as the model presented in Fig. 5.1, except that their displacement directions are different. The magnitude of the displacement of the movable boundary layer along the shear loading direction at each time step is given by $l_y\dot{\varepsilon}_{xy}\Delta t$ with l_y and Δt denoting the length of the simulation box in the y-direction and the simulation time step, respectively.

The modeling of the shear test via MD simulation is exemplified below by a study of the mechanical behaviors of a graphene nanosheet functionalized by hydrogen under a shear load (Kheirkhah et al., 2014). During the simulation, the initial structure of the functionalized graphene nanosheet is first relaxed in the NPT ensemble at zero pressure and 300 K. Afterward, the simulation is continued in the NVT ensemble, and a shear load is applied at a strain rate of 0.001 ps^{-1} using the displacement-control method. Specifically, a thin layer of C atoms on one side of the nanosheet is fixed, while a thin layer of C atoms on the other side is displaced by a constant value at each time step (as shown in the inset of Fig. 5.9A). The simulation results indicate that hydrogenation greatly degrades the mechanical properties of graphene (Fig. 5.9B). A small hydrogen coverage of 5% reduces the shear strength of the graphene by over 20%. At a full (100%) hydrogen coverage, the shear

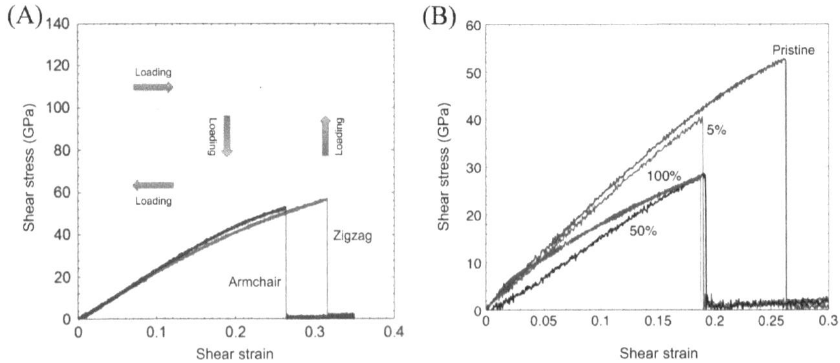

FIGURE 5.9 Stress–strain curves of (A) a pristine graphene nanosheet and (B) hydrogen-functionalized graphene under shear loading for various values of hydrogen coverage. Source: *Modified and reprinted with permission from a work by Kheirkhah et al. (2014).*

strength is reduced by over 50%. A detailed analysis indicates that the fracture of the functionalized graphene nanosheet initializes at the locations of the H atoms. In addition, wrinkles of graphene are induced during its shear deformation as shown in Fig. 5.10. The average amplitude of the wrinkles increases with both the hydrogen coverage and the shear strain.

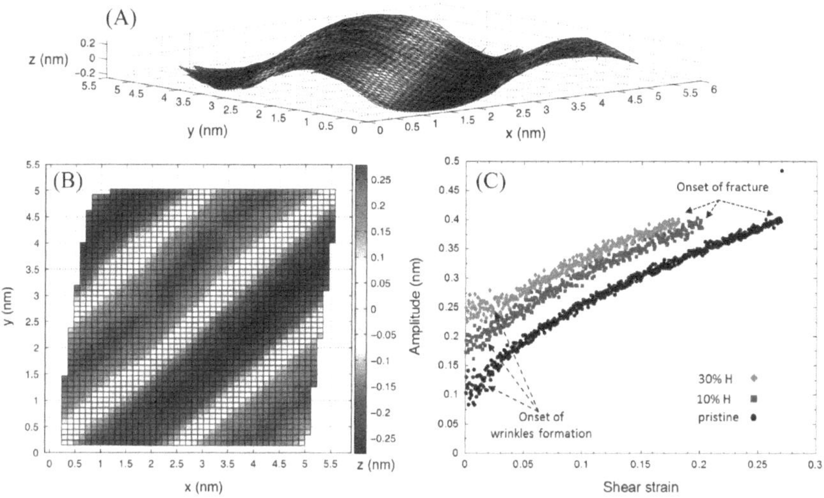

FIGURE 5.10 Formation of wrinkles in a functionalized graphene nanosheet under shear loading: (A) 3D contour plot of the displacements of the C atoms along the out-of-plane direction, (B) projection of the contour plot in the $x-y$ plane, and (C) dependence of the wrinkle amplitude on the shear strain and hydrogen coverage. Source: *Reprinted with permission from a work by Kheirkhah et al. (2014).*

5.3 Nanoindentation and nanoscratching tests

5.3.1 Nanoindentation tests

The nanoindentation test is a useful technique for characterizing various mechanical properties (e.g., hardness, elastic modulus, and viscoelasticity) of materials ranging from hard alloys to soft biomaterials. The parameters measured during the nanoindentation test include the loading force and the indentation depth.

In MD simulations, nanoindentation tests are usually modeled by forcing an indenter to penetrate a specimen. The indenter is commonly considered a rigid body during the simulation and may be designed to exhibit different geometries such as a cone or a half-sphere as shown in Fig. 5.11. In a manner similar to the procedure in an experimental nanoindentation test, an MD simulation model of the

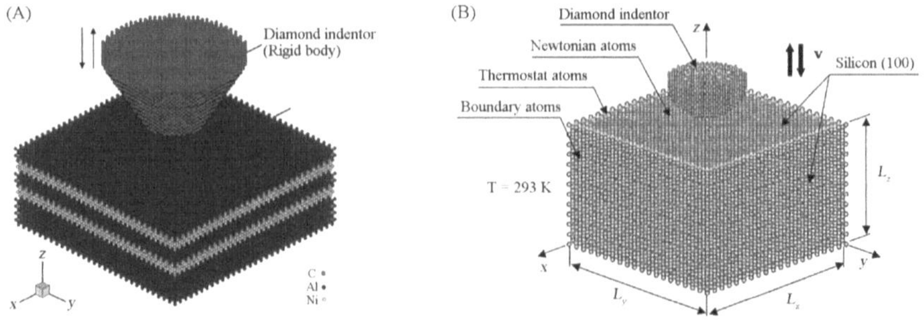

FIGURE 5.11 MD simulation models of the nanoindentation test, in which the indenters exhibit different geometries: (A) cone; (B) half-sphere. Source: *(A) Reprinted with permission from a work by Fang and Wu (2008); (B) reprinted with permission from a work by Cheong and Zhang (2000).*

nanoindentation test also includes a loading and unloading stage. The indenter moves toward the specimen at a preset velocity in a direction perpendicular to the specimen surface in the loading stage and in the opposite direction in the unloading stage. During the simulation, the velocity of the indenter and the simulation temperature are adjustable, which enables the study of their effects on the mechanical properties of the specimen.

In the simulation work of Fang and Wu (2008), a cone-shaped diamond indenter has been selected to model the nanoindentation test of a multilayer film composed of alternative aluminum and nickel thin layers (Fig. 5.11A). The conic angle of the indenter is 60°, and it has a spherical radius of 4 nm at the smaller end. PBCs are applied in the lateral directions. Two atomic layers at the bottom of the multilayer specimen are fixed in space during the simulation. The simulation system is first equilibrated at a preset temperature in the NVT ensemble for 100 ps before the nanoindentation process is initiated. In the loading stage of the indentation test, the indenter moves toward the specimen at a constant velocity of 100 m/s. When a preset indentation depth (i.e., penetration distance of the indenter into the specimen) is reached, the indenter is fixed in place for 15 ps to allow structural relaxation. The unloading process is subsequently initiated by relieving the contact between the specimen and the indenter. During the simulation, the loading force exerted on the indenter is calculated by summing up the forces exerted on each of its constituent atoms.

Through the simulation, the evolution of the indentation force can be investigated as shown in Fig. 5.12A. A jump-contact behavior can be observed during the loading process based on the evolution of the indentation force. Specifically, in the loading process, the loading force increases sharply. However, during the relaxation period, the loading force decreases rapidly because of stress relaxation of the specimen atoms near the indenter, thus exhibiting a prominent "jumping" behavior. As the indenter is withdrawn from the specimen in the unloading process, some specimen atoms adhere to the surface of the indenter (Fig. 5.12B). Separating these atoms from the specimen leads to a negative loading force. With an increasing indentation depth, the negative adhesion-induced loading force increases.

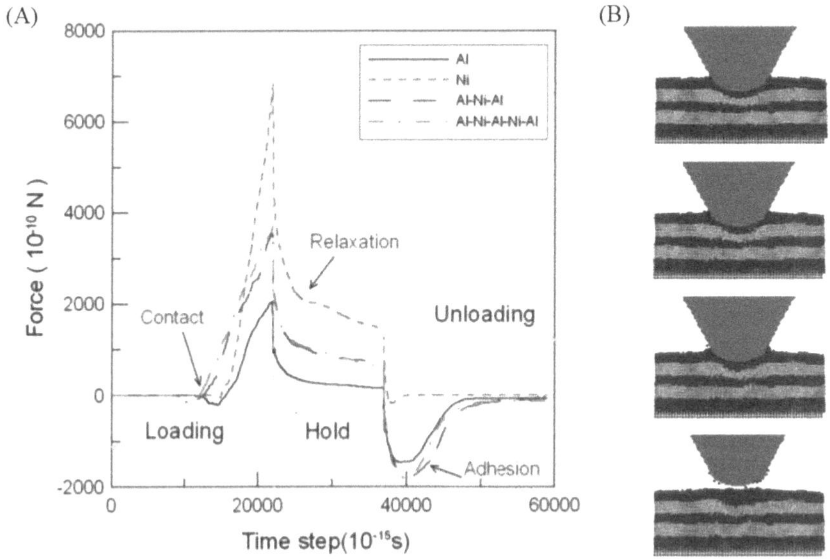

FIGURE 5.12 Loading–unloading process of the nanoindentation test of a multilayer film composed of alternative aluminum and nickel thin layers: (A) evolution of the loading force and (B) snapshots of the indentation process at a temperature of 300 K. Source: *Reprinted with permission from a work by Fang and Wu (2008).*

The nanoindentation test has been used to characterize the mechanical response of many different types of materials. For instance, MD simulations of nanoindentation have been performed on monocrystalline and nanotwinned tantalum nanofilms to unravel the effects of the twin boundary and twin thickness on their hardness and corresponding deformation mechanisms (Huang et al., 2017). In a simulation study conducted by Walsh et al. (2003) on both crystalline and amorphous silicon nitride nanofilms using a model containing 10 million atoms, a significant pile-up of the atoms from the film has been observed on its surface, which can be attributed to the pressure imposed by the indenter.

Additionally, nanoindentation can result in severe plastic deformation in regions near the indenter tip, and such plastic deformation can extend up to a significant distance beyond the contact area between the indenter and the film. Lu et al. (2015) have studied the deformation behaviors of single-crystal zirconium for different lattice orientations by conducting nanoindentation tests via MD simulation. The simulation results indicate that both elastic deformation and plastic deformation in the Zr crystal contribute to its hardness, with the plastic deformation playing a dominant role.

5.3.2 Nanoscratching tests

The nanoscratching test is used to measure the adhesion strength and failure modes of coatings and to characterize the nanoscale properties of bulk materials (Beake and Leggett, 2002). For instance, it can be used to evaluate the relationship between the mechanical properties of coatings and their adhesion strength to the corresponding substrates, which provides insight for improving the tribological performance of coatings. It can also be utilized to characterize the tribological behaviors of bulk materials such as human root canal dentin, bone cements formed by polymethyl methacrylate (PMMA), and other hip replacement components (Beake et al., 2013; Shokrieh et al., 2013). In experiments, nanoscratching testing is conducted by moving a tip under a ramped or constant load, often in a repetitive manner. Similarly to those of nanoindentation tests, the tips utilized in nanoscratching tests may be modified to exhibit different geometries for a variety of applications.

MD simulations can be employed to model nanoscratching tests. Compared with experiments, MD simulations are capable of correlating the normal load and friction forces with the structure evolution of materials under testing, and thus reveal their elastic and plastic deformation mechanisms. In MD simulations, the nanoscratching test is commonly conducted by moving a rigid tip on the specimen surface. The motion of the tip is typically controlled in two manners. In the first manner, the tip penetrates the specimen to a preset depth before sliding along its surface at a constant velocity and a fixed penetration depth. In the second manner, the tip penetrates the specimen at a constant vertical velocity without control of the penetration depth and simultaneously moves with a horizontal velocity. During the scratching process, forces exerted on the tip along the horizontal and vertical directions are measured to determine the mechanical response of the specimen.

The MD simulation of the nanoscratching test in the first tip-motion control manner is exemplified by a study of the frictional behavior of a multilayer graphene nanosheet (Zhang et al., 2015), as shown in Fig. 5.13. In this study, the simulation model consists of a hemispherical diamond tip with a diameter of 5 nm sliding against a four-layer graphene nanosheet. The tip is first indented toward the graphene nanosheet to a preset depth, followed by a structure relaxation process that lasts for 50 ps with a time step of 1 fs. Afterward, the tip moves at a fixed penetration depth along the graphene multilayer at a speed of 5 m/s. The atoms at the left and right boundaries of each graphene layer are fixed to restrict their movements with the tip. In addition, the tip is considered a rigid body to avoid possible atomic wear of its surface. It is found that the friction coefficient between the diamond tip and the graphene multilayer is extremely low when the scratch depth is smaller than 5.3 Å. However, when the scratch depth surpasses this value, the friction coefficient increases by at least 10 times, which is caused by cross-links formed between adjacent graphene layers.

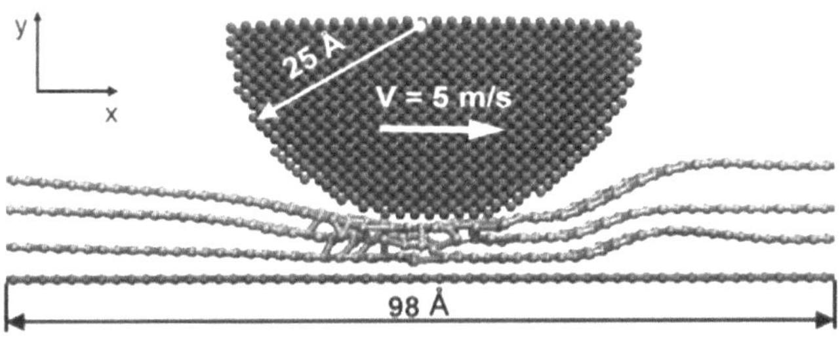

FIGURE 5.13 MD simulation model of the nanoscratching test of a multilayer graphene nanosheet using a diamond tip. Source: *Reprinted with permission from a work by Zhang et al. (2015).*

In a study by Bai et al. (2016c), an MD simulation has been conducted to assess the feasibility of employing graphene as a solid lubricant due to the frictionless characteristic of its surface. In such a simulation, a DLC film that is passivated by H atoms and covered by a graphene nanosheet is scratched by a hemispherical rigid diamond tip, as shown in Fig. 5.14. The atomic interactions are described by the AIREBO potential. During the scratching process, the diamond tip moves in both the horizontal and vertical directions at the same speed. Consequently, the indentation depth increases linearly with the simulation time. A bottom layer of the DLC film is defined and kept stationary to eliminate film motion during the simulation, and the temperature of the simulation system is kept at 300 K by coupling it with a Nose–Hoover thermostat. The simulation results demonstrate

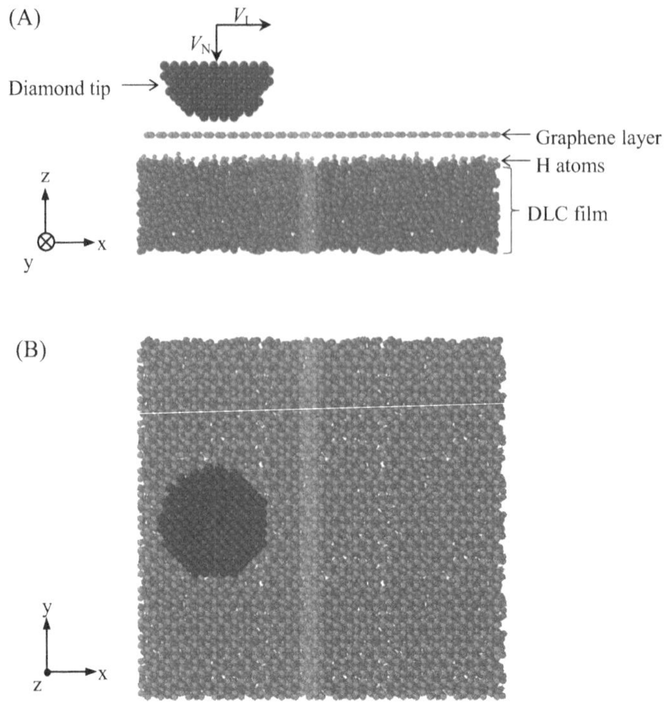

FIGURE 5.14 MD simulation model of a DLC film scratched by a diamond tip: (A) front view, and (B) top view. A graphene nanosheet is placed in between the tip and the DLC film as a solid lubricant. Source: *Reprinted with permission from a work by Bai et al. (2016c).*

that the friction force between the diamond tip and the DLC film is significantly lower than that obtained in the absence of the graphene lubricant. Meanwhile, the friction force increases significantly with the indentation depth because tribo-chemical reactions occur between the diamond tip and the graphene layer, and such reactions induce strong adhesion between them.

MD simulation has also been frequently applied to examine the mechanical behaviors of various bulk materials undergoing nano-scratching. For instance, the surface profile of a polystyrene specimen after being scratched by a hard tip is found to be strongly influenced by the arrangement of the constituent polystyrene chains (Du et al., 2013a). The deformation mechanisms of amorphous polystyrene undergoing scratching are found to be governed by two competing mechanisms: interchain sliding and intra-chain sliding (Du et al., 2013b). The nano-scratching test of crystalline copper demonstrates that scratching can greatly increase the energy of atoms in its subsurface, and thus lead to the creation of disordered structures (Chen et al., 2009).

5.4 Tribological behaviors

Friction and wear consume a huge amount of energy and are subjects of intense interest due to their ubiquitous manifestation in our daily lives and major roles in machinery degradation and power loss. The term "tribology" was coined in the 1950s to describe the aspects of materials science that include the studies of friction, wear, and lubrication. When materials are selected for fabricating mechanical components, it is essential to understand their tribological behaviors. Tribology also provides insight for the design of new mechanical systems with reduced wear and increased lifetime.

The tribological properties of materials have been extensively investigated on different scales ranging from the macroscale to the nanoscale. On the macroscale, there exist many well-established laws that govern different tribological phenomena. For example, Amontons's law of friction states that for two materials in contact, the friction force is proportional to the applied normal load and independent of the apparent contact area; Coulomb's law of friction states that the kinetic friction between a moving body and a surface is independent of the velocity of the moving body; the Archard equation expresses the wear volume of a material as a function of the applied load and the hardness of the material. In a nutshell, tribology on the macroscale possesses a highly robust framework and can provide invaluable guidance for the design of tribological components.

However, for microscopic and nanoscopic systems, the strong surface interaction and the large surface area to volume ratio induce a significant size effect that renders the macroscale tribological laws invalid, particularly at the nanoscale. The limitations of experimental techniques in investigating tribological phenomena on the micro/nanoscale have largely promoted the development of modeling methods for investigating nanoscale tribology. Among the many different types of theoretical approaches, MD simulation has been employed extensively for exploring the tribological properties of materials, which can be attributed to its well-developed algorithms and the utilization of potentials depicting highly realistic atomic interactions. The following section introduces the recent progress and simulation details of MD simulation studies in the fields of nanofriction, nanowear, and lubrication.

5.4.1 Nanofriction

There are two widely used MD simulation models for investigating nanofriction: the slider-block model and the tip-on-disk model.

In the slider-block model, two blocks slide against each other, and the contact interface between them is located within the $x-z$ plane, as illustrated in Fig. 5.15. A load perpendicular to the contact interface (along the y-direction) is applied, and a sliding velocity in the x-direction that is parallel to the interface is imparted. The upper and lower blocks are each divided into three layers along the loading direction. For both blocks, their outermost layer is considered a rigid body and thus named the rigid layer; the middle layer, known as the thermostatic layer, is kept at a constant temperature during the sliding process by coupling it to a thermostat to dissipate the thermal energy generated by friction; the innermost layer consists of Newtonian atoms that move because of the forces imposed on them by neighboring atoms according to the Newtonian law. It can be seen that the Newtonian layers from both blocks keep in contact during the simulation.

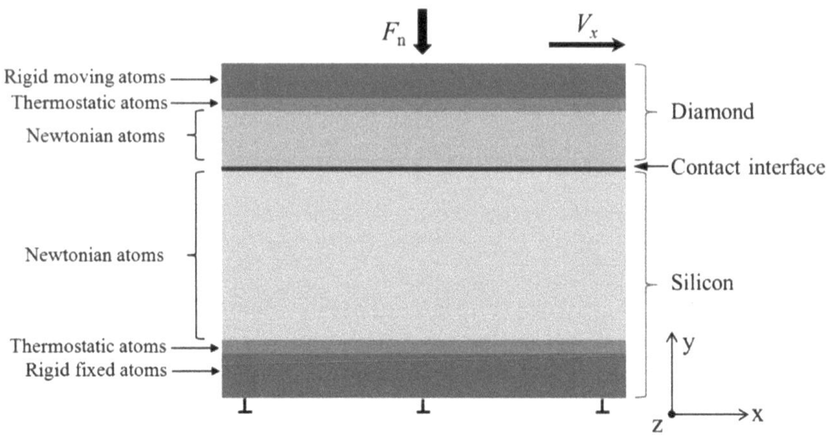

FIGURE 5.15 Typical MD simulation model for investigating nanofriction. Source: *Reprinted with permission from a work by Bai et al. (2015).*

The usual procedure for simulating nanofriction involves three steps. First, the structure relaxation of the simulated system is conducted via an energy minimization procedure, during which both the upper and lower rigid layers are fixed in place. Second, a normal load is applied by introducing an equal force on each atom in the upper rigid layer, followed by an equilibration procedure during which the normal load applied on the upper rigid layer is balanced by the force imposed on it by the remaining part of the upper block. Third, a sliding velocity v_x is given to the upper rigid layer, while the lower rigid layer remains fixed in place. During the sliding process, the friction force can be computed by summing up the lateral components of the forces contributed by each atom in the upper rigid layer.

The above computational model has been used to study nanofriction between diamond and silicon surfaces (Bai et al., 2015). In this study, the Tersoff potential is employed to describe the atomic interactions in the entire system. The simulation results indicate that the friction process involves two stages: the static friction stage and the kinetic friction stage. In the static friction stage, the contact interface between the diamond and silicon blocks is intact, and the friction force is determined by the elastic deformation of the blocks. Deformation is more prominent in the silicon block than in the diamond block because the former exhibits a lower mechanical strength. In the kinetic friction stage, the silicon block undergoes fracture, and friction is mainly caused by internal sliding within it.

Both the static and kinetic friction stages are illustrated by the atomic configuration of a thin slab of Si atoms as colored green in Fig. 5.16A. This thin slab of atoms serves as an inner marker that clearly illustrates the deformation of the silicon block. As the diamond block moves, the silicon block first undergoes deformation in a uniform manner, and the silicon–diamond interface remains intact without any sliding. The initially straight inner marker becomes inclined, which indicates that a

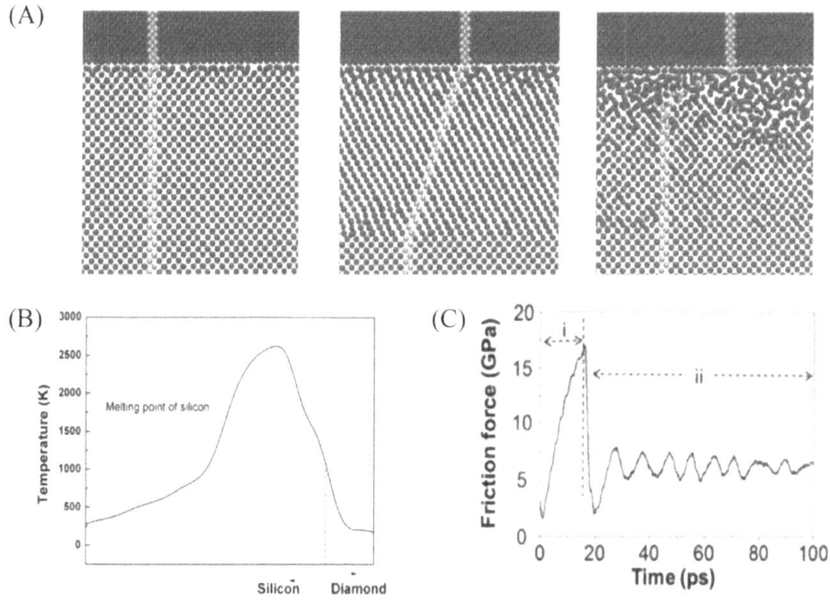

FIGURE 5.16 (A) Evolution of the interface between diamond and silicon during their relative sliding process; (B) temperature distribution along the loading direction during the fracture process of the silicon block; (C) evolution of the friction force with time. Source: *Reprinted with permission from a work by Bai et al. (2015).*

large elastic shear deformation is induced in the silicon block. Upon the elastic deformation reaching the elastic limit of silicon, fracture occurs at the subsurface of the silicon block in the form of atomic sliding. The elastic energy stored during the deformation process is released because of the fracture, which results in the relaxation of the silicon block, particularly for the region near the fracture that also experiences a local temperature increase. As the silicon block is relaxed, internal sliding and friction begin to occur within it, thereby initiating the kinetic friction stage.

During the fracture process, the temperature distribution along the loading direction can be obtained by dividing the Newtonian layers into many strips. The temperature of each strip is calculated by averaging the temperatures of its constituent atoms according to the relation between their kinetic energy and temperatures:

$$E_{\mathrm{avg}} = \frac{1}{N}\left[\frac{1}{2}\sum_{i=1}^{N} m_i(v_i - v_{\mathrm{com}})^2\right] = \frac{3}{2}k_{\mathrm{B}}T_{\mathrm{avg}}, \tag{5.9}$$

where E_{avg} denotes the average kinetic energy of atoms in a strip; N is the number of atoms in the strip; m_i and v_i denote the atomic mass and velocity of atom i, respectively; v_{com} denotes the velocity of the center of mass of the strip; k_{B} is the Boltzmann constant; T_{avg} denotes the average temperature of the strip. Fig. 5.16B shows that the temperature at the subsurface region of the silicon block exceeds the melting point of crystalline silicon and thus can result in its structural amorphization. Therefore, the mechanical strength of this region is greatly reduced, which facilitates internal sliding.

The internal sliding process can be further evidenced by the time evolution of the friction force shown in Fig. 5.16C. The maximum friction force in the static friction stage is much greater than that in the kinetic friction stage. Meanwhile, the friction force in the kinetic friction stage exhibits periodic fluctuations, which corresponds to the stick–slip friction behavior observed in experiments. For such a behavior, two surfaces in contact alternate between sticking to and sliding over each other. The stick–slip friction behavior observed for the diamond and silicon blocks is mainly caused by periodic deformation–relaxation of the silicon block.

The simulation model presented above has been used to explore the effects of various factors on the friction behaviors of materials (Bai et al., 2016a; Bai et al., 2016b). For instance, the effects of environmental hydrogen on the friction behaviors of DLC films have been studied (Bai et al., 2016b). In the study, an MD simulation model is developed by adding and randomly distributing H atoms between two DLC films. The temperature distribution, density distribution of different types of atoms, and graphitization of the films are calculated. It is observed that the H atoms can greatly reduce the friction forces between the DLC films. However, the mechanism of such a reduction relies on the average temperature rise of a small region near the friction interface. When the temperature rise is low, the friction reduction can be attributed to hydrogen passivation at the interfaces of the DLC films. When the temperature rise is high, the diffusion of H atoms into the DLC films and the graphitization of these films induce their structural degradation, thereby resulting in lower friction forces.

Kajita and Righi (2016) have employed ab initio MD simulation (AIMD) to investigate the effect of water on the lubricity of diamond coatings with Si dopants. The simulation model consists of two diamond slabs sliding against each other (Fig. 5.17). The surface of each diamond slab comprises dimers that give rise to alternating rows and trenches of sp^2- and sp^3-bonded C atoms. The Si dopants are incorporated into the diamond slabs by randomly substituting them for the C atoms to reach a doping density of ~8.3%. The water molecules are initially positioned in the configuration that is identified as the most stable one between the diamond slabs. Shear stresses and normal pressures are gradually applied to realize frictional sliding at the contact interface.

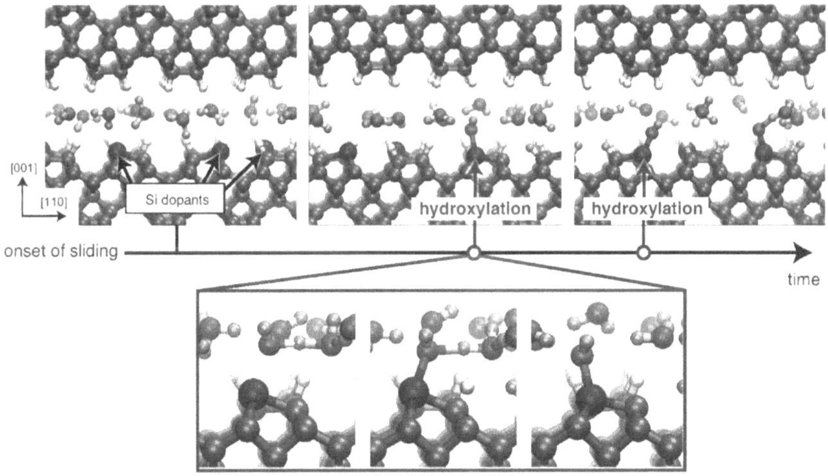

FIGURE 5.17 Snapshots (subsequence in time from left to right) acquired during the AIMD simulation of the sliding of two diamond slabs with Si dopants and intercalated with water molecules. The red and cyan spheres indicate the O and H atoms of water molecules, respectively; the white spheres indicate the H atoms that passivate the diamond surface; the blue spheres represent the Si dopants. Source: *Reprinted with permission from a work by Kajita and Righi (2016).*

During the AIMD simulation, the Car–Parrinello method (see Chapter 4: Advanced Ab Initio Molecular Dynamics and Coarse-Grained Molecular Dynamics) is used to determine the ionic and electronic degrees of freedom simultaneously, and a fictitious mass of 450 a.u. (1 a.u. = 9.109383×10^{-31} kg) is assigned to the electrons. The equations of motion for both the nuclei and electrons are integrated with the Verlet algorithm (see Chapter 1: Fundamentals of Classical Molecular Dynamics Simulation) with a time step of 0.12 fs. Two separate Nose–Hoover thermostats are applied to the nuclear and electronic subsystems, respectively, to guarantee that the adiabatic condition between these two subsystems is satisfied as required by the Car–Parrinello method. Under the adiabatic condition, the fictitious dynamics of the electrons instantaneously follow the nuclear motion. The nuclei in turn move according to an effective potential in classical mechanics determined by the electronic structure that is calculated by means of density functional theory with the Perdew–Burke–Ernzerhof approximation employed for the exchange–correlation functional (Perdew et al., 1996). In addition, during the calculation of the electronic structure, the ions are described by ultrasoft pseudopotentials (Vanderbilt, 1990), which are usually adopted to approximate the density of the core electrons to reduce computational costs. The overall electron density is obtained based on an electronic wavefunction that is expanded in a plane-wave basis set with a cutoff energy of 25 Ry (1 Ry = 13.605693 eV).

The simulation results indicate that the Si dopants at the diamond surfaces act as catalytic sites for surface hydroxylation (Fig. 5.17). The adsorbed hydroxyl groups on the diamond surface participate in the hydrogen bond networks of the water molecules that are confined at the sliding interface, thus anchoring a thin water film to the surface. This film prevents solid-on-solid contact and leads to ultralow friction.

Levita and Righi (2017) have conducted a similar study to investigate the effects of water intercalations on the friction behaviors of MoS_2 layers, which are an efficient solid lubricant, based on AIMD. The simulation results show that the intercalated water molecules can suppress the relative sliding of both pristine and defective MoS_2 layers. However, in the presence of normal loading and exposed edge terminations of the MoS_2 layers, water molecules are rapidly removed from the sliding interface and absorbed by the exposed edges either as OH/H segments or in their undissociated form. The results directly demonstrate that water molecules reduce the lubricating performance of MoS_2 layers even without any chemical oxidation occurring at the sliding interface.

Another important MD simulation model for the study of nanofriction employs a tip-on-disk type contact to imitate the friction measurement conducted using an atomic force microscope (AFM). This model is illustrated in Fig. 5.18, which is similar to the model in nanoscratching simulations (Fig. 5.13). The friction simulation is conducted by applying a perpendicular load to the tip and moving it laterally against the substrate surface at a constant velocity. The tip is commonly considered a rigid body to avoid deformation and wear. The specimen is divided into three layers: the fixed layer, the thermostat layer, and the Newtonian layer, in the same manner as the silicon block shown in Fig. 5.15. The friction force is calculated by summing up the tangential forces exerted on the tip atoms.

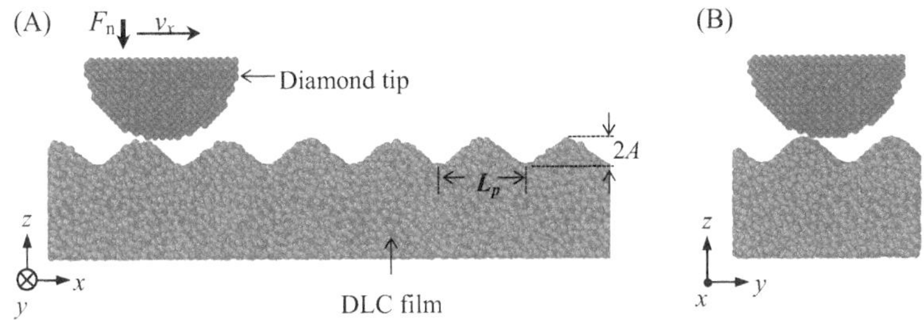

FIGURE 5.18 MD simulation model of a DLC film scratched by a diamond tip. A graphene nanosheet is placed in between the tip and the DLC film as a solid lubricant: (A) front view and (B) side view. Source: *Modified and reprinted with permission from a work by Bai et al. (2017b).*

Bai et al. (2017b) have employed this model to study the wear and friction between a diamond tip and DLC films with a rough surface. In this study, the rough surface profile of the DLC films is generated by reinventing the original flat DLC surface according to the function $Z = A\sin(2\pi x/L_p)\sin(2\pi y/L_p)$, where A denotes the amplitude, and L_p is the period length (Fig. 5.18). In the simulation, L_p is fixed, while A is varied to obtain different roughness values of the films. The simulation results indicate that the friction mechanism of the DLC films is highly dependent on the magnitude of the normal load. At a small load, the friction force is determined by the adhesion between the diamond tip and the film; this relation is in good agreement with the prediction by the Bowden–Tabor model, which is capable of accurately predicting the abrasive friction of iron on the nanoscale (Eder et al., 2015). However, at a large load, the friction force is reduced and deviates from the prediction by the Bowden–Tabor model.

The reduction of the friction force at high normal loads can be attributed to the diamond tip shifting a large number of C atoms of the film in front of or beneath it as indicated by the contour plot of the magnitudes of atomic displacements in Fig. 5.19. These moving C atoms form a transfer layer, which attaches itself to the diamond tip and isolates the tip from the rest of the film. Consequently, the sliding of the diamond tip against the DLC film is reduced to a relative motion between the

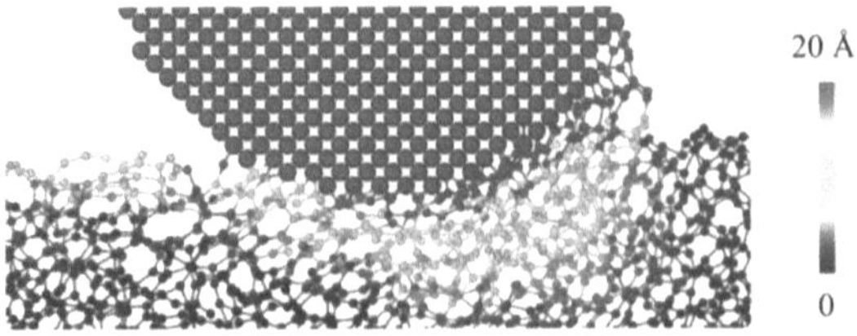

FIGURE 5.19 Contour plot of the magnitudes of atomic displacements of C atoms in a DLC film when a diamond tip slides against it. Source: *Reprinted with permission from a work by Bai et al. (2017b).*

transfer layer and the film. Since the sliding occurs in the form of internal friction within the DLC film, the friction force is thus determined by the rheological properties of this film. The rheological properties of amorphous solids are highly dependent on their microstructure and temperature. At a high normal load, the temperature rise of the DLC film due to friction is greatly increased, which further softens its microstructure and degrades its mechanical properties. Consequently, the DLC film is much more susceptible to shearing, thereby accounting for the greatly reduced frictional force at a high normal load.

2D materials are a class of materials with two dimensions beyond the nanoscale range (i.e., 1–100 nm) and the third dimension that is a single or a few atomic layers thick, and they have attracted immense attention due to their extraordinary mechanical, physical, and chemical properties. Investigating the nanofriction of 2D materials allows for the examination of their friction response to external loading and sliding and thus assists in promoting their applications. Based on MD simulations, the nanofriction of graphene has been extensively investigated, and it has been established that the nanofriction of graphene is highly dependent on many factors, such as the number of layers, chemical modifications, defects, external loading, and sliding velocity.

Ye et al. (2012) have conducted an MD simulation study on the friction between a diamond tip and a multilayer graphene nanosheet. The simulation results indicate that an increasing number of layers can result in decreased friction between graphene and the diamond tip. In addition, the nanofriction of graphene is largely enhanced when the graphene nanosheet is not flat but possesses wrinkles.

Li et al. (2014) have investigated the effects of fluorination of graphene on its friction behaviors. In their simulation, the atomic interactions within the graphene layers are described by the reactive force field (ReaxFF, see Chapter 8: Application of Molecular Dynamics Simulation in Other Problems) potential with the charge transfer effect being considered. The simulation results demonstrate that the fluorination of graphene can greatly increase the friction on its surface, which can be attributed to an increased corrugation of the potential energy surface resulting from the strong charge localization at the fluorine sites.

Liu et al. (2019) have studied the effects of the vacancy of graphene on its friction behaviors and have discovered that vacancies greatly increase friction on the graphene surface, which is mainly caused by the Ehrlich–Schwöebel barrier and the high chemical reactivity near the vacancy sites. Here, the Ehrlich–Schwöebel barrier refers to an energy barrier that an atom must overcome when it crosses a step edge in a material and experiences a different number of nearest neighbors.

5.4.2 Nanowear

Nanowear is one of the most common research topics in studies based on MD simulation. The MD simulation model for nanowear is the same as that for the tip-on-disk nanofriction. A key task of MD simulations of nanowear is to calculate the wear volume or the number of wear atoms in a system. In some studies, the wear atoms (also known as lost atoms or worn atoms) are identified by their displacements from their initial positions.

For instance, in the study on the nanowear of DLC films under a sliding diamond tip (Bai et al., 2017a), the wear atoms are defined as atoms that debond themselves from their nearest neighboring atoms. For such debonding to occur, the wear atoms must be displaced by a distance that is at least twice their equilibrium bond length. It is found that at a small normal load that is imposed by the diamond tip, the wear of the DLC film is purely abrasive, and the number of wear atoms can be predicted by the Archard equation which states that the volume of the removed debris due to wear is proportional to the work done by the friction force. However, at a large load, the number of wear atoms in the MD simulation deviates significantly from the prediction by the Archard equation, which could be attributed to the presence of both adhesive and abrasive wear. Besides the normal load, the wear mechanism of the DLC film also exhibit dependence on its surface roughness. As indicated by the work of Bai et al. (2017b), at a small surface roughness, the wear mechanism is adhesive wear. As the surface roughness increases, the wear mechanism changes from adhesive wear to abrasive wear, and the wear rate is reduced during the change although the friction force increases. During the abrasive wear after the change, the wear rate increases with a further increase in the surface roughness.

In the simulation study by Hu and Martini (2015) that investigates the effects of the roughness of a silicon surface on the nanowear of a copper substrate when they are sliding against each other, the Cu atoms exhibiting a displacement larger than the lattice constant of copper

(3.61 Å) are considered to be wear atoms. By this definition, wear atoms result from both plastic deformation and wear damage. It is observed that the number of wear atoms increases with the surface roughness, normal load, and sliding distance, and the corresponding wear mechanisms involve both abrasive and adhesive wear. In a study of the wear behavior of a γTi−Al alloy, where a hemispherical tip is modeled to slide on its surface, the wear atoms are defined as the ones that pile up on the initial surface at the end of the sliding (Xu et al., 2015). The simulation results indicate that the number of wear atoms is directly proportional to both the sliding distance and the penetration depth of the tip, which is consistent with the Archard equation.

The number of wear atoms during a wear process can also be calculated based on the sliding velocity of atoms. For instance, in a study of the abrasive wear of iron for which multiple particles are sliding against an iron substrate (Eder et al., 2015), the relative positions of these particles are fixed, and their rotation is forbidden, which corresponds to two-body wear or grinding with hard and fastened abrasives. The contact region between the particles and the substrate is divided into three zones according to the momentary advection velocities v of the atoms, as shown in Fig. 5.20. The wear atoms are defined as atoms for which $v > 0.9v_{max}$, where v_{max} denotes the maximum value of the advection velocity. Any atom satisfying the condition $v < 0.1v_{max}$ is labeled as part of the iron block, and those satisfying the condition $0.1v_{max} < v < 0.9v_{max}$ are defined as part of the shear zone.

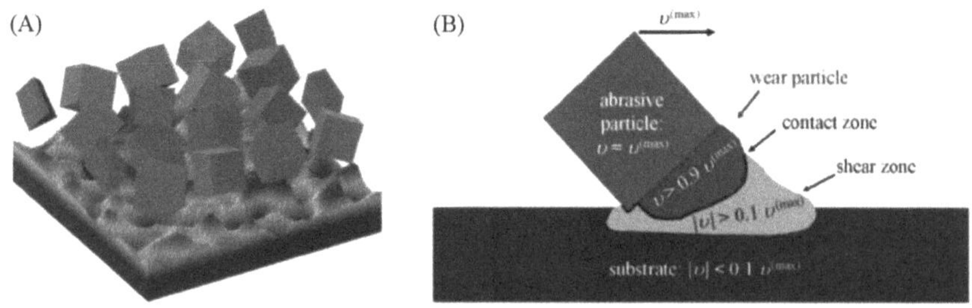

FIGURE 5.20 (A) MD simulation model for abrasive nanowear of iron under the sliding of multiple hard abrasive particles. (B) Schematic showing the definition of the wear atoms based on their advection velocity. Source: *Reprinted with permission from a work by Eder et al. (2015).*

The wear behaviors of various systems have been studied using MD simulation. For example, the wear of an AFM tip during its operation has been investigated by conducting MD simulations to study the wear loss of a tip in a tip-on-disk system (Hu et al., 2014). In another simulation study exploring the wear of a DLC tip sliding against a DLC substrate (Sha et al., 2013), the wear atoms of the tip are defined as atoms that leave it during the sliding process. The simulation demonstrates that the number of atoms lost from a tip is directly proportional to the sliding distance and the applied normal load, which is consistent with the Archard equation. While the friction force is determined by the number of atoms forming chemical bonds across the contact interface, the number of such bonds does not correlate with the wear process. Instead, the wear of the tip is mainly caused by the following processes: mass transport driven by the frictional (shear) traction across the contact interface that leads to the accumulation of mass at the trailing edge of the contact area and the subsequent loss of mass via a cluster detachment mechanism.

The wear behaviors of 2D materials such as graphene have also been investigated by MD simulations. The simulation of the wear of graphene typically involves sliding a tip against graphene layers, in a manner similar to that in the study of friction on the surface of graphene. Wang and Duan (2018) have observed that the adhesive wear of graphene occurs more frequently than abrasive wear, and the former is initiated by a rupture cavity caused by the dragging of the tip. Sandoz-Rosado et al. (2012) have studied the wear behavior of graphene supported by a substrate and observed that when the interaction between the sliding tip and the graphene layer is purely repulsive, the graphene layer never delaminates from the substrate. Meanwhile, a low substrate–graphene adhesion strength may reduce the overall damage to the graphene layer. The graphene layer is found to be able to withstand a normal load 8.5 times that for a DLC nanofilm when they possess similar thickness values. These results indicate that graphene holds promise as a solid lubricant.

5.4.3 Nanolubrication

Lubricants can significantly reduce the friction and wear of mechanical components and are thus extensively applied in industrial fields. With the rapid development of machinery and components on the micro- and nanoscale, the adhesion and friction at their contact interfaces present a high demand for lubricants. Since the framework of macroscale continuum mechanics is usually inapplicable on the nanoscale, the friction and lubrication properties of lubricants on the nanoscale cannot be explained by macroscale lubrication theories. Moreover, in experimental friction tests, it is difficult to observe nanoscale lubrication because of the in situ measurement of the structural changes of the lubricants. MD simulation is capable of directly characterizing the interactions between lubricants and contact surfaces and observing the structure evolution of the lubricants under various conditions, thereby providing a useful approach for investigating the lubrication mechanisms of lubricants on the nanoscale.

The MD simulation model for nanolubrication commonly consists of two relatively sliding blocks with a lubricant distributed between them.

The lubricants can be liquids or solids. Similarly to those in the MD simulation model for nanofriction, both the upper and lower blocks in the nanolubrication model are divided into three layers: the rigid layer, the thermostat layer, and the Newtonian layer, in the direction from the outermost boundary of the model to the contact interface. These layers have the same functions as their counterparts in the nanofriction model. The lubricant particles distributed between the blocks are also regarded as Newtonian, and thus can freely move under forces from their neighboring atoms. The type of potential energy function selected to describe the atomic interactions within each lubricant molecule depends on the material of the lubricant. Meanwhile, the intermolecular interactions among the lubricant molecules and the interactions between the lubricant molecules and the blocks are conveniently modeled by the Lennard–Jones (LJ) potential, in conjunction with the Coulombic potential for electrostatic forces.

Take the MD simulation of the lubrication behaviors of *n*-alkanes located between two iron oxide (FeO) blocks as an example. The chemical bonds within each *n*-alkane molecule and the nonbonded interactions among different *n*-alkane molecules are described by the L-OPLS–AA (long hydrocarbon optimized potentials for liquid simulations–all atoms) force field, which can accurately predict the behaviors of long alkanes and their phase transitions (Echeverri Restrepo et al., 2019). The atomic interactions within the FeO blocks are simplified by a harmonic spring potential. The interactions between the FeO blocks and the *n*-alkane molecules are characterized by a combination of the LJ potential and Coulombic potential.

The lubrication simulation includes three stages: the equilibrium stage, the compression stage, and the shear stage. In the equilibrium stage, a heating–quenching process is conducted for the *n*-alkane lubricant to ensure that a well-disordered structure is obtained. In the compression stage, a normal load is applied to the upper block by applying a constant normal force to the Fe atoms in the upper rigid layer while fixing the lower rigid layer in place. In the shear stage, a relative sliding velocity is assigned to the upper rigid layer with the lower rigid layer fixed. During the simulation, it can be observed that under a high normal load, the *n*-alkane molecules exhibit a strongly layered and long-ranged ordered structure and share similar behaviors with a solid-like film. On the contrary, at a high sliding velocity (i.e., a high shear rate), the *n*-alkane molecules exhibit a weakly layered and short-range ordered structure and behave more like a liquid lubricant.

Besides the *n*-alkanes discussed above, the nanolubrication simulation model has also been utilized to study the lubrication behaviors of other lubricants. For instance, Voeltzel et al. (2015) have studied the lubrication behavior of an ionic liquid containing $mmIm^+$ cations and

NTf_2^- anions between two FeO blocks. Martini et al. (2006) have simulated the viscous behavior of *n*-decane located between gold blocks. The lubrication performance of a perfluoropolyether Zdol film confined between two solid plates has also been investigated through MD simulation (Tanaka et al., 2003).

5.5 Interfacial effects in nanocomposites

Composite materials are combinations of two or more materials that result in products with enhanced qualities. Natural composite materials, such as mud and straw, have been utilized by man for thousands of years. Modern composite materials are typically fabricated by adding structural reinforcements such as fibers and particles to a matrix material. The most common composite material in contemporary infrastructure is steel-reinforced concrete. The constitutive steel rebars endow this composite material with high strength and stiffness. Without the reinforcement of the steel rebars, pure concrete cracks easily under external loading. Hence, a rebar-concrete composite exhibits enhanced mechanical properties compared with pure concrete. With the advancement of materials science, reinforcements such as carbon fibers, boron fibers, CNTs, and graphene are implemented in the fabrication of composite materials. Meanwhile, substances constituting the matrix of a composite extend to a vast variety of materials, including metals, polymers, and ceramics.

The enhanced mechanical properties of modern composite materials are attributed to the interactions between their constituent matrix and reinforcements, and the influence of the reinforcements on the deformation of the matrix. Hence, interfacial effects are critical for evaluating the deformation mechanisms of such composite materials. MD simulation has been utilized in many studies on composite materials as it not only allows a direct observation of the phenomena that occur at the interfaces of these composites but also provides a quantitative description of their interfacial behaviors.

5.5.1 Polymer-based nanocomposites

The addition of nanomaterials, such as CNTs, graphene, and nano-sized oxide or nitride particles, to polymer matrices can significantly improve their mechanical, electrical, and thermal properties. Because of the differences in physical and chemical properties between such nanofillers and a polymer matrix, a distinct interface commonly exists between them and plays a significant role in load bearing and heat and charge transfer. A comprehensive investigation of the characteristics of the constituent materials at their interfaces, which include molecular chain movements, atomic density distributions, and material compatibility, is essential for understanding the properties of these nanocomposites.

The commonly explored interfacial properties of polymer nanocomposites in MD simulations include interfacial structures, interfacial bonding strength, and molecular chain mobility.

Interfacial structures can be represented by the radial density distribution (RDD). In an MD simulation, the RDD of a nanocomposite can be obtained by dividing it into many slices and computing the mass density of each slice. As an example, the RDD of epoxy resin enforced by boron nitride nanosheets is presented in Fig. 5.21. The interfacial domain can be divided into three regions, namely, the compact region, the buffer region, and the normal region. The mass density decreases along the direction pointing away from the nanofillers: the compact region usually exhibits the highest strength but the lowest molecular chain mobility, while the normal region possesses the lowest strength but the highest chain mobility.

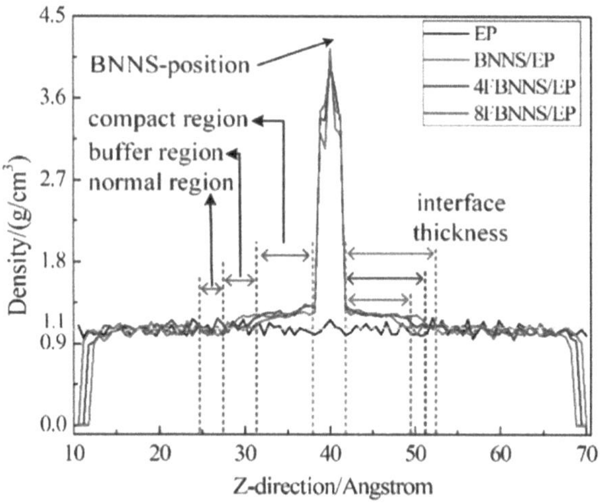

FIGURE 5.21 Radial density distribution of epoxy (EP) resin enforced by boron nitride nanosheets (BNNSs). 4FBNNS/EP and 8FBNNS/EP denote nanocomposites of which the BNNSs are functionalized with 4 and 8 functional groups, respectively. Source: *Reprinted with permission from a work by Li et al. (2019).*

The interfacial bonding strength between the polymer matrix and the nanofillers of a nanocomposite directly influences its capability in load bearing and heat and charge transfer. The interfacial bonding strength ΔE of a nanocomposite can be conveniently expressed as

$$\Delta E = E_{\text{total}} - (E_{\text{matrix}} + E_{\text{nanofillers}}), \tag{5.10}$$

where E_{total} denotes the total energy of the nanocomposite; E_{matrix} and $E_{\text{nanofillers}}$ refer to the energy of its matrix and nanofillers, respectively.

The energy of a system includes its kinetic energy and potential energy. In a molecular system, the potential energy can be further classified into intramolecular potential energy and intermolecular potential energy. For a polymer, the latter commonly includes the van der Waals interaction energy, electrostatic interaction energy, and energy contributed by hydrogen bonding.

From the calculated value of the interfacial bonding strength ΔE, the interfacial adhesion energy γ can be obtained by $\gamma = \Delta E/(2A)$, where A denotes the contact area between the nanofillers and the matrix. The adhesion energy is representative of the interaction energy per unit area and can thus be used to compare the adhesion strength of different interfaces.

The interfacial shearing strength is another parameter for characterizing the bonding strength between the nanofillers and the matrix of a nanocomposite, and it can be considered as the energy required to completely extract the nanofillers from the nanocomposite, as shown in Fig. 5.22. From a physics point of view, the interfacial shearing strength is equivalent to the interfacial bonding strength.

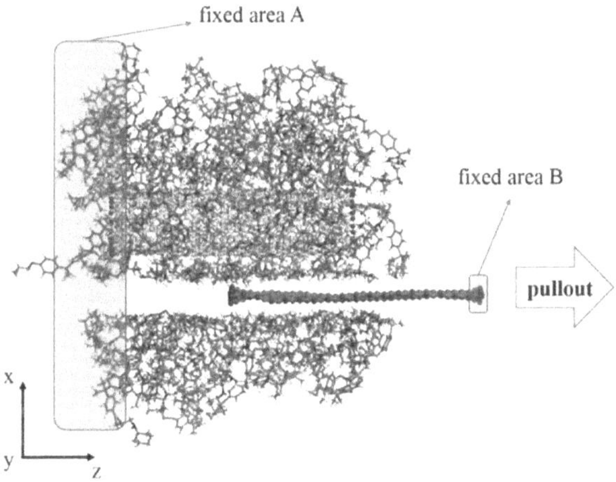

FIGURE 5.22 Schematic of a simulation model detailing the extraction of nanofillers from a nanocomposite. Source: *Reprinted with permission from a work by Sun et al. (2019).*

Molecular chain mobility is indicative of the behaviors of molecules exhibiting chain structures and is therefore capable of characterizing the interfacial compatibility between nanofillers and a matrix. Molecular chain mobility can be efficiently evaluated by the mean square displacement (MSD), the cohesive energy density ε_{coh}, and the free volume fraction f_{FVF}. A large MSD, a low cohesive density, and a high free volume fraction all correspond to high chain mobility. The cohesive energy density is defined as the cohesive energy per unit volume in the composite and can be calculated as follows:

$$\varepsilon_{\text{coh}} = \frac{E_{\text{coh}}}{V_{\text{m}}}, \tag{5.11}$$

$$E_{\text{coh}} = -\left(E_{\text{vdw}} + E_{\text{coul}} + E_{H\text{-bond}}\right), \tag{5.12}$$

where E_{coh} denotes the cohesive energy of the interface; V_{m} denotes the volume of the composite; E_{vdw}, E_{coul}, and $E_{H\text{-bond}}$ denote the energy contributions due to van der Waals interactions, electrostatic interactions, and hydrogen bonding, respectively. Usually, the incorporation of highly compatible nanofillers into a polymer matrix corresponds to a high cohesive energy density of the resultant composite due to the strong interactions between them.

The free volume of a composite refers to its internal space that is unoccupied by molecules. A large free volume fraction indicates that the molecules can move more easily and is thus closely related to the deformation behavior of the polymer matrix. The free volume fraction of a composite can be calculated by

$$f_{\text{FVF}} = \frac{V_{\text{free}} - V_{\text{vacuum}}}{V_{\text{occupy}} + V_{\text{free}} - V_{\text{vacuum}}}, \tag{5.13}$$

where V_{free} is the total free volume, V_{vacuum} is the sum of the free volumes of which each exceeds a critical size, and V_{occupy} is the total volume occupied by the nanofillers and matrix of the composite. High compatibility between the nanofillers and the matrix usually corresponds to a low free volume fraction due to the strong interactions between them.

Because of the strong interactions between the nanofillers and the polymer matrix at their interfaces, the properties of the matrix are greatly enhanced. MD simulation studies have been conducted intensively to explore the underlying mechanisms of such improvements.

Han and Elliott (2007) have studied the elastic properties of CNT/polymer nanocomposites constructed by embedding a single-wall CNT in each of two different amorphous polymer matrices PMMA and PmPV in the MD simulation model. The simulation results agree with the notion that CNTs can be potentially utilized to mechanically reinforce a polymer matrix, especially in the longitudinal directions of the incorporated CNTs. In addition, the effects of the alignment of the atomic structure of the CNT with that of the polymer matrix on the mechanical properties of the nanocomposites are nonnegligible if the interactions between the CNT and the polymer matrix are strong.

An MD simulation study conducted by Gersappe (2002) on polymers reinforced with nanofillers indicates that the nanofillers can significantly improve the toughness of polymers. Meanwhile, the capacity of the nanofillers inside the polymer matrix in dissipating energy is much more dependent on their mobility than their surface area, and thus their efficacy in enhancing the toughness of the matrix is mainly determined by their mobility within the matrix.

Li et al. (2016) have conducted MD simulations to explore the shearing behavior of a composite composed of a nitrile butadiene rubber (NBR) matrix reinforced by CNT nanofillers. The simulation results indicate that the CNTs are bonded to the NBR matrix and improve its resistance against shear deformation, resulting in a 60% increase in its shear modulus.

Adnan et al. (2007) have investigated the effects of the size of nanofillers on the elastic properties of polymer nanocomposites and concluded that these properties would be significantly enhanced with the reduction of the nanofiller size because of the densification of the polymer matrix near the nanofillers. Additionally, the total interaction energy at the nanofiller–matrix interfaces would experience a dramatic increase.

5.5.2 Metal-based nanocomposites

By incorporating nanofillers into metallic matrices, nanocomposites with highly enhanced mechanical properties can be obtained. In particular, the elastic and shear moduli and the strength of soft metals such as copper and aluminum can be significantly increased by distributing nanofillers such as CNTs and graphene at their grain boundaries.

Zhang et al. (2020) have explored the effects of grain boundary (GB) engineering on the mechanical properties of a graphene/copper nanocomposite via MD simulation. In the study, the model of the nanocomposite is constructed by placing graphene nanoflakes at the GBs of a nanocrystalline copper cube, as shown in Fig. 5.23. The average size of the copper grains is ~5.4 nm, and the grain orientations are randomized. The EAM (short for embedded atom method) potential is employed to describe the interactions among the copper atoms, while the AIREBO potential is implemented to model the interactions among the C atoms in graphene. The interaction between the Cu and C atoms is characterized by the LJ potential. A standard tensile test utilizing the rescaling method (discussed in Section 5.2) is conducted to capture the tensile response of the nanocomposite. The simulation results demonstrate that the graphene nanoflakes can tremendously enhance the mechanical strength of the copper matrix by accumulating dislocations at the GBs and inducing a pronounced strain hardening effect. These results indicate that GB engineering is an effective strategy for tailoring the mechanical properties of nanocrystalline copper.

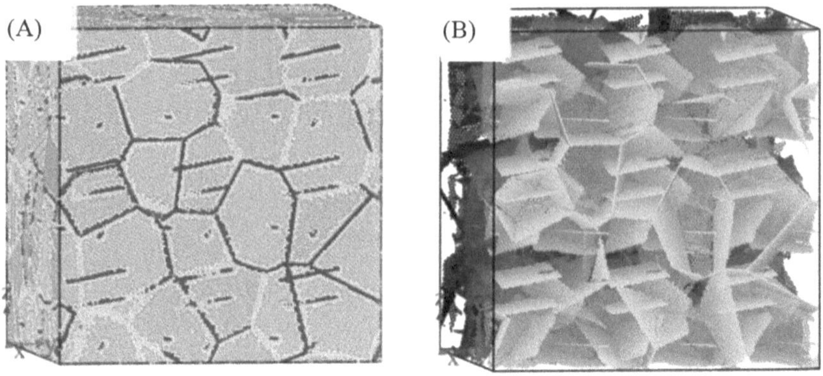

FIGURE 5.23 MD simulation model for studying the mechanical properties of a graphene/copper nanocomposite: (A) atomic configuration of the overall nanocomposite, and (B) atomic configuration of the graphene nanoflakes distributed at the GBs of the nanocrystalline copper cube. In (A), the green and purple regions illustrate the copper matrix and the graphene nanoflakes, respectively; the gray regions denote the Cu atoms at the GBs. Source: *Reprinted with permission from a work by Zhang et al. (2020).*

In a similar study, Cong and Lee (2018) have studied the mechanical properties of aluminum-based nanocomposites with boron nitride nanotubes (BNNTs) as the nanofillers. In the simulation model, a BNNT is embedded in an aluminum block with PBCs applied in the x-, y-, and z-directions. The atomic interactions within a BNNT are described by the Tersoff potential, and those within the aluminum matrix are modeled by the EAM potential. Meanwhile, the interactions between the constituent atoms of the BNNT and the Al atoms are described by an LJ potential. A uniaxial tensile test employing the rescaling method is conducted by applying a tensile load in the direction of the BNNT axis at a strain rate of $0.0005\ \mathrm{ps}^{-1}$. It is found that the BNNT plays a significant role in bearing external loads even though it constitutes only a small volume fraction of the composite. In addition, increasing the volume fraction of the BNNT is determined to be more efficient than increasing its diameter in further enhancing the tensile strength of the nanocomposite.

The above-mentioned low-dimensional materials, such as CNTs and graphene, have also been applied to metallic glasses with amorphous structures to reinforce their mechanical behaviors. For instance, Rezaei et al. (2016) have studied the mechanical behavior of a $Cu_{64}Zr_{36}$ metallic glass nanocomposite reinforced by CNTs. Both tensile and compressive tests are conducted on the composite with a strain rate of $10^8\ \mathrm{s}^{-1}$ to characterize its mechanical behaviors. The simulation results indicate that long CNTs are able to significantly increase the elastic modulus, yield strength, and yield strain of the metallic glass, while short CNTs contribute negligibly to its strength. In general, the enhancement of the metallic glass by long CNTs can be attributed to the fact that such CNTs are able to absorb a considerable amount of energy from an external load, and they can also influence the plastic deformation of the metallic glass by arresting the formation of shear bands.

5.6 Defect effects

Defects are commonly present in solids and can highly influence their performances in practical applications. According to their dimensions, defects can be categorized into four types, namely 0D (e.g., vacancies), 1D (e.g., disclinations and dislocations), 2D (e.g., GBs and interphase boundaries), and 3D defects (e.g., voids and cracks). MD simulation serves as a useful tool to directly evaluate the effects of various types of defects on the mechanical properties of solids and their microstructural evolution during deformation.

Based on MD simulation, Zhou et al. (2006) have investigated the effect of a negative wedge disclination in a cylindrical bicrystalline nickel nanowire on its intergranular crack formation at approximately absolute zero temperature using the EAM potential. They have proposed a method to introduce a negative wedge disclination in a cylinder on the atomistic level.

A bicrystalline cylinder consisting of two face-centered cubic (FCC) nickel crystals with a common [001] axis is first formed to exhibit a symmetric tilt GB with the reciprocal coincidence site density $\Sigma = 5$, lattice plane (310), and misorientation angle $\theta_1 = 53.13^\circ$ (Fig. 5.24A). Note that the atomic structures of GBs are built based on a disclination-structural unit model for FCC metals developed by Gertsman et al. (1989). Similarly, another bicrystalline cylinder is also formed to exhibit a symmetric tilt GB with the reciprocal coincidence site density $\Sigma = 5$, lattice plane (210), and misorientation angle $\theta_2 = 36.87^\circ$. Each bicrystalline cylinder is then cut along its {110} lattice planes, and the two resulting wedges are brought into contact at a common origin (Fig. 5.24B). These two wedges overlap because their {110} lattice planes are misoriented by an angle of $(\theta_1 - \theta_2)/2$. To remove the overlap and match the lattice planes, each constituent atom of the two wedges is displaced by a certain amount. The displacement is calculated based on the isotropic elasticity theory for wedge disclinations, and therefore a cylinder containing a wedge disclination of strength $\omega = (\theta_1 - \theta_2)$ is formed (Fig. 5.24C). The strength of the disclination can be modified by varying the misorientation angle θ_2 of the left bicrystal.

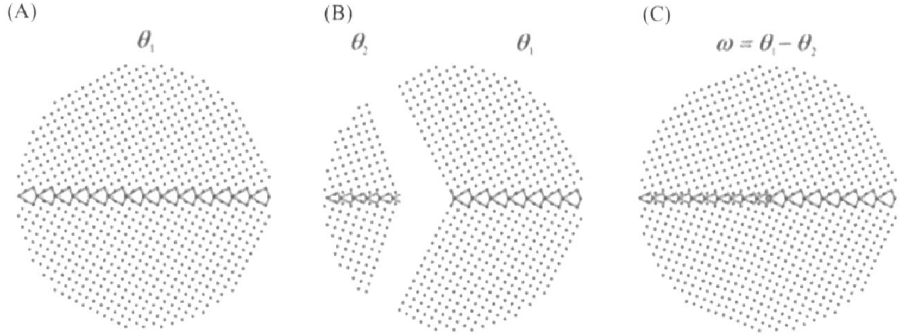

FIGURE 5.24 Schematic of the construction of the atomic configuration of a bicrystaline nickel nanowire containing a negative wedge disclination of strength ω. (A) An fcc bicrystal with [001] $\theta_1 = 53.13°$ tilt GB. (B) Two fcc bicrystals with [001] $\theta_2 = 36.87°$ (left) and [001] $\theta_1 = 53.13°$ (right) tilt GBs, respectively. (C) A bicrystalline nanowire containing an $\omega = 16.26°$ wedge disclination, constructed from the two bicrystals in (B), along the central axis. Source: *Reprinted with permission from a work by Zhou et al. (2006).*

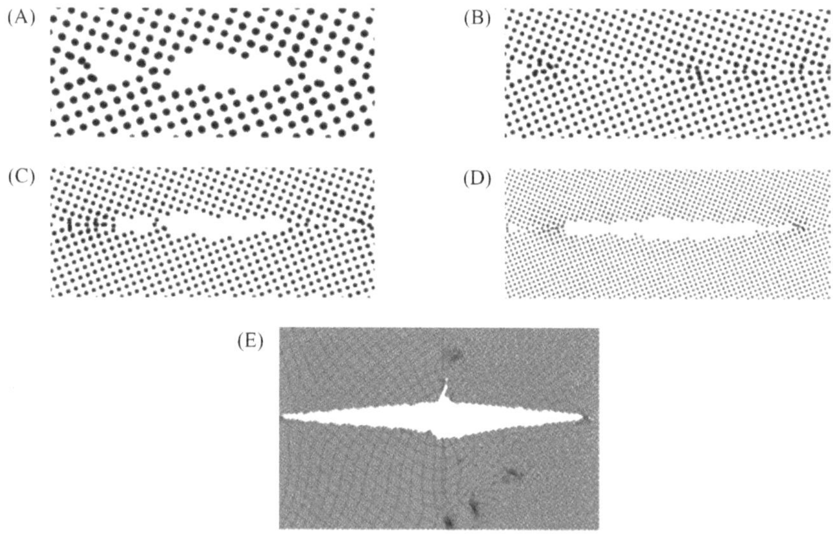

FIGURE 5.25 Crack nucleation near the cores of disclinations with strength values of (A) 9.99°, (B) 10.39°, (C) 11.06°, (D) 13.16°, and (E) 22.61°. Atoms on all (002) planes are represented by black circles. Source: *Reprinted with permission from a work by Zhou et al. (2006).*

The MD simulation results in Fig. 5.25 indicate that the structural stability of the disclinations highly depends on their strength. It is evident that the low-strength disclinations remain stable with a nonbroken core, which does not differ significantly from the initial structure. When the strength of the disclination increases, a greater disorder is observed in its core. Beyond a certain value of the strength, the disclination is no longer stable, and a crack opens up along the GBs on both of its sides. Fig. 5.25 shows the sequence of relaxed structures obtained with an initial crack-free structure of the disclinated cylindrical bicrystal for an increasing disclination strength. The critical disclination strength ω_c at which the disclination initiates a crack cannot be exactly determined from the atomistic computations. For $\omega = 9.99°$, a very small crack is seen, while for $\omega = 10.39°$, no definite crack is observed, and at $\omega = 11.06°$, a crack appears again. In other words, the crack nucleation exhibits a nonmonotonic behavior.

Detailed analysis indicates that a further increase of the disclination strength beyond its critical value leads to an increase in the length of the crack after structure relaxation. The dependence of the stable crack length on the disclination strength is presented in Fig. 5.26 and compared with that predicted by the continuum mechanics model. It can be observed that when the disclination strength exceeds its critical value by a small amount, the stable crack length obtained from the MD simulation and that from the continuum mechanics model agree well with each other. When the disclination strength is much greater than its critical value, the stable crack length obtained from the MD simulation is ~15% smaller than the value obtained from the continuum mechanics model. Such a discrepancy becomes larger as ω increases, which can be attributed to the fact that surface tension, which is captured in the MD simulation, is not taken into account in the continuum mechanics model. This surface tension induces a bulk compression, which is nonuniform as a result of crack perturbation, and such nonuniform bulk compression exerts a greater influence on longer crack lengths than on shorter ones.

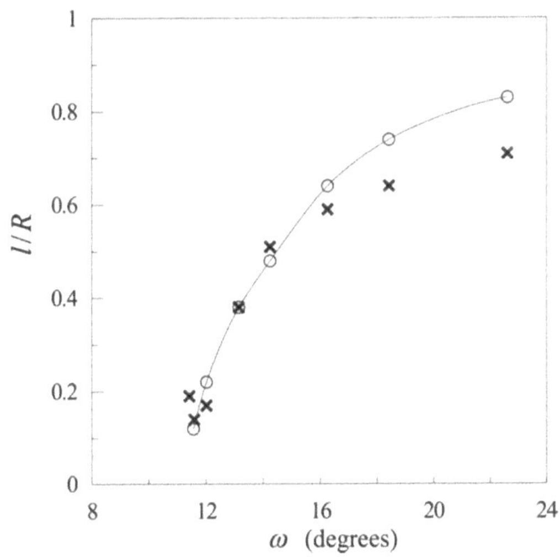

FIGURE 5.26 Dependence of the stable crack length on the disclination strength as calculated by MD simulation (*cross marks*) and the continuum mechanics model (*open circles*). Source: *Reprinted with permission from a work by Zhou et al. (2006).*

The stability of wedge disclinations in hexagonal close-packed (HCP) titanium nanowires at approximately absolute zero temperature has been studied by Wu et al. (2006). Similarly to the construction of wedge disclinations in the FCC nickel nanowires, the wedge disclinations in the HCP titanium nanowires are obtained based on the disclination – structural unit model for HCP metals proposed by Wu et al. (2004). The disclinations in the HCP titanium nanowires also become unstable when their strength exceeds a certain critical value. Moreover, it has been discovered that the nature of the relaxation mechanisms of unstable disclinations depends on the GB structures. For low, intermediate, and high values of the disclination strength ω, the relaxation mechanisms correspond to basal plane cleavage, a combination of basal plane cleavage and new grain nucleation, and prismatic plane cleavage, respectively.

Zhou et al. (2007) have further analyzed the effects of temperature and size on the stability of wedge disclinations in bicrystalline titanium nanowires. The simulation results indicate that for given values of the nanowire diameter D and temperature T, there exists a critical disclination strength ω_c above which a crack initiates from the disclination core when the bicrystalline structure is relaxed. However, depending on the values of D and T, a structural transformation may precede or follow the crack nucleation when the disclination strength increases and approaches ω_c.

The simulation results of the structural transformation of the disclinated bicrystalline titanium nanowires during structural relaxation are summarized in Fig. 5.27 by a relaxation mechanism map in which the critical temperature T_c is plotted against the nanowire diameter D on the logarithmic scale. In region I above the solid straight line, new grain nucleation always occurs before crack nucleation at ω_c, while in region II below the solid line, crack nucleation occurs before grain nucleation. In addition, above the dashed line, new grain nucleation can occur in the bicrystalline nanowire in the absence of a disclination.

Fig. 5.27 also provides a global perspective of the coupling between the size and temperature effects. In particular, $\log(T_c)$ decreases linearly with $\log(D)$ for the transition in the operating sequence of relaxation mechanisms (indicated by the solid line and hollow triangles) and in the spontaneous grain nucleation (indicated by the dashed line and hollow circles). At room temperature and above, crack nucleation is always preceded by grain nucleation for all the diameters considered. In the region between the solid line and the dashed line, new grain nucleation can only occur in the presence of a disclination, and it will be followed by crack nucleation if the disclination strength corresponds to its critical value.

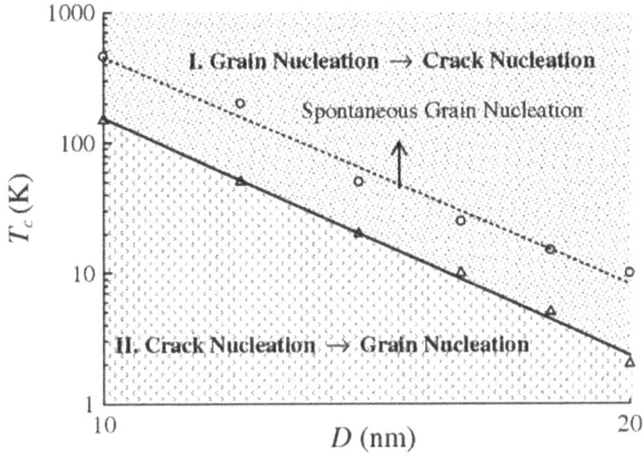

FIGURE 5.27 Relaxation mechanism map of nanowires containing $(11\bar{2}6)$ tilt GBs. The regions above and below the solid line correspond to domains with different sequences of operating mechanisms. Above the dashed line, spontaneous grain nucleation occurs. Source: *Reprinted with permission from a work by Zhou et al. (2007).*

Further simulations have been conducted for disclinated bicrystalline systems involving other HCP metals that share similar GB structures with the aforementioned disclinated bicrystalline titanium nanowires. A total of 16 HCP metals have been considered, including yttrium, ruthenium, and titanium. Crack nucleation has been observed in all the disclinated bicrystalline systems, seven of which exhibit amorphization, and several of them display grain nucleation.

Based on the studies on the disclination at the GBs of bicrystals, Wu et al. (2007) have further explored the stability of a disclination triple junction in a tricrystalline nickel nanowire with an FCC lattice structure. To construct the triple junction, three bicrystalline cylinders, each of which possesses a symmetric GB with a [001] tilt axis and misorientation θ, are first constructed according to the structural unit model. Afterward, these three bicrystalline cylinders are cut along their {110} planes to form three circular wedges, which are then brought into contact at a common origin, as shown in Fig. 5.28A. Each circular wedge subtends an angle α at the center of the nanowire to be formed, and the angle α is related to the misorientation angle by the equation $\alpha = 180^\circ - \theta$. Generally, the three wedges overlap because their {110} lattice planes are misoriented. To remove the overlap and match the lattice planes of the three wedges, each of their constitutive atoms is displaced by a vector calculated based on the isotropic elasticity theory for wedge disclinations, similarly to the aforementioned case of negative wedge disclinations in bicrystalline nickel nanowires. The strength

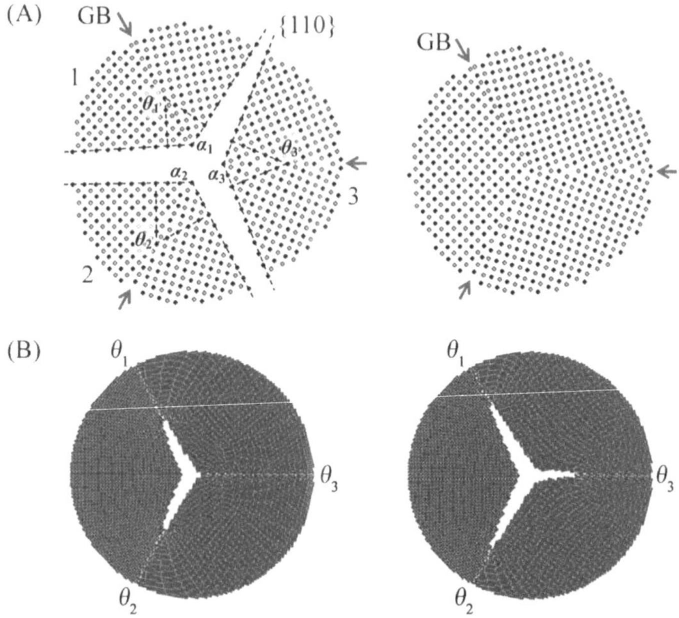

FIGURE 5.28 (A) Schematic of the construction of a disclination triple junction. (B) Atomistic configurations in the central region (20 nm diameter) of a 40 nm diameter nickel nanowire after structure relaxation for two different disclination strength values of $\omega = 21.71^\circ$ and $\omega = 24.11^\circ$. Source: *Modified and reprinted with permission from a work by Wu et al. (2007).*

ω of the as-constructed disclination can be calculated in terms of the misorientation angles as $\omega = \theta_1 + \theta_2 + \theta_3 - 180^\circ$, and it can be adjusted by fixing any two of the three misorientation angles and varying the third angle.

Based on the simulation results, it has been found that for the disclination triple junctions, there exists a critical disclination strength beyond which the disclinations relax via crack nucleation in the manner of GB cleavage. In addition, the critical strength demonstrates dependence on the diameter of the nanowire that can be described by an approximate relation $\omega_c \propto D^{-0.6}$. Moreover, an increase in the disclination strength by a few degrees above the critical value results in a significant increase in the crack volume fraction.

Zhou et al. (2008b) have further conducted the MD simulations of disclination triple junctions in tricrystalline cobalt nanowires exhibiting an HCP lattice structure. The results indicate that complex relaxed structures can occur, including cracks, cavities, distorted GBs, dislocations, twin bands, diffuse dislocation walls, and amorphized zones. The realized structures depend on the nanowire size, temperature, disclination strength, and GB structures. Moreover, a critical disclination strength ω_c exists, below which the disclination is stable and above which it relaxes via competing mechanisms. The critical disclination strength demonstrates a strong size effect at all temperatures.

When the disclination strength ω is low but higher than ω_c, structural relaxation is mainly realized through crack nucleation and GB distortion. As ω increases further, the disclinations are relaxed not only by crack nucleation and GB distortion but also by the formation of a twin band and a dislocation wall, as shown in Fig. 5.29. In this case, the trifurcated crack is short but possesses a large opening displacement compared with its counterpart at a lower value of ω. Moreover, it has been discovered that with an increasing temperature, the modes of structural relaxation change from crack nucleation and GB distortion to cavitation and amorphization.

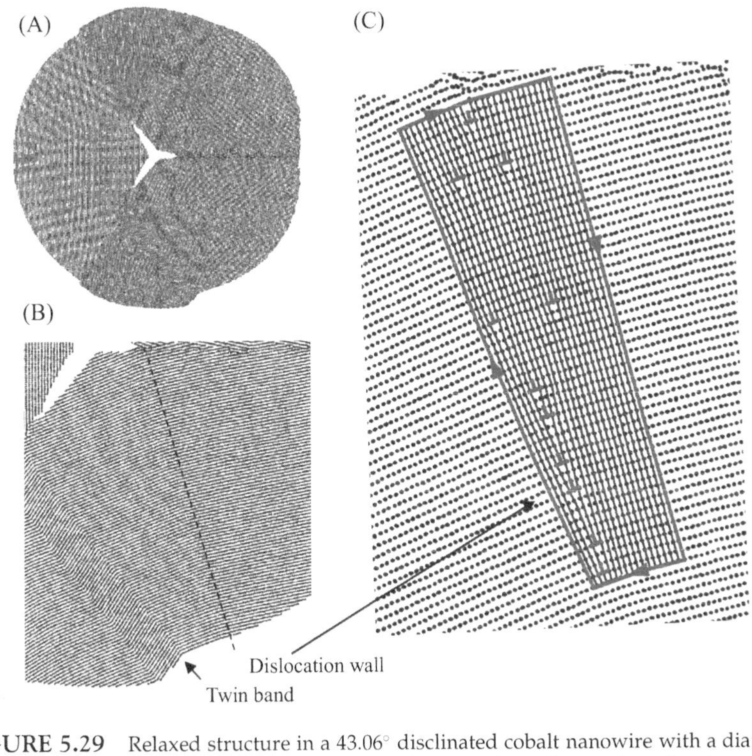

FIGURE 5.29 Relaxed structure in a 43.06° disclinated cobalt nanowire with a diameter of 40 nm at 300 K. (A) View of the nanowire showing the trifurcated crack, zigzag GBs, steps on the nanowire surface, twin bands, and dislocation walls within the grains. (B) Magnified view of a twin band and a dislocation wall near the lower-left GB. (C) Further magnification of the diffuse dislocation wall with a misorientation of ~7.5°. Source: *Modified and reprinted with permission from a work by Zhou et al. (2008b).*

The critical disclination strength ω_c exhibits linear dependence on the diameter of the nanowire in the double logarithmic scale for all the considered diameters in the range of 20–50 nm. When this linear relationship is extrapolated for diameters in the range of 200–1000 nm, the critical disclination strength ω_c is in good agreement with the experimental data for ultrafine-grained metals.

In addition to the investigation on the structural instability of different disclinations in nanowires with different lattice structures, Zhou et al. (2008a) have studied the effects of disclinations on the mechanical properties of nanofilms through MD simulation. In the simulation, the tensile strength of a bicrystalline nickel nanofilm with a wedge disclination is tested. The disclination is located in the middle of a symmetrical tilt GB that terminates at right angles to the free surfaces. It has been discovered that the tensile strength of the bicrystal increases with the disclination strength ω up to a value of $10-12^\circ$ before decreasing rapidly as ω increases further. The maximum tensile strength of the bicrystal can reach up to $\sim 80\%$ of the theoretical tensile strength of an ideal film without any disclinations or defects. The decrease in the tensile strength for large values of ω can be attributed to the site of crack nucleation shifting from the GB extremity to the disclination core. Moreover, it has been shown that a change in the GB misorientation causes a variance of 30% to the tensile strength, which suggests a possible approach to enhance the strength of nanofilms by introducing appropriate GB misorientations and disclination strength.

Aside from the above-discussed simulation studies on the effects of 1D defects on the mechanical behaviors of nanowires, MD simulation studies have been conducted to investigate the effects of 2D defects on the mechanical behaviors of materials. As a common type of 2D defect, GBs can significantly influence the mechanical response of polycrystalline solids. For instance, Lin et al. (2015b) have investigated the effects of GBs on the tensile response of bicrystalline silicon nanofilms. Both twist and tilt GBs have been considered. Their simulation results indicate that the fracture strength of bicrystalline silicon nanofilms containing either a twist GB or a tilt GB is lower than that of single-crystalline silicon nanofilms, which can be attributed to the GBs inducing changes in the crystallographic orientations and increasing the resistance to dislocation sliding.

During severe plastic deformation or solidification processes, GBs may attract impurity atoms causing the formation of their GB segregation, which can significantly alter the properties of the corresponding material. Segregation of certain impurity atoms in the GBs of metallic bulk materials can enhance their mechanical strength, particularly for metallic materials with nanoscale grains, of which the GB networks are dense, and the effects of segregation are prominent.

Babicheva et al. (2015) have investigated the influence of segregation of different metal atoms including Cu, Ti, Mg, Fe, Co, and Pb on the plastic deformation of nanocrystalline aluminum via MD simulation. The simulation model is a cubic computational cell with an edge length of 42 nm that contains a total of ~5 million atoms. The average size of the grains is taken to be 9 nm. Segregation atoms are introduced to the GBs to replace some Al atoms there. The coordination number of an Al atom is used to determine if it is within a GB; typically, the coordination number of Al atoms inside a grain exhibiting an FCC lattice structure is 12, but Al atoms in GBs possess a smaller coordination number of 10 or 11. These Al atoms with relatively small coordination numbers in the GBs of the nanocrystalline system are replaced by atoms of one of the above-mentioned metals, resulting in the formation of their respective GB segregation. The EAM potential is adopted to describe the interactions among the atoms, and the system temperature is kept at 300 K using the Nose–Hoover thermostat. During the simulation, PBCs are applied along all the x-, y-, and z-directions. The deformation of the system is realized by gradually increasing the shear strain component γ_{xy} in the $x-y$ plane while maintaining a constant hydrostatic pressure at 6 GPa using the MTTK barostat. The shear stress components τ_{xz} and τ_{yz} are kept at zero during the deformation process.

The simulation results demonstrate that GB segregation involving different elements exhibits diverse evolution patterns during shear deformation, as shown in Fig. 5.30. The impurity atoms of the Al–Co, Al–Mg, and Al–Pb alloys remain in their respective GBs, while those of the other alloys do not. Moreover, it has been discovered that both the segregation of Fe atoms and that of Co atoms in the GBs of the aluminum nanocrystalline system enhance its

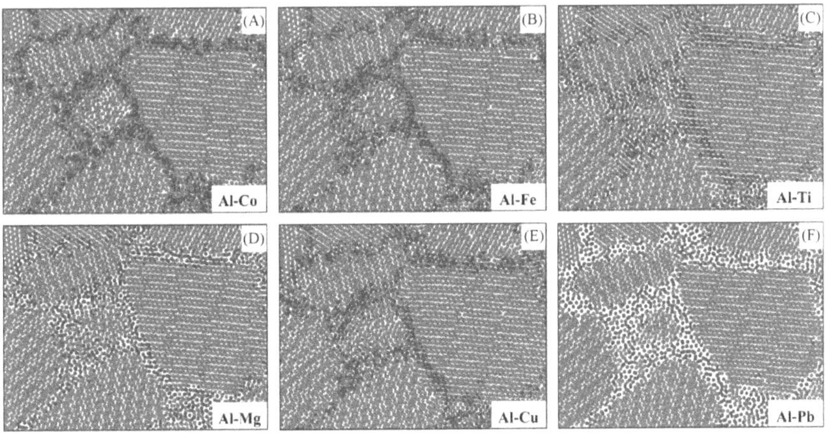

FIGURE 5.30 Atomic structures of nanocrystalline aluminum alloys with impurity atoms subjected to a shear strain of $\gamma_{xy} = 0.06$: (A) Al–Co, (B) Al–Fe, (C) Al–Ti, (D) Al–Mg, (E) Al–Cu, and (F) Al–Pb. The red spheres represent Al atoms, while the blue spheres represent impurity elements. Source: *Reprinted with permission from a work by Babicheva et al. (2015).*

mechanical strength. Meanwhile, the segregation of Cu atoms and that of Pb atoms have opposite effects, while the effect of segregation of both Mg and Ti atoms on the mechanical strength of the nanocrystalline system is negligible.

Babicheva et al. (2016) have also examined the deformation mechanisms of binary aluminum alloys (Al–Co and Al–Mg) with GB segregation and compared them with that of a binary aluminum solid solution in which impurity atoms are distributed randomly within the aluminum matrix. The results indicate that the distribution of the alloying elements (Co and Mg) significantly influences the deformation behaviors of the aluminum alloys. In particular, the presence of Mg atoms in the GBs results in plastic deformation of the aluminum alloys, which is manifested through three mechanisms: GB sliding, GB migration, and grain growth. However, the presence of Mg atoms as randomly distributed solutes has a negligible effect on the deformation mechanisms of the aluminum alloys, and they display the same deformation behaviors (i.e., GB sliding and GB migration) as pure aluminum. Meanwhile, the segregation of Co atoms in the GBs can effectively inhibit GB sliding and GB migration, thereby greatly increasing the mechanical strength of the aluminum alloys. In contrast, Co atoms that are randomly distributed as solutes lower the mechanical strength of the aluminum alloys as a result of structural amorphization that can easily initiate plastic flow.

Babicheva et al. (2017) have further investigated the effect of GB segregation on the cyclic tensile deformation of an aluminum bicrystalline system via MD simulation. Their simulation model consists of two aluminum grains separated by an asymmetric tilt GB. The GB segregation is formed by replacing the two rows of Al atoms at the GB with Co or Ti atoms. To expedite the occurrence of failure at the GB, a microcrack is generated at the GB located in the middle of the model. During the cyclic tensile deformation along the direction normal to the GB plane, the minimum tensile stress is set at zero, while the maximum tensile stress is set at 85% of the ultimate strength of the bicrystalline system that is calculated during a quasi-static uniaxial tensile test. It has been discovered that while pure aluminum experiences plastic deformation in the form of perfect dislocation sliding followed by twining, the addition of Co or Ti atoms in the GBs can decrease plastic flow by inhibiting dislocation sliding and twining. Moreover, the segregation of the Co or Ti atoms in the GBs can promote the splitting of perfect dislocations into Shockley partials in the aluminum bicrystalline system. Consequently, the rate of propagation of cracks in the system is lowered. It has been demonstrated that GB segregations can effectively enhance the fatigue properties of nanocrystalline aluminum.

Voids are common 3D defects in materials that are formed during their fabrication stages. Understanding the influence of these defects on the structural stability and fracture mechanisms of materials is important for optimizing their fabrication procedures and expanding the purviews of their applications.

A typical MD simulation study on the mechanical behaviors of materials containing voids is concerning the effect of voids on the interfacial fracture mechanism of a bilayer system (Lin et al., 2015a). This system consists of an amorphous Si_3N_4 (a-Si_3N_4) layer that is positioned on top of a silicon layer. In the simulation, two models are implemented. A small void is created in the silicon layer near the interface for the first model and exactly at the interface for the second model. To simulate the mode-I fracture and mode-II fracture in such a bilayer structure, tensile loading and shear loading are applied to the bilayer system, respectively. The simulation results show that under tensile loading at a temperature of 300 K, the bilayer system with a void within its silicon layer exhibits the initiation of cracks at the void, and the subsequent propagation of the cracks toward the interface eventually results in fracture of the bilayer structure. For the bilayer system with a void at its interface, fracture occurs at the void. Under shear loading, the temperature and the position of the void significantly influence the magnitude of the critical stress corresponding to interfacial sliding, but they barely affect the shear deformation process.

Besides the effects of voids on the mechanical behaviors of materials, MD simulation has also been applied to study the nucleation, growth, and coalescence of voids in materials. For instance, Rawat and Chaturvedi (2020) have investigated the evolution of voids in single-crystalline iron under uniaxial, biaxial, and triaxial tensile loading conditions. The simulation results indicate that the average void growth rate (the increase of the average void volume per increment of the tensile strain) is the highest for the uniaxial case, while it is the lowest for the triaxial case. In addition, the average void spacing of separated voids is the highest for the uniaxial case where the void density is the lowest, while it is the lowest for the triaxial case where the void density is the highest. Seppälä et al. (2005) have studied the coalescence of voids during the dynamic fracture of ductile metals. The interactions among voids have been quantified through the rate of reduction of the intervoid distance, the correlated directional growth of the voids, and the correlated shape evolution of the voids. Based on these quantifications, it has been identified that the critical intervoid ligament distance marking the onset of coalescence of two voids is close to their radii and independent of their initial separation distance.

In the above-mentioned two studies, the voids are embedded in single-crystalline materials without the consideration of interactions between voids and GBs. Zhang et al. (2019) have addressed this issue by investigating the interactions between nanovoids and migrating GBs in bicrystalline copper. It has been found that the interactions between GBs and voids can lead to the dissolution of voids. Besides GBs, many other factors, including temperature, grain size, strain rate, and chemical environment, can affect the nucleation, growth, and coalescence of voids. For instance, Rawat and Chaturvedi (2021) have shown that the nucleation rate of voids is highly accelerated as the temperature increases. Chandler et al. (2008) have demonstrated that H atoms that are trapped at the GBs of bicrystalline nickel reduce the critical stress and strain for void nucleation. In addition, the effect of the trapped H atoms on the void nucleation only depends on the hydrogen concentration, regardless of the GB misorientations. Since the fracture of materials usually occurs through the nucleation, growth, and coalescence of voids, further studies on the evolution of voids and its influencing factors are crucial for a better understanding of the mechanical properties of materials and their applications.

References

Adnan, A., Sun, C.T., Mahfuz, H., 2007. A molecular dynamics simulation study to investigate the effect of filler size on elastic properties of polymer nanocomposites. Compos. Sci. Technol. 67, 348–356.

Babicheva, R.I., Dmitriev, S.V., Bachurin, D.V., Srikanth, N., Zhang, Y., Kok, S.W., et al., 2017. Effect of grain boundary segregation of Co or Ti on cyclic deformation of aluminium bi-crystals. Int. J. Fatigue 102, 270–281.

Babicheva, R.I., Dmitriev, S.V., Bai, L., Zhang, Y., Kok, S.W., Kang, G., et al., 2016. Effect of grain boundary segregation on the deformation mechanisms and mechanical properties of nanocrystalline binary aluminum alloys. Comput. Mater. Sci. 117, 445–454.

Babicheva, R.I., Dmitriev, S.V., Zhang, Y., Kok, S.W., Srikanth, N., Liu, B., et al., 2015. Effect of grain boundary segregations of Fe, Co, Cu, Ti, Mg and Pb on small plastic deformation of nanocrystalline Al. Computat. Mater. Sci. 98, 410–416.

Bai, L., Sha, Z.-D., Srikanth, N., Pei, Q.-X., Wang, X., Srolovitz, D.J., et al., 2015. Friction between silicon and diamond at the nanoscale. J. Phys. D Appl. Phys. 48, 1–9.

Bai, L., Srikanth, N., Kang, G., Zhou, K., 2016a. Influence of third particle on the tribological behaviors of diamond-like carbon films. Sci. Rep. 6, 1–11.

Bai, L., Srikanth, N., Korznikova, E.A., Baimova, J.A., Dmitriev, S.V., Zhou, K., 2017a. Wear and friction between smooth or rough diamond-like carbon films and diamond tips. Wear 372, 12–20.

Bai, L., Srikanth, N., Korznikova, E.A., Baimova, J.A., Dmitriev, S.V., Zhou, K., 2017b. Wear and friction between smooth or rough diamond-like carbon films and diamond tips. Wear 372, 12–20.

Bai, L., Srikanth, N., Wu, H., Liu, F., Liu, B., Zhou, K., 2016b. Effect of environmental hydrogen atoms on the tribological behaviors of diamond-like carbon films. Tribol. Int. 99, 258–266.

Bai, L., Srikanth, N., Wu, H., Liu, Y., Liu, B., Zhou, K., 2016d. Investigation on tensile behaviors of diamond-like carbon films. J. Non-Cryst. Solids 443, 8–16.

Bai, L., Srikanth, N., Zhao, B., Liu, B., Liu, Z., Zhou, K., 2016c. Lubrication mechanisms of graphene for DLC films scratched by a diamond tip. J. Phys. D Appl. Phys. 49, 1–11.
Bai, L., Sun, P.-P., Liu, B., Liu, Z., Zhou, K., 2018. Mechanical behaviors of T-carbon: a molecular dynamics study. Carbon N. Y. 138, 357–362.
Beake, B.D., Leggett, G.J., 2002. Nanoindentation and nanoscratch testing of uniaxially and biaxially drawn poly (ethylene terephthalate) film. Polymer (Guildf) 43, 319–327.
Beake, B.D., Harris, A.J., Liskiewicz, T.W., 2013. Review of recent progress in nanoscratch testing. Tribol.-Mater., Surf. Interfaces 7, 87–96.
Born, M., Huang, K., 1954. Dynamical Theory of Crystal Lattices. Clarendon Press, Oxford.
Chandler, M.Q., Horstemeyer, M.F., Baskes, M.I., Wagner, G.J., Gullett, P.M., Jelinek, B., 2008. Hydrogen effects on nanovoid nucleation at nickel grain boundaries. Acta Mater. 56, 619–631.
Chen, J., Liang, Y., Chen, M., Bai, Q., Tang, Y., 2009. A study of the subsurface damaged layers in nanoscratching. Int. J. Abras. Technol. 2, 368–381.
Cheong, W.C.D., Zhang, L.C., 2000. Molecular dynamics simulation of phase transformations in silicon monocrystals due to nano-indentation. Nanotechnology 11, 173–180.
Cong, Z., Lee, S., 2018. Study of mechanical behavior of BNNT-reinforced aluminum composites using molecular dynamics simulations. Compos. Struct. 194, 80–86.
Du, K., Tang, Y., Zhang, J., Sun, T., 2013a. Molecular dynamics modelling and simulation of mechanical nanoscratching of polystyrene. Int. J. Nanomanufact. 9, 98–107.
Du, K., Tang, Y., Zhang, J., Xu, F., Yan, Y., Sun, T., 2013b. Velocity-dependent nanoscratching of amorphous polystyrene. Curr. Nanosci. 9, 153–158.
Echeverri Restrepo, S., van Eijk, M.C.P., Ewen, J.P., 2019. Behaviour of n-alkanes confined between iron oxide surfaces at high pressure and shear rate: a nonequilibrium molecular dynamics study. Tribol. Int. 137, 420–432.
Eder, S.J., Feldbauer, G., Bianchi, D., Cihak-Bayr, U., Betz, G., Vernes, A., 2015. Applicability of macroscopic wear and friction laws on the atomic length scale. Phys. Rev. Lett. 115, 025502-1–025502-5.
Fang, T.-H., Wu, H.-H., 2008. Molecular dynamics simulations on nanoindentation mechanisms of multilayered films. Comput. Mater. Sci. 43, 785–790.
Gersappe, D., 2002. Molecular mechanisms of failure in polymer nanocomposites. Phys. Rev. Lett. 89, 058301-1–058301-4.
Gertsman, V.Y., Nazarov, A.A., Romanov, A.E., Valiev, R.Z., Vladimirov, V.I., 1989. Dislocation-structural unit model of grain boundaries. Philos. Magaz. A 59, 1113.
Han, Y., Elliott, J., 2007. Molecular dynamics simulations of the elastic properties of polymer/carbon nanotube composites. Comput. Mater. Sci. 39, 315–323.
Hu, X., Martini, A., 2015. Atomistic simulation of the effect of roughness on nanoscale wear. Comput. Mater. Sci. 102, 208–212.
Hu, X., Tourek, C.J., Ye, Z., Sundararajan, S., Martini, A., 2014. Structural and chemical evolution of the near-apex region of an atomic force microscope tip subject to sliding. Tribol. Lett. 53, 181–187.
Huang, C., Peng, X., Fu, T., Chen, X., Xiang, H., Li, Q., et al., 2017. Molecular dynamics simulation of BCC Ta with coherent twin boundaries under nanoindentation. Mater. Sci. Eng.: A 700, 609–616.
Kajita, S., Righi, M.C., 2016. A fundamental mechanism for carbon-film lubricity identified by means of ab initio molecular dynamics. Carbon N. Y. 103, 193–199.
Kang, Z., Li, M., Tang, Q., 2010. Buckling behavior of carbon nanotube-based intramolecular junctions under compression: molecular dynamics simulation and finite element analysis. Comput. Mater. Sci. 50, 253–259.
Kheirkhah, A.H., Iranizad, E.S., Raeisi, M., Rajabpour, A., 2014. Mechanical properties of hydrogen functionalized graphene under shear deformation: a molecular dynamics study. Solid State Commun. 177, 98–102.

Levita, G., Righi, M.C., 2017. Effects of water intercalation and tribochemistry on MoS2 lubricity: an ab initio molecular dynamics investigation. Chemphyschem 18, 1475–1480.

Li, J., Chen, J., Zhu, M., Song, H., Zhang, H., 2019. Interfacial characteristics of boron nitride nanosheet/epoxy resin nanocomposites: a molecular dynamics simulation. Appl. Sci. 9, 1–14.

Li, Q., Liu, X.-Z., Kim, S.-P., Shenoy, V.B., Sheehan, P.E., Robinson, J.T., et al., 2014. Fluorination of graphene enhances friction due to increased corrugation. Nano Lett. 14, 5212–5217.

Li, Y., Wang, S., Arash, B., Wang, Q., 2016. A study on tribology of nitrile-butadiene rubber composites by incorporation of carbon nanotubes: molecular dynamics simulations. Carbon N. Y. 100, 145–150.

Lin, P., Babicheva, R.I., Xue, M., Zhang, H., Xu, H., Liu, B., et al., 2015a. Effects of temperature and voids on the interfacial fracture of Si/a-Si3N4 bilayer systems. Phys. Status Solidi B-Basic Solid State Phys. 252, 2013–2019.

Lin, P., Babicheva, R.I., Xue, M., Zhang, H., Xu, H., Liu, B., et al., 2015b. Tensile response of bi-crystalline Si nanofilms with twist and tilt grain boundaries. Comput. Mater. Sci. 96, 295–299.

Liu, J., Qi, Y., Li, Q., Duan, T., Yue, W., Vadakkepatt, A., et al., 2019. Vacancy-controlled friction on 2D materials: roughness, flexibility, and chemical reactions. Carbon N. Y. 142, 363–372.

Liu, Q., Shen, S., 2012. On the large-strain plasticity of silicon nanowires: effects of axial orientation and surface. Int. J. Plast. 38, 146–158.

Liu, W., Ruan, H., Zhang, L., 2013. On the plasticity event in metallic glass. Philos. Magaz. Lett. 93, 158–165.

Lu, Z., Chernatynskiy, A., Noordhoek, M.J., Sinnott, S.B., Phillpot, S.R., 2015. Nanoindentation of Zr by molecular dynamics simulation. J. Nuclear Mater. 467, 742–757.

Martini, A., Liu, Y., Snurr, R.Q., Wang, Q.J., 2006. Molecular dynamics characterization of thin film viscosity for EHL simulation. Tribol. Lett. 21, 217–225.

Perdew, J.P., Burke, K., Ernzerhof, M., 1996. Generalized gradient approximation made simple. Phys. Rev. Lett. 77, 3865–3868.

Rawat, S., Chaturvedi, S., 2020. Evolution of voids in single-crystal iron under uniaxial, biaxial and triaxial loading conditions. Philos. Magaz. 100, 2068–2090.

Rawat, S., Chaturvedi, S., 2021. Effect of temperature on the evolution dynamics of voids in dynamic fracture of single crystal iron: a molecular dynamics study. Philos. Magaz. 101, 657–672.

Rezaei, R., Shariati, M., Tavakoli-Anbaran, H., Deng, C., 2016. Mechanical characteristics of CNT-reinforced metallic glass nanocomposites by molecular dynamics simulations. Comput. Mater. Sci. 119, 19–26.

Sandoz-Rosado, E.J., Tertuliano, O.A., Terrell, E.J., 2012. An atomistic study of the abrasive wear and failure of graphene sheets when used as a solid lubricant and a comparison to diamond-like-carbon coatings. Carbon N. Y. 50, 4078–4084.

Seppälä, E.T., Belak, J., Rudd, R.E., 2005. Three-dimensional molecular dynamics simulations of void coalescence during dynamic fracture of ductile metals. Phys. Rev. B 71, 064112.

Sha, Z., Qu, S., Liu, Z., Wang, T., Gao, H., 2015. Cyclic deformation in metallic glasses. Nano Lett. 15, 7010–7015.

Sha, Z.-D., Sorkin, V., Branicio, P.S., Pei, Q.-X., Zhang, Y.-W., Srolovitz, D.J., 2013. Large-scale molecular dynamics simulations of wear in diamond-like carbon at the nanoscale. Appl. Phys. Lett. 103, 073118-1–073118-5.

Shen, S., Atluri, S.N., 2004. Atomic-level stress calculation and continuum-molecular system equivalence. Comp. Model. Eng. Sci. 6, 91–104.

Shokrieh, M., Hosseinkhani, M., Naimi-Jamal, M., Tourani, H., 2013. Nanoindentation and nanoscratch investigations on graphene-based nanocomposites. Poly. Test. 32, 45–51.

Sun, S., Chen, S., Weng, X., Shan, F., Hu, S., 2019. Effect of carbon nanotube addition on the interfacial adhesion between graphene and epoxy: a molecular dynamics simulation. Polymers 11, 1–12.

Tanaka, K., Kato, T., Matsumoto, Y., 2003. Molecular dynamics simulation of vibrational friction force due to molecular deformation in confined lubricant film. J. Tribol.—Transact. ASME 125, 587–591.

Tang, C., Guo, W., Chen, C., 2009. Molecular dynamics simulation of tensile elongation of carbon nanotubes: temperature and size effects. Phys. Rev. B 79, 155436-1–155436-9.

Vanderbilt, D., 1990. Soft self-consistent pseudopotentials in a generalized eigenvalue formalism. Phys. Rev. B 41, 7892–7895.

Voeltzel, N., Giuliani, A., Fillot, N., Vergne, P., Joly, L., 2015. Nanolubrication by ionic liquids: molecular dynamics simulations reveal an anomalous effective rheology. Phys. Chem. Chem. Phys. 17, 23226–23235.

Walsh, P., Omeltchenko, A., Kalia, R.K., Nakano, A., Vashishta, P., Saini, S., 2003. Nanoindentation of silicon nitride: a multimillion-atom molecular dynamics study. Appl. Phys. Lett. 82, 118–120.

Wang, L., Duan, F., 2018. Nanoscale wear mechanisms of few-layer graphene sheets induced by interfacial adhesion. Tribol. Int. 123, 266–272.

Wu, M.S., Nazarov, A.A., Zhou, K., 2004. Misorientation dependence of the energy of [1100] symmetrical tilt boundaries in hcp metals: prediction by the disclination-structural unit model. Philos. Magaz. 84, 785–806.

Wu, M.S., Zhou, K., Nazarov, A.A., 2006. Stability and relaxation mechanisms of a wedge disclination in an HCP bicrystalline nanowire. Model. Simul. Mater. Sci. Eng. 14, 647–661.

Wu, M.S., Zhou, K., Nazarov, A.A., 2007. Crack nucleation at disclinated triple junctions. Phys. Rev. B 76, 134105-1–134105-6.

Xu, S., Wan, Q., Sha, Z., Liu, Z., 2015. Molecular dynamics simulations of nano-indentation and wear of the gamma Ti-Al alloy. Comput. Mater. Sci. 110, 247–253.

Xu, W., Kim, W.K., 2019. Molecular dynamics simulation of the uniaxial tensile test of silicon nanowires using the MEAM potential. Mech. Mater. 137, 1–12.

Ye, Z., Tang, C., Dong, Y., Martini, A., 2012. Role of wrinkle height in friction variation with number of graphene layers. J. Appl. Phys. 112, 116102-1–116102-3.

Yuan, F., Huang, L., 2012. Molecular dynamics simulation of amorphous silica under uniaxial tension: from bulk to nanowire. J. Non-Crystal. Solids 358, 3481–3487.

Zhang, L., Shibuta, Y., Lu, C., Huang, X., 2019. Interaction between nano-voids and migrating grain boundary by molecular dynamics simulation. Acta Mater. 173, 206–224.

Zhang, Q., Diao, D., Kubo, M., 2015. Nanoscratching of multi-layer graphene by molecular dynamics simulations. Tribol. Int. 88, 85–88.

Zhang, S., Huang, P., Wang, F., 2020. Graphene-boundary strengthening mechanism in Cu/graphene nanocomposites: a molecular dynamics simulation. Mater. Design 190, 1–11.

Zhou, K., Nazarov, A.A., Wu, M.S., 2006. Continuum and atomistic studies of a disclinated crack in a bicrystalline nanowire. Phys. Rev. B 73, 045410-1–045410-11.

Zhou, K., Nazarov, A.A., Wu, M.S., 2007. Competing relaxation mechanisms in a disclinated nanowire: temperature and size effects. Phys. Rev. Lett. 98, 035501-1–035501-4.

Zhou, K., Nazarov, A.A., Wu, M.S., 2008a. Atomistic simulations of the tensile strength of a disclinated bicrystalline nanofilm. Philos. Magaz. 88, 3181–3191.

Zhou, K., Wu, M.S., Nazarov, A.A., 2008b. Relaxation of a disclinated tricrystalline nanowire. Acta Mater. 56, 5828–5836.

CHAPTER

6

Application of molecular dynamics simulation in thermal problems

This chapter focuses on the applications of molecular dynamics (MD) simulation in thermal problems. It begins with the two most commonly used MD simulation methods for calculating the thermal conductivity of a solid: the direct method and the Green–Kubo method. Important technical details of both methods are highlighted. Afterward, the application of MD simulation in investigating interfacial thermal transport is demonstrated with two case studies. Finally, MD simulations of thermal rectification effects are introduced.

6.1 Demand for understanding the thermal properties of nanomaterials

The rapid development of the synthesis and processing of nanomaterials and their wide applications in electronic and energy industries have resulted in an ever-increasing demand for a greater scientific understanding of their thermal properties. This demand is mainly driven by the following factors.

First, the continuous miniaturization and concomitant increase in the component density of electronic devices place ever-increasing thermal loads on them. Under Moore's law, the sizes of individual electronic device components decrease exponentially every year. The removal of heat from electronic devices is of crucial importance to both their performance and function, and thus thermal design has become an integral part of their development. Moreover, as electronic devices approach the nanoscale, their thermal transport behaviors demonstrate distinct features from those observed on the macroscale. These unique features

DOI: https://doi.org/10.1016/B978-0-12-816419-8.00011-8

should also be properly understood and incorporated into the complete electrical–thermal design of electronic devices.

Second, the increasing demand for energy calls for practical solutions that provide novel and efficient approaches to energy generation, consumption, and recycling. An attractive solution is to utilize newly developed nanotechnologies and nanomaterials to enhance heat management in energy-associated applications. In some cases, such as nanoelectronics, nanomaterials with high thermal conductivity are required to render thermal dissipation the highest possible efficiency. In other cases, such as thermoelectric devices and thermal barrier coatings, nanomaterials or nanoparticles with low thermal conductivity are desired to improve the thermal conversion or preservation involved.

Third, an understanding of the thermal properties of nanomaterials can also lead to the discovery of novel nanotechnologies. Traditionally, phonons are considered the main heat carriers, and electrons are considered the main electricity carriers. Recently, it has been established that phonons can also be manipulated to transfer and process information (Wang and Li, 2008a). More importantly, various types of thermal logic devices have been theoretically proposed or even experimentally fabricated, such as the thermal rectifier (Hu et al., 2006; Xie et al., 2012), thermal transistor (Lo et al., 2008), thermal logic gate (Wang and Li, 2007), and thermal memory cell (Wang and Li, 2008b). Based on these basic thermal components, it is possible to construct thermal circuits similar to their electronic counterparts. The key working mechanism of these thermal components is the thermal rectification effect that arises from a high dependence of the heat transport behavior of a material on the direction of an applied temperature gradient. The thermal rectification effect is analogous to the phenomenon observed in an electronic rectifier for which the electrical gradient direction determines the electric transport behavior. Compared with bulk materials, nanomaterials, in particular some nanocomposites, exhibit such a thermal rectification effect and possess large rectifying ratios, and therefore are highly promising for applications in thermal devices.

MD simulation is ideal for investigating the thermal properties of nanomaterials because it is capable of studying the detailed atomic structures under thermal loads by comprehensively analyzing the kinetic behaviors and interactions of individual atoms. Hence, a detailed description of the relationship between the microstructure of nanomaterials and their properties can be obtained, and the most important aspects of their thermal properties can be identified. This chapter aims to provide a solid foundation for the simulation of the thermal properties of nanomaterials with a focus on thermal transfer problems.

6.2 Molecular dynamics simulation methods for thermal conductivity calculation

Thermal conductivity relates the heat flux to the temperature gradient via Fourier's law as

$$J = -\lambda \nabla T, \tag{6.1}$$

where J denotes the heat flux vector, λ denotes the second-order thermal conductivity tensor, and the Nabla operator $\nabla T = (\partial T/\partial x, \partial T/\partial y, \partial T/\partial z)$ denotes the temperature gradient in the Cartesian coordinate system. For many simple applications, the thermal conductivity in a particular direction is of interest, and the scalar form of the Fourier's law in this direction can be adopted to investigate the thermal conductivity. For example, the thermal conductivity λ_{zz} in the z-direction can be calculated by $J_z = -\lambda_{zz}\partial T/\partial z$. Experimentally, the thermal conductivity of a material is typically obtained by measuring the temperature gradient resulting from the application of a heat flux. In MD simulations, the thermal conductivity of a system can be computed using either a nonequilibrium MD (NEMD) method or an equilibrium MD (EMD) method. The two most commonly applied methods for computing thermal conductivity are the "direct method" and the Green–Kubo method. The direct method is an NEMD method that relies on imposing a temperature gradient across the simulation cell and is therefore analogous to the experimental measurement process. In contrast, the Green–Kubo method is an EMD method that utilizes fluctuations in the heat flux to compute thermal conductivity based on the fluctuation–dissipation theorem. For a homogeneous material, both the direct method and the Green–Kubo method can be adopted to predict its thermal conductivity. However, for an inhomogeneous material, such as a composite material or a material containing interfaces, the Green–Kubo method is inapplicable. This is because the Green–Kubo method assumes that the spatial variation of an applied heat current must be equivalent to that of the temperature gradient. This condition cannot be satisfied in an inhomogeneous system. In this section, the direct method and the Green–Kubo method are discussed in detail with their respective strengths and weaknesses highlighted.

6.2.1 Direct method

The direct method for computing the thermal conductivity of a system is illustrated in Fig. 6.1. A rectangular simulation box with a dimension of $2L$ in the z-direction is presented as an example, and periodic boundary conditions (PBCs) are applied in all directions. Initially, the

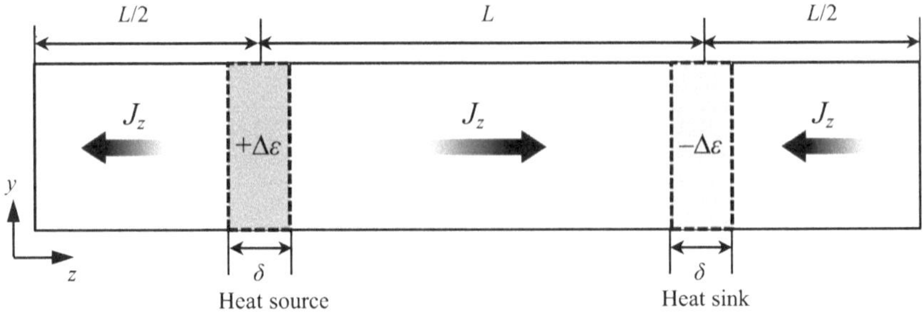

FIGURE 6.1 Schematic representation of a simulation cell for calculating thermal conductivity using the direct method.

MD system is equilibrated and relaxed in the NPT (constant number of species, pressure, and temperature) ensemble at a preset temperature and zero pressure. Upon reaching the equilibrium state, the simulation is continued in the NVE (constant number of species, pressure, and total energy) ensemble to keep the total energy of the system conserved. A constant heat flux J_z (assuming that the thermal conductivity along the z-direction is measured here) is then imposed on the system at each time step with a duration of Δt. This heat flux is achieved by adding a small amount of heat $\Delta\varepsilon$ into a region located at $z = 3L/2$ with a thickness of δ, which renders this region a heat source. Similarly, $\Delta\varepsilon$ is removed from another region that is located at $z = L/2$ with the same thickness, and this region is referred to as a heat sink. After a transient period with a sufficient duration, a stable temperature gradient $\partial T/\partial z$ is established along the heat flux direction.

Depending on the system under investigation, the transient period required for the system to reach a steady temperature profile can reach up to a few nanoseconds. Hence, a relatively long simulation time is required for calculating thermal conductivity using the direct method, and the data generated over a long time in the initial stage of the simulation is generally discarded to avoid the incorporation of data corresponding to the transient period into the calculation of the temperature gradient. In a practical MD simulation, the duration of this transient period is often unknown beforehand. As a result, more data may be unnecessarily discarded during the calculation of the temperature gradient.

The thermal conductivity λ_{zz} in the z-direction can be obtained from Fourier's law, for which J_z is given by $J_z = \Delta\varepsilon/(2A\Delta t)$, i.e., $\lambda_{zz} = -\Delta\varepsilon/\left[2A\Delta t(\partial T/\partial z)\right]$, where A denotes the cross-section area of the system. Here, the factor of 2 accounts for the heat flux from the heat source dissipating uniformly in both directions.

Due to the symmetric setup of the simulation system, the temperature gradient is calculated by averaging its values obtained from both sides of the heat source. This averaging is helpful in reducing the inherent statistical errors in MD simulations but doubles the computational cost. For large MD systems, an alternative simulation setup shown in Fig. 6.2 would be preferable. In this setup, the system is halved in size along the direction of the heat flux, and the PBC is only applied in the y-direction. To prevent lateral drifting and rotation of the entire system, both ends of the simulation cell are fixed by freezing the movements of the particles inside a thin slab. The heat source and heat sink regions are set to be adjacent to either of these two fixed regions. From Figs. 6.1 and 6.2, it is evident that the presence of a heat source and heat sink results in a heat flux that flows in a well-defined direction. Note that to obtain the thermal conductivity along different lattice directions, entirely new simulations with different simulation cells must be performed.

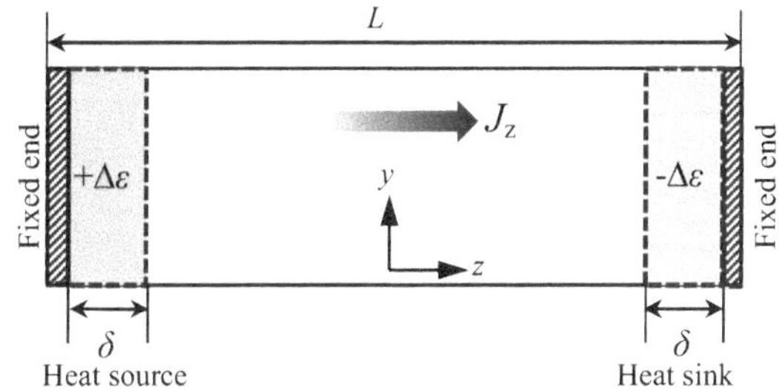

FIGURE 6.2 MD simulation cell for the direct method with a reduced system size and fixed boundaries.

Generally, the addition of energy to the heat source and the extraction of energy from the heat sink are both realized by rescaling the velocities of the particles within it (Jund and Jullien, 1999). Nevertheless, to avoid an artificial drift of the kinetic energy of the particles, velocity rescaling has to be performed with the total momentum of the system being conserved. For particle i in the heat source or the heat sink, its velocity v_i at each time step is modified according to the following equation:

$$v'_i = v_C + \alpha(v_i - v_C), \tag{6.2}$$

where v_C is the velocity of the center of mass of all the particles in the heat source or the heat sink, and the factor α relates to the energy $\Delta\varepsilon$ added or extracted at each time step *via* the relation:

$$\alpha = \sqrt{1 \pm \frac{\Delta\varepsilon}{E_k^R}}. \tag{6.3}$$

The positive and negative signs depend on whether particle i is in the heat source or heat sink. The relative kinetic energy of the system E_k^R is given by

$$E_k^R = \frac{1}{2}\sum_i m_i v_i \cdot v_i + \frac{1}{2}\sum_i m_i v_C \cdot v_C. \tag{6.4}$$

For an MD system without any translational movement, the velocities of the center of mass of the heat source and the heat sink are zero, and Eqs. (6.2) and (6.4) can be simplified by setting $v_C = 0$.

Alternatively, the addition/extraction of energy $\Delta\varepsilon$ can be realized through a velocity exchanging algorithm (Müller-Plathe, 1997). Specifically, the velocity vector of an atom in the heat sink is exchanged with that of an atom in the heat source at a preset frequency in such a manner that the temperature increases in the heat source but decreases in the heat sink. These two selected atoms correspond to the atom with the lowest kinetic energy in the heat source and the atom possessing the highest kinetic energy in the heat sink. Generally, the temperature difference between the heat source and heat sink is small, and there is always overlap between the distributions of the kinetic energy of the atoms in the heat source and the atoms in the heat sink, so the "hottest" atom in the heat sink always possesses a higher kinetic energy than the "coolest" atom in the heat source. Consequently, a heat flux is guaranteed to exist with a direction from the heat source to the heat sink, which leads to a resultant temperature gradient in the intervening region.

Upon reaching a steady state, the artificial energy transfer from the heat sink to the heat source induced by the velocity exchange is precisely balanced by the heat flux in the opposite direction induced by the temperature difference developed between the heat source and heat sink. The resultant heat flux is explicitly determined by summing up all the energy transported by each velocity exchange. Therefore, given that the particles in the heat source and the heat sink have the same atomic mass m, the thermal conductivity λ_{zz} can be calculated by

$$\lambda_{zz} = \frac{\sum_{N_{\text{transfer}}} \frac{m}{2}\left(v_{\text{hi}}^2 - v_{\text{lo}}^2\right)}{2tA\langle \partial T/\partial z\rangle}, \tag{6.5}$$

where N_{transfer} is the total number of velocity exchange events that occur over time t; v_{hi} denotes the highest atomic velocity in the heat sink, and v_{lo} is the lowest atomic velocity in the heat source. Differing from the velocity rescaling method in which the energy addition/extraction $\Delta\varepsilon$ is kept constant, the amount of energy transferred between the heat source and the heat sink varies with time for the velocity exchange approach. Hence, the temperature gradient also exhibits fluctuations, but its averaged value over time is a constant. Consequently, $\langle \partial T/\partial z \rangle$, the averaged value of the temperature gradient over time, is used in Eq. (6.5) instead of its instantaneous value.

One advantage of the velocity exchange approach is that the total linear momentum, total kinetic energy, and total energy of the MD system can be conserved. However, since the velocity exchange is only performed between two atoms, the resultant energy addition/subtraction $\Delta\varepsilon$ is small, and thus the temperature gradient will be small, particularly for large systems with low thermal conductivity. Consequently, a significantly long simulation time is required for obtaining a well-converged temperature gradient through time-averaging that can eliminate the impact of the noise produced by local temperature fluctuations. Meanwhile, the velocity exchange approach is highly sensitive to the energy exchange frequency. At a high frequency, the energy added into the heat source cannot be dissipated timely and accumulates in it, leading to a constant increase in its temperature. Similarly, the energy extracted from the heat sink is not timely compensated for, leading to a constant decrease in its temperature. Consequently, the temperature difference between the heat source and heat sink becomes increasingly large. Thus, selecting an appropriate exchanging frequency is crucial in achieving a well-converged temperature gradient in a short simulation time.

Another approach for generating heat flux in the simulation models shown in Figs. 6.1 and 6.2 is by setting the heat source at a temperature $T + \Delta T$ and the heat sink at a lower temperature $T - \Delta T$ by coupling the heat source and heat sink to two separate thermostats. Here, T is the system temperature. Upon reaching a steady state, the temperature difference between the heat source and the heat sink establishes a steady temperature gradient in the simulation system. To maintain the steady temperature gradient, energy is supplied to (extracted from) the heat source (sink) by coupling it with a thermostat to prevent its continuous cooling (heating) due to the flow of heat. The total amount of energy supplied to the heat source or subtracted from the heat sink increases linearly with time, and the slope of this linear relation gives the heat flux.

To calculate the steady temperature gradient, the simulation cell is divided into K thin slabs of equal thickness in the heat flux direction. The instantaneous local temperature of the kth slab is calculated by

$$T(z_k) = \frac{1}{3n_k k_B}\sum_{i}^{n_k} m_i v_i^2, \tag{6.6}$$

where z_k denotes the position of the center of the slab. The summation extends over n_k atoms in the kth slab; k_B is the Boltzmann constant (1.380649×10^{-23} J K^{-1}); the mass and velocity of atom i are denoted by m_i and v_i, respectively. To eliminate the noise in temperature caused by local thermal fluctuations, averaging over time needs to be performed for the temperature of each thin slab by

$$\langle T(z_k)\rangle_M = \frac{1}{M}\sum_{m=0}^{M-1} T_{N-m}(z_k). \tag{6.7}$$

Here, $\langle T(z_k)\rangle_M$ is the temperature of the kth slab averaged over the final M time steps of the simulation, and $T_{N-m}(z_k)$ is the instantaneous temperature for the $(N-m)$th time step.

Fig. 6.3 shows the steady temperature profile obtained in the MD simulation for calculating the thermal conductivity of crystalline Si at 500 K using the direct method. A rectangular simulation box with a dimension of $4 \times 4 \times 288$ lattice units is adopted. The symmetric simulation model in Fig. 6.1 is used, and the Stillinger–Weber (SW) potential (see Chapter 2: Potential Energy Functions) is adopted to describe the interatomic interactions. The Newtonian equations of motion of the atoms are integrated with a time step of 0.55 fs. The velocity rescaling method with a heat addition/extraction rate of $\Delta\varepsilon = 5 \times 10^{-4}$ eV per time step is employed to generate a heat flux of 1.06×10^{-4} eV nm^{-2}.

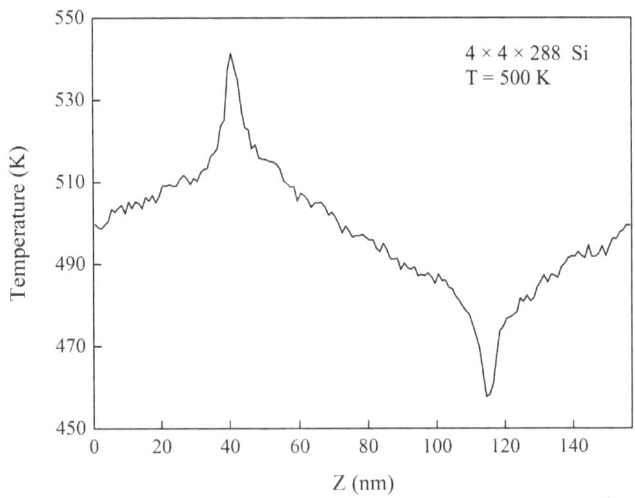

FIGURE 6.3 Typical steady-state temperature profile for calculating the thermal conductivity of a rectangular bar composed of crystalline Si with a dimension of 4 × 4 × 288 lattice units using the simulation cell presented in Fig. 6.1.

Two noticeable features of the temperature profile can be observed in Fig. 6.3. Within 6 nm from the heat source or heat sink, a nonlinear region can be observed in the temperature profile, which is attributed to enhanced phonon scattering caused by the artificial velocity rescaling of the atoms in the heat source and the heat sink. This nonlinearity is prominent in good thermal conductors but may disappear in poor ones. In the intermediate region (typically a distance of 0.2 L away from the heat source or the heat sink), the temperature profile exhibits a nearly linear pattern, based on which a temperature gradient can be obtained to calculate thermal conductivity. The gradients measured for the two opposite heat flux directions are similar, with values of 0.31 K nm^{-1} and 0.32 K nm^{-1}, respectively. Typically, the gradients fitted to the two linear regions differ by less than 15%, and such a difference is usually caused by the unequal heat fluxes in the two directions. This difference is due to imperfect symmetries in the atomic structure, dynamic behavior of the atoms, and/or statistical inaccuracies of data, which are issues independent of the geometrical symmetry of the simulation cell.

The above discussion provides the fundamental knowledge for applying the direct method to calculate thermal conductivity. There are still several critical "practical issues" that must be clarified. The first pertains to the calculation of temperature. In MD simulations, the temperature of a system is generally calculated by Eqs. (1.51) and (1.52) based on the classical equipartition theorem, which is only valid at temperatures much higher than the Debye temperature θ_D of the system, which is given by $\theta_D = \hbar\omega_D/k_B$. Here, $\hbar$ is the reduced Planck constant (1.054572×10^{-34} J s), and ω_D denotes the Debye frequency that corresponds to the highest phonon frequency allowed in the Debye model. According to quantum mechanics, the energy spectrum of a quantum oscillator (e.g., an atom in a solid) is given by a series of discretized energy levels $(n_e + 1/2)\hbar\omega$, with n_e denoting the phonon occupation number and ω denoting the phonon frequency. When the thermal energy k_BT is much greater than the energy difference $\hbar\omega$ between two adjacent quantum energy levels, i.e., $k_BT > > \hbar\omega$, the quantum characteristic of the oscillator can be ignored, and it can be approximated as a classical oscillator. Under this condition, the possible energy states of the classical oscillator can be regarded as continuous and occupied with the same probability, and the classical equipartition theorem is also valid. On the other hand, in the low-temperature regime in which k_BT is comparable to or smaller than $\hbar\omega$, it is important to consider the quantum characteristic of the oscillator by considering its different quantum occupation numbers for each energy state as described by the Boltzmann distribution.

In MD simulation, each atom in a solid can be approximated as an oscillator. When the temperature of a system is lower than or comparable to its Debye temperature, the quantum characteristic of the atoms usually needs to be considered by performing a quantum correction to the system temperature. The quantum correction is performed by imposing the requirement that the mean kinetic energy of the classical

MD system be equal to that of the actual quantum system. The energy of the former can be expressed as

$$k_B T_{MD} = \int d\omega D(\omega)\hbar\omega\left(n_e + \frac{1}{2}\right), \tag{6.8}$$

where $D(\omega)$ is the phonon density of states, and n_e is defined by

$$n_e = \frac{1}{\exp\left(\frac{\hbar\omega}{k_B T}\right) - 1}. \tag{6.9}$$

Based on this equation, the MD temperature T_{MD} is related to the quantum temperature T, and temperature remapping can be performed. Since the thermal conductivity is implicitly associated with the temperature gradient, a rescaling of the thermal conductivity is also required and given by

$$J_z = \lambda_{zz}\left(\frac{\partial T}{\partial z}\right) = \lambda_{zz}\left(\frac{\partial T}{\partial T_{MD}}\right)\left(\frac{\partial T_{MD}}{\partial z}\right). \tag{6.10}$$

Fig. 6.4A depicts a typical relationship between the classical temperature T_{MD} obtained in an MD simulation using Eqs. (1.51) and (1.52), and

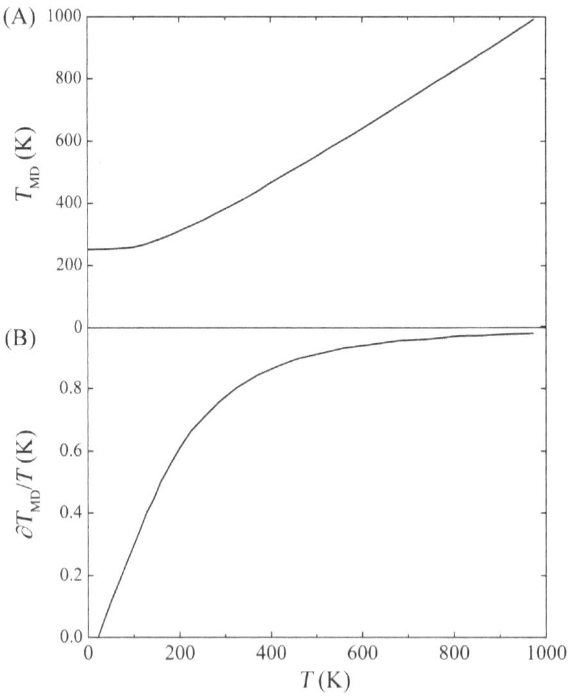

FIGURE 6.4 (A) Example of the relationship between the classical temperature T_{MD} and the quantum temperature as expressed in Eq. (6.8). (B) The corresponding relation between $\partial T_{MD}/\partial T$ and the quantum temperature T.

the quantum temperature determined by Eq. (6.8). The first-order derivative of the curve in Fig. 6.4A produces $\partial T_{MD}/\partial T$ that is shown in Fig. 6.4B. It is observed that the derivative $\partial T_{MD}/\partial T$ decreases rapidly with T in the low-temperature regime, and therefore the thermal conductivity obtained after the quantum correction becomes increasingly lower than its uncorrected counterpart as T decreases. This outcome is expected because the phonon occupation number of a material decreases with temperature. Here, a phonon is a quantum mechanical description of a collectively excited vibration mode of atoms or molecules of a crystal lattice.

Materials with a high hardness usually have a high Debye frequency ω_D because of the stiff bonds between their constituent atoms. Hence, a high Debye temperature θ_D can be expected, and a quantum correction of temperature is usually necessary. For example, diamond has high hardness, and its Debye temperature is as high as ~1800 K. Hence, a quantum correction is usually performed to calculate the thermal conductivity of diamond at low temperatures. However, when the focus of a study is on the effects of various factors on the thermal conductivity of a material instead of giving an accurate prediction of its absolute value, the quantum correction can usually be omitted in MD simulation practices.

The second practical issue of the direct method is the finite-size effect. To understand this effect, some basic knowledge of the mean free path (MFP) of phonons is first introduced. In metals, thermal energy is primarily carried by electrons, whereas in semiconductors and insulators, thermal energy is transferred via the propagation of phonons. When phonons propagate within a material, they are dissipated in the form of scattering, either by themselves or by geometrical defects or constraints within the material including point defects, dislocations, grain boundaries, surfaces, and interfaces. By treating phonons as pseudo-particles traveling through a solid, the basic kinetic theory can be applied to obtain the expression for the thermal conductivity $\lambda = C_V v_P l/3$ (assume the thermal conductivity is isotropic here). Here, C_V is the constant volume heat capacity; v_P is the group velocity of the phonons that can be approximated by the speed of sound within the solid in most cases; l is the length of the MFP of the phonons, which represents the characteristic length between two successive phonon scattering events.

The heat capacity and the phonon group velocity can be conveniently determined; they are essentially independent of the structure of a system as it approaches the nanoscale. The length l of the phonon MFP of a material is highly dependent on its atomic structure and is related to the phonon relaxation time τ and group velocity v as $l = \tau v$. The relaxation time τ can be approximated based on Matthiessen's rule, which states that the total phonon scattering rate (reciprocal of the relaxation time) is the sum of all independent scattering processes, i.e., $\tau^{-1} = \sum \tau_i^{-1}$, where i indexes all scattering processes including phonon–phonon scattering and phonon scatterings caused by point defects, grain boundaries, and interfaces. For instance, the total phonon scattering rate for a polycrystalline nanomaterial can be computed as $\tau^{-1} = \tau_{\mathrm{b}}^{-1} + \tau_{\mathrm{im}}^{-1} + \tau_{\mathrm{g}}^{-1} + \tau_{\mathrm{s}}^{-1}$, for which $\tau_{\mathrm{b}} = l_{\mathrm{b}}/v$, $\tau_{\mathrm{g}} = d/v$, and $\tau_{\mathrm{s}} = L/v$. Here, τ_b denotes the intrinsic relaxation time of the corresponding single-crystal bulk material, τ_{im} is the relaxation time in the presence of impurities, τ_{g} is the relaxation time under the effect of grain boundary scattering, τ_{s} is the relaxation time under the effect of surface scattering, l_{b} is the intrinsic length of the phonon MFP of the bulk material, d is the mean grain size, and L is the length of the material in the direction of the heat flux.

The finite-size effect of the simulation model in Fig. 6.1 is caused by the boundaries of the heat source and heat sink. Since the atomic dynamics in the heat source and heat sink are artificially altered during the simulation to generate a heat flux, their boundaries cause phonon scattering. In a situation in which the intrinsic length l_{b} of the phonon MFP exceeds the dimensions of a system, which is known as the Casimir limit, phonons propagate within the system in a ballistic manner and seldom experience phonon–phonon scattering if no defects are present. Phonon scattering only occurs at the boundaries of the heat source and heat sink, and the relaxation time τ_{s} is the dominant factor that determines the length l of the phonon MFP. Hence, l is bounded by the size of the simulation system and will increase if the system dimensions are increased. Consequently, the thermal conductivity λ will increase with the system size according to the kinetic theory of thermal conductivity.

Typical lengths of the phonon MFP in metals and semiconductors at room temperature range from 10 nm to 100 nm. To obtain the intrinsic thermal conductivity of a material that is independent of the simulation system size, the dimensions of the system should be much larger than the length of the phonon MFP, which is unfortunately impractical for some materials. For instance, the length of the phonon MFP in graphene is over 700 nm, and an MD simulation with cell dimensions on the order of micrometers is computationally costly. Therefore, the thermal conductivity of a bulk material or an infinite system is usually obtained by extrapolating the results of a set of simulations with different system sizes based on Matthiessen's rule. Take the calculation of the thermal conductivity of single-crystalline Si using the model in Fig. 6.1 as an example, and the length of the phonon MFP l can be approximated by

$$\frac{1}{l} = \frac{1}{l_{\mathrm{b}}} + \frac{2}{L}. \tag{6.11}$$

The factor of 2 accounts for the fact that when phonons travel along the direction of the heat flux, the average distance between two consecutive scattering events is taken to be $L/2$. Note that $L/2$ is only a rough approximation of the average distance of phonon scattering caused by the boundaries of the heat source and heat sink. This approximation does not affect the bulk-phase thermal conductivity predicted by the linear extrapolation method.

Based on the classical equipartition theorem, the specific heat capacity C_V of a solid can be approximated by $3k_B\rho_N/2$ with ρ_N denoting the number density of its constituent atoms. Recall that $\lambda = C_V v_P l/3$, and the linear relationship between $1/\lambda$ and $1/L$ can thus be given by

$$\frac{1}{\lambda} = \frac{2}{k_B v_P \rho_N}\left(\frac{1}{l_b} + \frac{2}{L}\right). \tag{6.12}$$

Eq. (6.12) suggests that a plot of $1/\lambda$ with respect to $1/L$ should be linear, and the thermal conductivity of an infinitely long system can be obtained by extrapolating this linear relationship to $L \to \infty$. For cases in which the phonon group velocity v_p and the intrinsic length l_b of the phonon MFP in a bulk material are not known beforehand, a more general form of Eq. (6.12) can be adopted where $1/\lambda = A + B/L$. The value of λ for an infinitely long system ($L \to \infty$) is obtained as $1/A$. Fig. 6.5 shows the relationship between the parameters $1/\lambda$ and $1/L$ for single-crystalline Si at 500 K and 1000 K, and a linear relationship can be clearly observed at both temperatures.

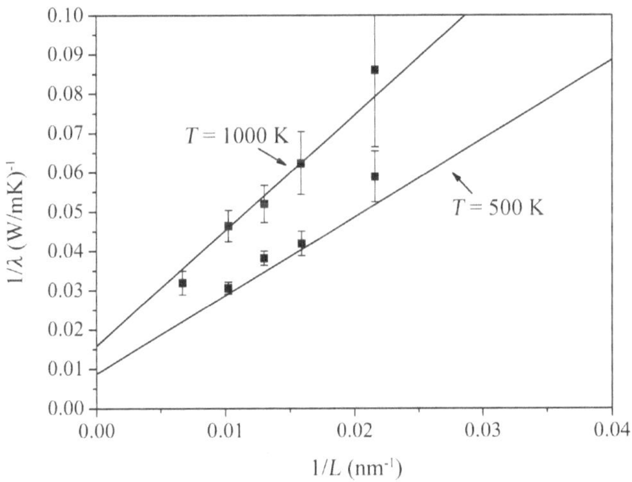

FIGURE 6.5 Relationship between $1/\lambda$ and $1/L$ for single-crystalline Si at 500 K and 1000 K. The simulation model presented in Fig. 6.1 is adopted, and the SW potential is used to describe the interatomic interactions. Source: *Reprinted and modified with permission from a work by Schelling et al. (2002).*

Note that the linear extrapolation method can only predict the thermal conductivity of a bulk material accurately when the dimensions of all the simulation cells in a set of simulations exceed a critical value. When this requirement is not fulfilled, the predicted bulk thermal conductivity is usually an underestimate of the true value. By developing an analytical model, Sellan et al. (2010) proposed a method to determine this critical value of the cell dimensions. In their method, an upper limit λ_{∞}^{max} is first set for the bulk thermal conductivity, and it can be predicted by the linear extrapolating method as

$$\lambda_{\infty}^{max} = \frac{Lk_B\rho_N(C_{11}^{1/2} + 2C_{44}^{1/2})}{18\rho^{1/2}}, \tag{6.13}$$

where ρ is the mass density, and C_{11} and C_{44} are the elastic constants of the material. If the bulk thermal conductivity λ estimated by the linear extrapolation procedure is greater than λ_{∞}^{max}, such a prediction will be deemed inaccurate. The dimensions of the simulation cells should thus be increased, and the simulation is repeated until $\lambda \leq \lambda_{\infty}^{max}$.

A drawback of the linear extrapolation method is that high-quality data must be obtained for simulations with relatively small simulation cells. This requirement is because small errors in the data points used for the linear plot can result in large errors in the predictions made by the linear extrapolation method, particularly when the intrinsic bulk thermal conductivity of the system is large. It can be seen from Fig. 6.5 that for bulk Si, the error of $1/\lambda$ increases rapidly as the simulation cell length L decreases. Therefore, for relatively small systems, MD simulations must be performed for a sufficiently long duration to achieve a smooth temperature profile that minimizes the errors in the linear fitting of the temperature gradient. For more technical details of minimizing linear fitting errors, one may refer to a work by Howell (2011).

Another practical issue of the direct method pertains to how the temperature difference between the heat source and the heat sink is determined. On the one hand, a large temperature difference is helpful in achieving a large temperature gradient that corresponds to lower statistical errors by minimizing the impact of local thermal fluctuations. On the other hand, a small temperature difference is desirable to circumvent nonlinearity in the temperature profile across the entire system because of the temperature dependence of thermal conductivity. When the heat flux in the direct method is generated by coupling the heat source and heat sink with two separate thermostats, a temperature difference in the range of 10–50 K is preferable depending on the system size and the thermal conductivity of the material in question. When the direct method is applied by directly imposing a constant heat flux via velocity rescaling, a value on the order of 1.0×10^{-4} eV nm^{-2} would be a good choice.

6.2.2 Green–Kubo method

The Green–Kubo method of calculating the thermal conductivity relies on relating the equilibrium heat flux autocorrelation function (HFACF) to the thermal conductivity via the Green–Kubo expression

$$\lambda_{\mu v}(\tau_{\mathrm{m}}) = \frac{1}{\Omega k_{\mathrm{B}} T^2} \int_0^{\tau_{\mathrm{m}}} \langle J_\mu(\tau) J_v(0) \rangle \mathrm{d}\tau, \tag{6.14}$$

where Ω is the system volume, T is the system temperature, J_μ and J_v are the heat fluxes in the μ and ν directions, respectively, and the angular brackets denote an ensemble average that is replaced by a time average in MD simulations according to the ergodic hypothesis. As the upper integral limit τ_{m} increases and exceeds the time required for the HFACF to decay to zero, the integral in Eq. (6.14) converges and provides the thermal conductivity of a material. Since MD simulations are performed in discrete time steps, the integral in Eq. (6.14) can be replaced with a summation as

$$\lambda_{\mu v}(\tau_M) = \frac{\Delta t}{\Omega k_{\mathrm{B}} T^2} \sum_{m=1}^{M} (N-m)^{-1} \sum_{n=1}^{N-m} J_\mu(m+n) J_v(n), \tag{6.15}$$

where τ_M is given by $M\Delta t$. To obtain a good statistical average, the summation limit M should be substantially smaller than the total number of simulation time steps N. The expression $J_\mu(m+n)$ is the μth component of the heat flux at the $(m+n)$th time step. The heat flux vector J is defined as

$$J(t) = \frac{\mathrm{d}}{\mathrm{d}t} \sum_i r_i(t) e_i(t). \tag{6.16}$$

Here, $r_i(t)$ is the time-dependent position of the ith atom, and $e_i(t)$ is its site energy that consists of both a kinetic energy component and a potential energy component as

$$e_i(t) = \frac{1}{2} m_i v_i \cdot v_i + e_i^{\mathrm{Pot}}. \tag{6.17}$$

For a two-body or pair potential $V_2(r_{ij})$, it is reasonable to divide the potential energy evenly between the two atoms in this pair, i.e.,

$e_i^{\mathrm{Pot}} = \sum_{j,j\neq i} V_2(r_{ij})/2$. By computing the time derivative in Eq. (6.16), J can be further expressed as

$$J(t) = \sum_i v_i e_i + \sum_i \boldsymbol{r}_i \left[\sum_j v_j \cdot \frac{\partial e_i}{\partial \boldsymbol{r}_j} + \sum_j \boldsymbol{F}_j \cdot \frac{\partial e_i}{\partial \boldsymbol{p}_j} \right], \tag{6.18}$$

where $\boldsymbol{F}_j$ is the force imposed on the jth atom due to its interaction with all the other atoms. To reach the above equation, $e_i(t)$ is considered as a function of the positions $\boldsymbol{r}_j$ and momenta $\boldsymbol{p}_j$ of all the atoms in the system. By labeling the force exerted on the jth atom by the ith atom as $\boldsymbol{F}_{ji}$, which can be calculated by $\boldsymbol{F}_{ji} = -\partial V_2(r_{ij})/\partial \boldsymbol{r}_j$ or $\boldsymbol{F}_{ji} = -2\partial e_i/\partial \boldsymbol{r}_j$, Eq. (6.18) can be rewritten as

$$J(t) = \sum_i v_i e_i + \frac{1}{2} \sum_{\substack{i,j \\ i\neq j}} -\left(v_j \cdot \boldsymbol{F}_{ji}\right) \boldsymbol{r}_i + \frac{1}{2} \sum_i (\boldsymbol{F}_i \cdot v_i)\, \boldsymbol{r}_i, \tag{6.19}$$

which is equivalent to

$$J(t) = \sum_i v_i e_i + \frac{1}{2} \sum_{\substack{i,j \\ i\neq j}} -\left(v_j \cdot \boldsymbol{F}_{ji}\right) \boldsymbol{r}_i + \frac{1}{2} \sum_{\substack{i,j \\ i\neq j}} \left(\boldsymbol{F}_{ij} \cdot v_i\right) \boldsymbol{r}_i, \tag{6.20}$$

by considering the fact $\boldsymbol{F}_i = \sum_{j,j\neq i} \boldsymbol{F}_{ij}$. After interchanging the indices i and j of the second term on the right, Eq. (6.20) can be recast as

$$J(t) = \sum_i v_i e_i + \frac{1}{2} \sum_{\substack{i,j \\ i\neq j}} \left(\boldsymbol{F}_{ij} \cdot v_i\right) \boldsymbol{r}_{ij}. \tag{6.21}$$

The first term in Eq. (6.21) is the convection term that describes atomic diffusion and is expected to be significant only at high temperatures near the melting point of the material. The second term describes the interactions between neighboring atoms.

An important issue associated with the Green–Kubo method is the lack of a precise definition of the local site energy $e_i(t)$ when the interatomic interactions are described by a many-body potential. For the heat flux given by Eq. (6.21), it is assumed that the pair potential energy is divided equally between two atoms. However, this assumption is invalid for a many-body potential. By taking the SW potential as an example [Eqs. (2.38)–(2.43)], the potential energy term of the site energy e_i^{Pot} can be defined as

$$e_i^{\mathrm{Pot}} = \frac{1}{2} \sum_{j,j\neq i} V_2(r_{ij}) + \frac{1}{6} \sum_{jk,j\neq k\neq i} V_3(\boldsymbol{r}_i, \boldsymbol{r}_j, \boldsymbol{r}_k). \tag{6.22}$$

The first term is the pair potential term and is defined as half of $V_2(r_{ij})$ by invoking the assumption that it is divided equally between two atoms. The second term is the three-body potential energy term obtained by assuming that $V_3(\boldsymbol{r}_i, \boldsymbol{r}_j, \boldsymbol{r}_k)$ can also be distributed evenly among the three atoms contributing to the three-body interactions. It is apparent that by summing e_i^{Pot} over i, the potential energy of the whole system is obtained. The definition of e_i^{Pot} in the above equation corresponds to a heat flux defined as

$$J(t) = \sum_i v_i e_i + \frac{1}{2} \sum_{\substack{i,j \\ i \neq j}} (\boldsymbol{F}_{ij} \cdot v_i)\, \boldsymbol{r}_{ij} + \frac{1}{6} \sum_{\substack{ijk \\ i \neq j \neq k}} (\boldsymbol{r}_{ij} + \boldsymbol{r}_{ik})(\boldsymbol{F}_{ijk} \cdot v_i), \quad (6.23)$$

where $\boldsymbol{F}_{ijk}$ is the force on atom i due to the three-body interactions between atoms i, j, and k and is given by

$$\boldsymbol{F}_{ijk} = -\nabla_i V_3(\boldsymbol{r}_i, \boldsymbol{r}_j, \boldsymbol{r}_k). \quad (6.24)$$

Another commonly adopted definition of e_i^{Pot} is to assign the three-body potential energy $V_3(\boldsymbol{r}_i, \boldsymbol{r}_j, \boldsymbol{r}_k)$ entirely to atom i that forms a bond angle with atoms j and k (with atom i staying at the vertex of the angle), which results in a definition for the local site energy as

$$e_i^{\mathrm{Pot}} = \frac{1}{2} \sum_{j, j \neq i} V_2(r_{ij}) + \frac{1}{2} \varepsilon \sum_{jk, j \neq k \neq i} h(r_{ij}, r_{ik}, \theta_{jik}), \quad (6.25)$$

where ε is a scaling factor and $h(r_{ij}, r_{ik}, \theta_{jik})$ is a functional term adopted to describe the three-body interactions (see the SW potential in Chapter 2: Potential Energy Functions). The definition of e_i^{Pot} in Eq. (6.25) leads to an alternative expression for the heat flux as

$$J(t) = \sum_i v_i e_i + \frac{1}{2} \sum_{\substack{i,j \\ i \neq j}} (\boldsymbol{F}_{ij} \cdot v_i)\, \boldsymbol{r}_{ij} + \sum_{\substack{ijk \\ i \neq j \neq k}} \boldsymbol{r}_{ij} \left[\boldsymbol{F}_j(ijk) \cdot v_j\right], \quad (6.26)$$

where the three-body force term $\boldsymbol{F}_j(ijk)$ is given by

$$\boldsymbol{F}_j(ijk) = -\varepsilon \nabla_j h(r_{ij}, r_{ik}, \theta_{jik}). \quad (6.27)$$

Although an exact definition of the local site energy is not available, it has been demonstrated that the distribution of the potential energy among interacting atoms does not significantly impact the simulation results because the temperature gradient created in a material varies on a length scale much larger than the interatomic distance (Lee et al., 1991).

One advantage of the Green–Kubo method is that the entire thermal conductivity tensor of the material in question can be obtained with only a single simulation, and therefore potential anisotropic properties pertaining to the thermal conductivity can be studied. However, it is commonly acknowledged that the Green–Kubo method requires a long simulation time to achieve the convergence of the thermal conductivity. A large number of MD steps N must be used in Eq. (6.15) for accurate statistical averaging to obtain reliable results. In addition, the number of steps M should be sufficiently large to ensure that the autocorrelation function decays to zero over a time interval $M\Delta t$. Fig. 6.6 shows typical HFACFs for calculating the thermal conductivity of Si at 500 K and 1000 K using the SW potential. Here, the HFACFs are normalized based on their values at $t=0$ and calculated by

$$\frac{\langle \mathbf{J}(\tau)\mathbf{J}(0)\rangle}{\langle \mathbf{J}(0)\mathbf{J}(0)\rangle} = \frac{\langle J_x(\tau)J_x(0)\rangle + \langle J_y(\tau)J_y(0)\rangle + \langle J_z(\tau)J_z(0)\rangle}{\langle J_x(0)J_x(0)\rangle + \langle J_y(0)J_y(0)\rangle + \langle J_z(0)J_z(0)\rangle}. \tag{6.28}$$

It is evident from the inset of Fig. 6.6A that in the first 0.1 ps, the HFACF undergoes a rapid decay, which is believed to be related to the high-frequency optical phonon modes that contribute minimally to the thermal conductivity of the material because they couple only weakly to the low-frequency acoustic modes (Che et al., 2000). Afterward, the HFACF exhibits a much lower decay rate that lasts for ~100 ps.

Over a longer time, the autocorrelation function is dominated by noise (indicated by the fluctuations illustrated in gray in Fig. 6.6A), whose amplitude $W(\tau)$ can be estimated by the root-mean-square of the HFACF as (Howell, 2012)

$$W(\tau) = \sqrt{\frac{\langle J_x(\tau)J_y(0)\rangle^2 + \langle J_y(\tau)J_z(0)\rangle^2 + \langle J_z(\tau)J_x(0)\rangle^2}{3}}. \tag{6.29}$$

Since the cross-correlation function equals zero in the thermodynamic limit, $W(\tau)$ is composed solely of noise. If t_{signal} is defined as the time at which the HFACF possesses a magnitude equivalent to that of W, it is observed that the HFACF shows prominent oscillations superposed on the decay for $t > t_{\text{signal}}$; t_{signal} is ~52 ps at 1000 K and ~160 ps at 500 K.

Therefore, to obtain a well-converged value of the thermal conductivity, the upper limit τ_{m} of the integral in Eq. (6.14) should at least be greater than t_{signal} that is determined by the approach discussed above. This requirement generally leads to a total simulation time of over 1 ns and even tens of nanoseconds in some cases. To reduce the total

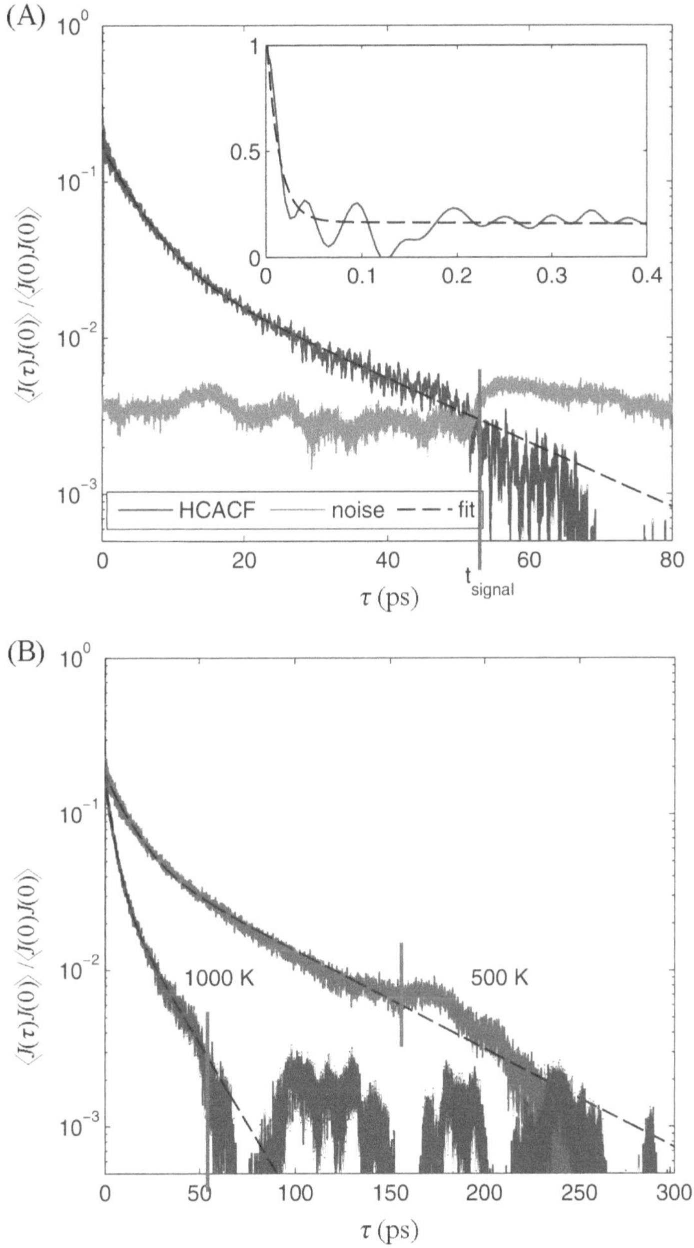

FIGURE 6.6 Representative HFACFs for Si normalized by their values at $t = 0$ and plotted on a logarithmic y-axis. The dashed lines are fitted using Eq. (6.30). (A) HFACF corresponding to Si at 1000 K with its initial decay behavior shown in the inset. (B) Comparison of the HFACFs corresponding to Si at 500 K and 1000 K. Source: *Modified and reprinted with permission from a work by Howell (2012).*

simulation time, the simulation results for the HFACF $\langle J(\tau)J(0)\rangle$ in a short simulation are often assumed to decay exponentially and can be fitted as an exponential function of τ_m, which is then integrated via Eq. (6.14). Generally, the exponential fitting is performed for the values of the HFACF obtained in the initial phases of the simulation in which its oscillations are small, and a good statistical averaging is relatively easy to achieve. To capture the characteristics of the decay of $\langle J(\tau)J(0)\rangle$ more accurately, it was recommended by Howell (2012) that the simulation results of $\langle J(\tau)J(0)\rangle$ be fitted as the sum of multiple exponential functions instead of a single one. For instance, the exponential fitting of $\langle J(\tau)J(0)\rangle$ delineated by the dashed line in Fig. 6.6 is the sum of three exponential functions:

$$\langle J(\tau)J(0)\rangle = \sum_{i=1}^{3} A_i \exp\left(-\frac{\tau}{\tau_i}\right), \tag{6.30}$$

where A_i and τ_i are the fitting constants. Fig. 6.7 shows the accumulative thermal conductivity for different upper integration limits τ_m calculated using the HFACFs presented in Fig. 6.6. The blue and green lines represent the results obtained by direct numerical integration via Eq. (6.15), whereas the black lines represent the results calculated based on the integration of the fitted exponential function in Eq. (6.30). Five separate simulations are conducted for both Si at 1000 K and Si at 500 K. Each of the former simulations lasts for a duration of $D = 12$ ns (illustrated by thin blue lines), and each of the latter simulations lasts for $D = 24$ ns (illustrated by thin green lines). The average value of the results obtained from the five repeating simulations is then computed to reduce statistical errors and is equivalent to the result obtained based on the simulation with a duration of $5D$ with $D = 12$ ns or $D = 24$ ns.

It is observed from Fig. 6.7 that up to $\tau_m = t_{signal}$, the values of the thermal conductivity λ obtained by the direct numerical integration and those by the integration of the fitted exponential function are identical. In addition, both increase with the upper integration limit τ_m and begin to converge at $\tau_m = t_{signal}$. However, due to the noise present in the HFACF (Fig. 6.6A), the values of λ obtained by both integration methods begin to deviate from each other. This discrepancy that occurs when $\tau_m > t_{signal}$ leads to an uncertainty issue of the Green–Kubo method in predicting λ. Fortunately, the contribution of the HFACF to the total thermal conductivity when $\tau_m > t_{signal}$ is typically less than 5%, and thus the value of λ predicted at $\tau_m = t_{signal}$ can be safely adopted as a good estimation of the true thermal conductivity of the system.

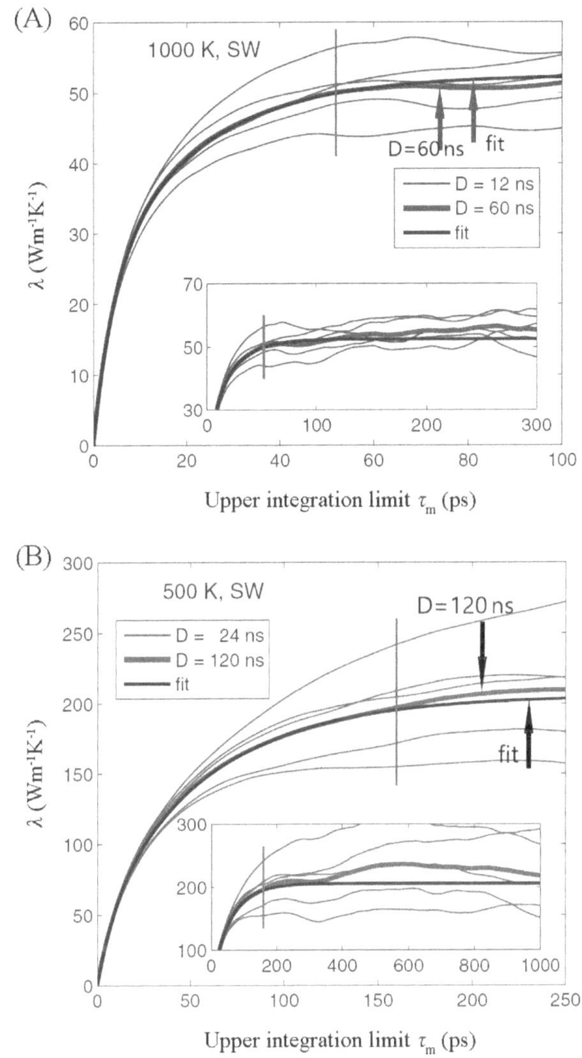

FIGURE 6.7 Accumulative thermal conductivity as a function of the upper integration limit τ_m in Eq. (6.14) at (A) 1000 K and (B) 500 K for the HCACFs shown in Fig. 6.6. Each of the numerical integrals of five separate simulations lasts for a duration $D = 12$ ns for Si at 1000 K (thin blue lines) and $D = 24$ ns for Si at 500 K (thin green lines), and their mean values (thick blue and green lines) are compared with the integral of the triple exponential fitting function of $\langle J(\tau)J(0)\rangle$ in Eq. (6.30). The vertical lines indicate the positions where $\tau_m = t_{\text{signal}}$. The insets present the long-time behavior of the thermal conductivity of the systems that is dominated by noise. Source: *Modified and reprinted with permission from a work by Howell (2012).*

It can be rigorously proven that the relative error in such an estimation of λ is bounded by $2\sqrt{\tau_{m,max}/D}$, where $\tau_{m,max}$ is the largest numerically meaningful upper limit for the integral in Eq. (6.14), which refers to the maximum correlation time of the heat flux $\boldsymbol{J}(\tau)$ and can be reasonably set to be t_{signal}. For a simulation duration $D = 12$ ns and $t_{signal} = 52$ ps, the relative error for the value of λ calculated for Si at 1000 K is estimated to be $\sim 4\%$.

Besides the exponential fitting approach, another approach to reduce the computational load of the Green–Kubo method is to apply Fourier transformations to the heat flux, and the limit of the following expression at $\omega \to 0$ is taken subsequently, which is given by

$$\lambda_{\mu v}(\omega) = \frac{1}{\Omega k_B T^2} J_\mu(\omega) J_v^*(\omega). \tag{6.31}$$

Here, $J_\mu(\omega)$ is the μth component of the heat flux operator $\boldsymbol{J}(\omega)$ that is expressed as

$$\boldsymbol{J}(\omega) = \int_0^\infty \mathrm{d}t \boldsymbol{J}(t) \exp(\mathrm{i}\omega t), \tag{6.32}$$

and $J_v^*(\omega)$ denotes the complex conjugate of $J_v(\omega)$. Note that both the exponential fitting approach and the Fourier transformation approach rely on the assumption that the HFACF decays in an exponential manner, which leads to different estimations of the thermal conductivity, compared with the direct numerical integration method. It is commonly established that to obtain the most reliable estimation of the thermal conductivity of a system using the Green–Kubo method, the direct numerical integration method should be adopted, although an exceedingly long simulation time may be required.

Before closing this section, it is worth noticing that as with the direct method discussed in Section 6.2.1, the Green–Kubo method also suffers from the finite-size effect. However, the finite-size effect for the Green–Kubo method is much less significant than that for the direct method. Che et al. (2000) has attributed the finite-size effect of the Green–Kubo method to the application of PBCs. Under PBCs, a phonon may pass the same point in space several times without being scattered when the system dimension is much smaller than the intrinsic phonon MFP. Since the system may retain some dynamic information during the passage of the phonon, artificial correlations may exist in the HFACF. In this case, the correlation function may not be reliable for a simulation time longer than the average time required for a phonon to travel across the simulation cell.

6.3 Molecular dynamics simulation of interfacial thermal transport

6.3.1 Interfacial thermal transport models

The significance of interfacial thermal transport has been attracting the interest of researchers ever since the thermal boundary resistance R_K between a solid and superfluid helium was first reported by Kapitza (1941). When the discovery was made, it was originally speculated that such resistance was caused by certain special properties of superfluid helium. It was later established that the Kapitza resistance exists at the interface between any pair of dissimilar materials. The Kapitza thermal resistance can be intuitively interpreted as a result of an interface creating an interruption in a regular crystalline lattice in which phonons propagate. As phonons encounter the interface, they either pass through the interface or are reflected. In either case, phonon scattering occurs, and thermal resistance is often generated.

On the macroscale, phonon transmission at an interface can be understood through two theoretical models: the acoustic mismatch model and the diffuse mismatch model. For an interface between dissimilar materials, their different mass densities and the different speeds of sound propagating through them result in a mismatch in their acoustic impedance, which is analogous to the mismatch in the refractive indices of two optically different materials. The effect of the acoustic impedance on phonon transmission is captured in the acoustic mismatch model, which assumes that no phonon scattering occurs at the interface, and the transmission of phonons across the interface only depends on the mismatch in the acoustic impedance of the two materials. In contrast, the diffuse mismatch model assumes that phonons are scattered diffusively by the interface, and their transmission coefficients are only determined by the difference in the phonon densities of states across the interface. Both models can be applied to predict the phonon transmission coefficient and thus the interfacial thermal transport behavior of the system. However, such predictions are solely based on the bulk properties of the materials forming the interface, and the effects of a detailed atomic structure of the interface are ignored.

To effectively describe the effects of a detailed interface structure, the interfacial thermal transport behaviors have been increasingly investigated through MD simulation due to its simple implementation and low requirements for prior assumptions of heat transport at the interface. The most straightforward method to investigate interfacial thermal transport behaviors is similar to the direct method for calculating thermal conductivity, as discussed previously in Section 6.2.1.

The MD simulation model presented in Fig. 6.2 can be adopted, and an interface is implemented at its center (Fig. 6.8). When a heat flux J is imposed in the direction perpendicular to the interface, a steady sharp temperature drop ΔT_{In} can be observed at the interface after some time. The interfacial thermal conductance G, which is the reciprocal of the interface thermal resistance R_K, is determined by the relationship:

$$J = G\Delta T_{\mathrm{In}}. \tag{6.33}$$

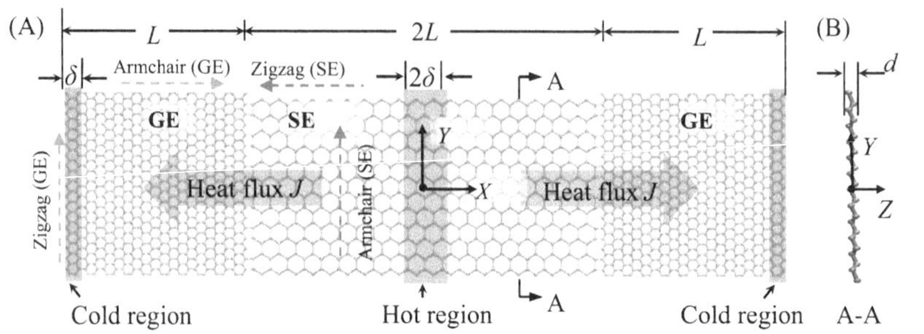

FIGURE 6.8 (A) Atomic structure of a hybrid GE/SE monolayer and (B) local buckling of SE. Source: *Modified and reprinted with permission from a work by Liu et al. (2014a).*

For cases in which the interface is formed between a thin film and a substrate or between two thin films, the direct method is inapplicable because of the small thickness of the thin film. Consequently, the thermal relaxation method is usually adopted. In this method, an initial temperature difference ΔT is implemented across the interface, and the system is subsequently relaxed thermally to allow heat transport to occur. As heat is transported across the interface from the region with a higher temperature to the region with a lower temperature, the temperature difference ΔT decays exponentially with time. The rate of decay is then adopted as a critical feature to characterize the interfacial thermal transport behavior of the system. To better illustrate the above two methods, two case studies are presented in the following section. In the first case, the interfacial thermal transport behavior of a silicene/graphene monolayer heterostructure is investigated by employing the direct method (Liu et al., 2014a), while in the second case, the interfacial thermal transport behavior of a silicene/graphene bilayer heterostructure is explored via the thermal relaxation method (Liu et al., 2014b).

6.3.2 Interfacial thermal transport of a silicene/graphene hybrid monolayer

Low-dimensional nanomaterials have been attracting tremendous interest for the past few decades because of their anisotropic and dimension-tunable properties. They have been recognized for their potential applications in next-generation nanoelectronics, offering new functionalities that were once restricted by the limitations of conventional Si and compound semiconductors. Two-dimensional (2D) nanomaterials, characterized by a physical thickness in the order of nanometers, stand out among low-dimensional structures and have become some of the most prominent materials in nanotechnology. Consequently, in the past decade, much attention has been diverted to 2D nanomaterials, which has been partially motivated by the successful preparation of free-standing graphene (GE) (Novoselov et al., 2004), the first 2D atomic crystal available.

GE can be regarded as a single layer of carbon atoms arranged into a honeycomb-lattice structure. Pristine GE is considered a semimetal or zero-gap semiconductor, and the electrons and holes in GE behave like zero-mass Dirac Fermions (Novoselov et al., 2004). Experimental results from transport measurements have indicated that GE has remarkably high electron mobility at room temperature with reported values of over 15,000 $cm^2\,V^{-1}\,s^{-1}$ (Geim and Novoselov, 2007). Both theoretical and experimental studies have revealed that single-layer GE has an ultrahigh thermal conductivity of 3000–5000 $W\,m^{-1}\,K^{-1}$ (Ghosh et al., 2008; Rajabpour and Vaez Allaei, 2012). Moreover, GE is one of the strongest materials known to exist. Measurements have shown that GE possesses an intrinsic strength of 130 GPa and a tensile modulus (stiffness) of 1 TPa, which is 200 times greater than that of steel (Lee et al., 2008).

Due to the novel and supreme physical and mechanical properties of GE, it has become the subject of intensive research since its successful fabrication via mechanical exfoliation in 2004. The success in GE preparation and its wide range of application prospects have sparked broad interest in the search for novel 2D nanomaterials that are analogous to GE. The fact that Si lies in the same column of the periodic table as C has therefore evoked considerable attempts to produce a Si analog of GE, which is known as silicene (SE). Similarly to GE, SE is also composed of hexagonal rings and exhibits remarkable electronic and magnetic properties (Cahangirov et al., 2009). Moreover, SE is a gapless semimetal, and its charge carriers behave as zero-mass Dirac Fermions with ultrahigh Fermi velocities, which are comparable to those of GE (De Padova et al., 2012; Vogt et al., 2012).

In parallel to the search for new GE-analogous nanomaterials, research efforts have also been devoted to the design of composite nanomaterials by integrating various types of 2D nanomaterials with GE to form hybrid heterostructures. By modifying the combining sequences or patterns of the constituent materials, heterostructures with fascinating properties can be achieved. For instance, when GE and SE are combined to form a hybrid bilayer, it has been discovered that the doping carrier concentration of SE and GE can be modulated by altering their interfacial spacing, which leads to the formation of tunable p–n junctions in their hybrid structure (Hu et al., 2013). The optical absorption rate of the heterostructure is also superior to those of individual SE and GE monolayers, and it has been suggested that the properties of Dirac electrons in both GE and SE can be preserved within the heterostructure. With a proper choice of geometrical parameters for its constituent GE and SE monolayers, it will be possible to engineer the heterostructure and shift its Fermi level close to the Dirac points (Matthes et al., 2012).

To take advantage of such GE/SE heterostructures in advanced nanodevices, their thermal transport behaviors should be understood. On the nanoscale, Joule heating induced by an electronic current can cause heat to be spatially localized, which may lead to thermal hot spots. This problem will be more prominent for SE than GE and bulk Si as SE has a much lower thermal conductivity. In this section, an MD simulation to explore the interfacial thermal transport of a GE/SE monolayer heterostructure is presented. The effects of system size, temperature, external strain, and the magnitude of heat flux on the interfacial thermal conductance are discussed.

6.3.2.1 Simulation model

SE has an atomic structure with intrinsic local buckling, and not all the Si atoms are located on the same plane, as shown in Fig. 6.8. In contrast, GE possesses a planar structure. The in-plane projection of the bond length of SE (2.30 Å) is much larger than the bond length of GE (1.42 Å). This difference renders it difficult to construct a lattice-matched interface between GE and SE along the same chirality. Since the lattice constant of GE (2.46 Å) is only ~7% greater than the in-plane projection of the bond length of SE, it is possible to form an interface by connecting the zigzag edge of GE with the armchair edge of SE. The atomic configuration of a hybrid GE/SE monolayer is shown in Fig. 6.8. One SE monolayer of length $2L$ is inserted between two adjacent GE monolayers of length L. The armchair (zigzag) and zigzag (armchair) edges of the GE (SE) monolayer are orientated along the x- and y-directions, respectively. Initially, all the C and Si atoms are placed in a honeycomb-lattice structure with the lattice constants of GE and SE set at 2.46 Å and 3.98 Å, respectively. To capture the buckling feature of SE, the Si atoms in the hybrid GE/SE monolayer are initially displaced in an out-of-plane manner with a buckling distance of 0.44 Å (Fig. 6.8B). The interface is established by elongating the SE monolayer by ~7% along the y-direction to match the lattice of the GE monolayer. The distance between the C and Si atoms at the interface is set at 1.80 Å, which is the Si–C bond length of SiC.

To calculate the interfacial thermal conductance of the hybrid GE/SE monolayer, a constant heat flux J is imposed on it and maintained for several time steps to develop a steady temperature profile. For a hybrid system composed of an interface, a temperature drop ΔT_{In} at the interface is usually formed, which allows for the measurement of the interfacial thermal conductance or the Kapitza conductance as $G = J/\Delta T_{\text{In}}$.

The thermal conductivity λ of a layered structure can be calculated using a mixture rule-based expression $\lambda^{-1} = \sum f_i \lambda_i^{-1}$, where f_i and λ_i are the volume fraction and thermal conductivity of the ith component, respectively. Based on this expression, the thermal conductivity λ_{H} of the hybrid GE/SE monolayer can be calculated by the following equation with its interfaces taken into account (Zhou et al., 2007):

$$\lambda_{\text{H}}^{-1} = \frac{1}{2\lambda_{\text{GE}}} + \frac{1}{2\lambda_{\text{SE}}} + \frac{1}{2LG}. \tag{6.34}$$

Here, λ_{GE} and λ_{SE} are the thermal conductivity of the GE and SE sheets, respectively, and they can be calculated from Fourier's law $J = -\lambda \nabla T$. The effect of the interface on the thermal transport behavior of the hybrid GE/SE monolayer can be estimated by the ratio $r = (2GL)^{-1}/\lambda_{\text{H}}^{-1}$. Since the reciprocal of the thermal conductivity of a material is its thermal resistivity, r can be interpreted as the ratio of the effective thermal resistivity of the interface to the thermal resistivity of the hybrid GE/SE monolayer. A large value of r represents a great impedance in the thermal transport within the hybrid monolayer due to its interface(s).

To simplify the calculation of r, the temperature gradients ∇T in the GE and SE monolayers are approximated by $-\Delta T_{\text{GE}}/L$ and $-\Delta T_{\text{SE}}/L$, respectively, where ΔT_{GE} and ΔT_{SE} denote temperature drops formed along the directions of the heat flux in the GE and SE monolayers, respectively. By simply replacing G with $J/\Delta T_{\text{In}}$, λ_{GE} with $JL/\Delta T_{\text{GE}}$, and λ_{SE} with $JL/\Delta T_{\text{SE}}$, the ratio r can be expressed as $r = \Delta T_{\text{In}}/(\Delta T_{\text{In}} + \Delta T_{\text{GE}} + \Delta T_{\text{SE}})$.

PBCs are applied in the x- and y-directions. The interactions between the C–C atoms and C–Si atoms are described by the Tersoff potential with an optimized set of parameters (Lindsay and Broido, 2010) to better represent the lattice dynamic properties of GE. The interactions between the Si–Si atoms are described by the SW potential with a specifically developed set of parameters (Zhang et al., 2014) for the SE monolayer. The direct method presented in Fig. 6.2 is adopted, and the time step Δt is set at 0.5 fs. The only difference between the simulation model in Fig. 6.2 and this model is that in this case study, both the heat flux direction from the SE monolayer to the GE monolayer and that from GE to SE are considered. When J flows from SE to GE, it is denoted by J^+, while J^- denotes a reverse heat flux.

6.3.2.2 *Interfacial thermal conductance*

For an imposed heat flux J, the temperature distribution along the length direction of the hybrid GE/SE monolayer is exemplified by the case of $L = 17$ nm, as shown in Fig. 6.9. A dramatic temperature drop ΔT_{In} occurs across the GE/SE interfaces. Within the individual GE and SE monolayers, their temperature distributions show a linear pattern except for the region in the middle of the SE monolayer in which phonon scattering at the heat source gives rise to a certain level of nonlinearity. Compared with ΔT_{In}, ΔT_{GE} and ΔT_{SE} are small and r has a value of 0.7. This observation indicates that the interfaces are the dominant resistance for the thermal transport in the hybrid GE/SE monolayer, thereby increasing the interfacial thermal conductance is important for improving the thermal conduction ability of the hybrid monolayer.

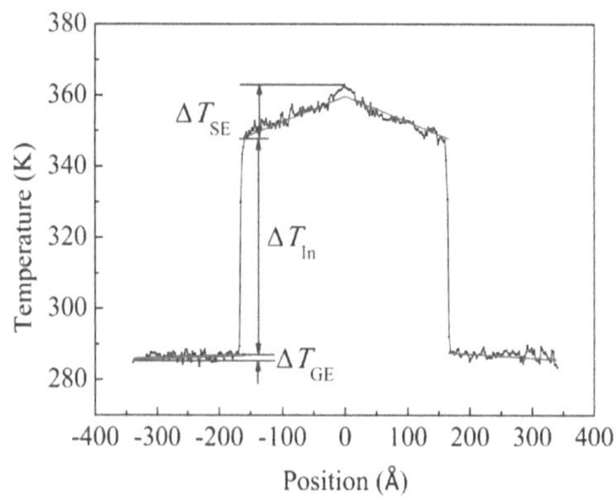

FIGURE 6.9 Typical temperature profile of the hybrid GE/SE monolayer along the length direction. Source: *Reprinted with permission from a work by Liu et al. (2014a).*

Fig. 6.10 shows the dependence of the thermal conductance G of the system on its length L for both J^+ and J^-. It is observed that G increases monotonically with L and reaches a saturated value of ~250 MW m^{-2} K^{-1} when $L = 40$ nm. In contrast, the thermal resistance ratio r decreases with L, indicating that the dominant thermal transport resistance due to the interface is weakened as the length of the monolayer increases. The thermal conductance G of the GE/SE interface has the magnitude of the same order as those of the interfaces within Si-based hybrid systems such as carbon nanotube (CNT)/Si (~300 MW m^{-2} K^{-1}) (Diao et al., 2008), Si/SiO_2

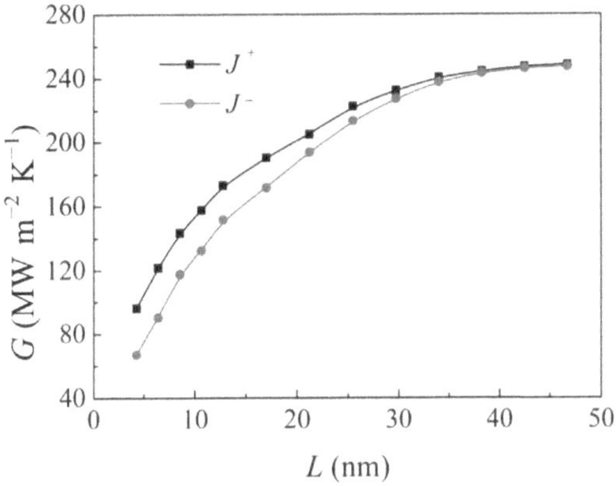

FIGURE 6.10 Dependence of thermal conductance G of the system on its length L for the heat fluxes J^+ and J^-. Source: *Modified and reprinted with permission from a work by Liu et al. (2014a).*

(~235 MW m^{-2} K^{-1}) (Chen et al., 2012), and Si/Ge (~330 MW m^{-2} K^{-1}) (Samvedi and Tomar, 2009), but it is one order of magnitude larger than those of multilayered GE nanostructures such as GE/phenolic resin (~20 MW m^{-2} K^{-1}) (Luo and Lloyd, 2012) and GE/SiC (~50 MW m^{-2} K^{-1}) (Xu and Buehler, 2012). The high interfacial thermal conductance of the hybrid GE/SE monolayer is due to its strong interfacial covalent bonds, in contrast to the weak van der Waals forces at the interfaces of multilayered GE nanostructures.

The length dependence of the interfacial thermal conductance arises when the system has a small size, especially when it is smaller than the length of the phonon MFP. As the system size increases, phonon modes with longer wavelengths are excited. These phonon modes can easily transmit across the interface without inelastic scattering, and thus contribute additionally to the interfacial thermal conduction.

To further understand the length dependence of the interfacial thermal conductance of the system and the role of long-wavelength phonons in its thermal transport behavior, the phonon spectra $P(\omega)$ of the GE and SE monolayers are explored and presented in Fig. 6.11A. Here, $P(\omega)$ is calculated by

$$P(\omega) = \frac{1}{\sqrt{2\pi}} \int_0^{\infty} e^{i\omega t} \langle \sum_{j=1}^{N} v_j(t) v_j(0) \rangle d\omega, \tag{6.35}$$

where ω and $v_j(t)$ denote the angular frequency and velocity of atom j at time t, respectively. The ensemble average, denoted by the expression with the angular brackets, is substituted with a time average, which is performed over 50 ps after the system has reached the equilibrium state. During the process of obtaining the time average, sample

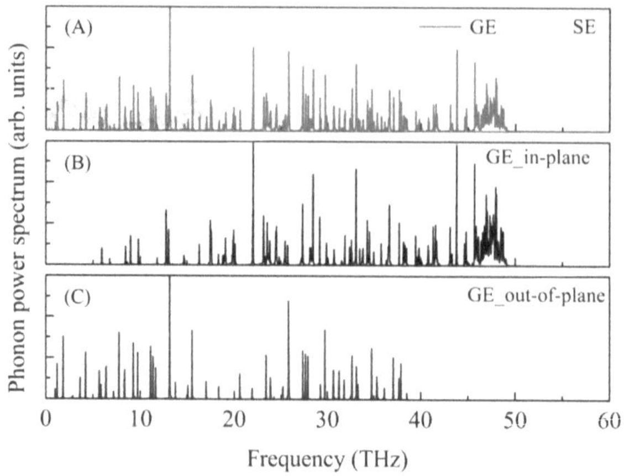

FIGURE 6.11 (A) Total phonon spectra of GE and SE and the decomposition of the GE spectrum into (B) in-plane and (B) out-of-plane components. Source: *Reprinted with permission from a work by Liu et al. (2014a).*

velocities from the simulation are recorded every 5 fs. Since GE is highly anisotropic, its spectrum can be further decomposed into in-plane and out-of-plane components, as presented in Fig. 6.11B and C, respectively.

It is observed that most of the overlaps between the phonon spectra of the GE and SE monolayers are located in the low-frequency range of 1–15 THz, and the out-of-plane component of the GE spectrum contributes to most of these overlaps. A similar coupling phenomenon has also been observed between the phonons in polymer and the low-frequency out-of-plane phonons in GE, which has been identified as the most important channel for thermal transport across the interface between polymer and GE (Luo and Lloyd, 2012).

Due to the elongation of the system, more low-frequency phonons are available in the GE monolayer to couple with the phonons in the SE monolayer, thus resulting in a larger value of G. However, in the extremely low-frequency region ($<\sim$0.5 THz), the phonon spectrum power of SE is weak, suggesting that fewer phonons are excited in this region. Hence, when the length of the system L further increases from 40 nm, the newly introduced long-wavelength phonons in the GE monolayer possess frequencies lower than 0.5 THz, but insufficient phonons in SE are available to couple with them. Consequently, a saturated value of G is reached as L increases.

6.3.2.3 *Temperature effect*

Nanodevices usually work at high temperatures due to the existence of hot spots and their high heat densities. Therefore, the study of the effect of temperature on the interfacial thermal conductance of a system is vital to understanding the performance of nanodevices. A system with a length $L = 17$ nm is chosen as an example. Fig. 6.12A shows that its interfacial thermal conductance G increases monotonically with temperature T, which is a phenomenon that has also been observed for other hybrid systems such as CNT/Si (Diao et al., 2008), SiO_2/Si (Chen et al., 2012), and Si/Ge (Samvedi and Tomar, 2009). This temperature dependence is in direct contrast to that of λ_{GE} and λ_{SE} observed in the individual GE or SE monolayer. Hence, it can be predicted that the ratio of the thermal resistivity at the interface to that of the entire hybrid monolayer, denoted by r, decreases with T, which is further validated from the simulation results presented in Fig. 6.12B.

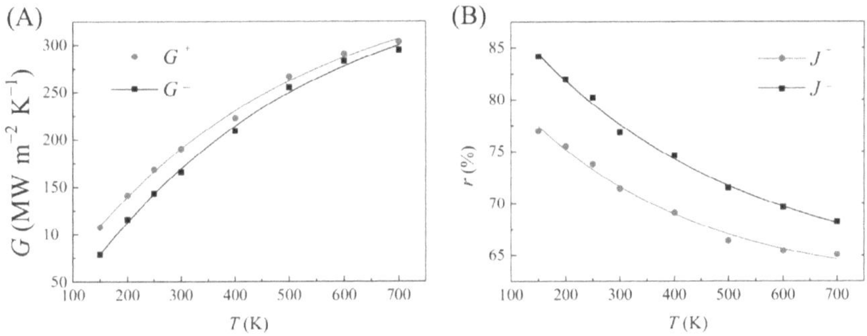

FIGURE 6.12 Temperature dependence of (A) the interfacial thermal conductance G and (B) the thermal resistance ratio r for two opposite directions of the heat flux. Source: *Modified and reprinted with permission from a work by Liu et al. (2014a).*

When T increases up to 700 K, the values of r for both J^+ and J^- still remain above 0.65, which indicates that the thermal resistance of the entire hybrid monolayer is dominated by its interfacial thermal resistance even at high temperatures. As a result, the overall thermal resistance of the hybrid monolayer decreases as T increases. This feature is advantageous for heat dissipation within GE/SE monolayers, particularly at high temperatures.

The temperature dependence of the interfacial thermal conductance of the system can be explained as follows. At low temperatures, fewer phonons, in particular phonons with high frequencies, are excited to participate in thermal transport within the system. Therefore, the coupling between phonons across the interface is reduced, leading to a smaller value of G. As T increases, more high-frequency phonons are available to participate in thermal transport. Meanwhile, higher temperatures also bring about enhanced inelastic interface phonon scattering, which leads to an increase in both the anharmonicity of the atomic interactions and the phonon transmission coefficients of the interfaces within the system. Hence, G increases with T.

6.3.2.4 Strain effect

The thermal transport behaviors of nanomaterials can be significantly modified by applying an external strain (Li et al., 2010). In the initial model of the GE/SE hybrid monolayer, SE is elongated by 7% along the interfacial direction to match the lattice of GE in a strain-free state. To study the strain effect, the monolayer is compressed along its interfacial direction until the tensile strain in the SE monolayer is reduced to zero. As the in-plane tensile strain in the SE monolayer reduces, the value of G increases monotonically (Fig. 6.13), and the values of G^+ and G^- converge toward each other. Here, G^+ and G^- denote the interfacial thermal conductance of the system corresponding to J^+ and J^-, respectively.

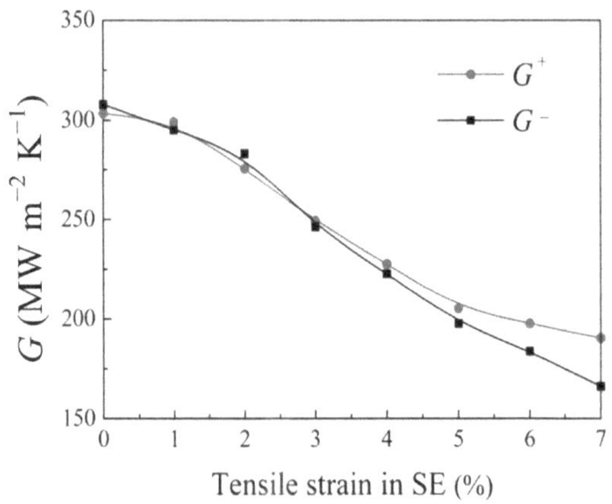

FIGURE 6.13 Dependence of the interfacial thermal conductance G on the tensile strain in the SE monolayer along its interfacial direction. Source: *Modified and reprinted with permission from a work by Liu et al. (2014a).*

The effect of compression on the interfacial thermal conductance G can be explained by the broadening of the phonon spectrum of the SE monolayer, as shown in Fig. 6.14. With a reduction in the initial tensile strain in the SE monolayer, its phonon spectrum shifts slightly toward higher frequencies. Moreover, the peak located at around 2 THz is broadened and split into two new peaks, which gives rise to more phonons with frequencies in the range of 1–7 THz. On the other hand, since the in-plane stiffness of the GE monolayer under compression is almost zero, its phonon spectrum is unaffected by the compression applied. Therefore, the mismatch between the phonon spectra of the GE and SE monolayers is reduced, especially in the low-frequency range, and phonons across the interfaces are better coupled, thereby leading to a larger value of G.

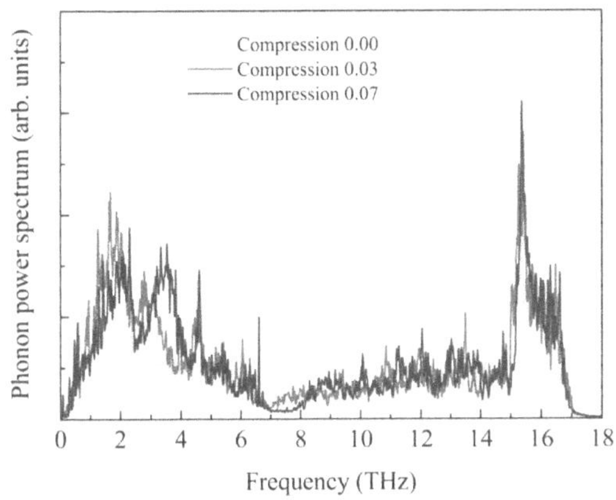

FIGURE 6.14 Phonon spectra of Si atoms under different tensile strains in SE. Source: *Reprinted with permission from a work by Liu et al. (2014a).*

6.3.2.5 Heat flux effect

For the simulation discussed above, the heat flux J is kept constant at 10.5 GW m^{-2}. In practical applications, the heat flux in nanodevices is inclined to change under different working conditions. Thus, the effect of changes in the heat flux J on the interfacial thermal conductance G of the system should be investigated.

For the interfacial thermal transport of the GE/SE hybrid monolayer, a critical value of $J = 42$ GW m^{-2} is established, below which G is virtually independent of J, and the thermal transport in the hybrid GE/SE monolayer is mostly dominated by Fourier heat conduction. When $J > 42$ GW m^{-2}, low-frequency kinetic waves are excited in the SE monolayer and subsequently propagate across the entire hybrid monolayer. Similar kinetic waves have also been observed in CNTs and GE nanoribbons under high heat fluxes (Zhang et al., 2012; Zheng et al., 2014), and it has been proposed that such waves provide an additional channel for non-Fourier heat conduction (Zhang et al., 2012). The critical value of the heat flux for hybrid monolayers is much smaller than those for CNTs ($\sim$250 GW m^{-2}) (Zhang et al., 2012) and GE nanoribbons ($\sim$400 GW m^{-2}) (Zheng et al., 2014).

Fig. 6.15 presents the phonon spectra of the GE and SE monolayers for $J = 52.5$ GW m^{-2} and demonstrates the excitation of low-frequency kinetic waves for $J > 42$ GW m^{-2}. A sharp peak is present at 0.65 THz for both the phonon spectra of the GE and SE monolayers and has been determined to originate from their out-of-plane components. The average out-of-plane displacements of all the C and Si atoms for $J > 42$ GW m^{-2} are also significantly larger than those for $J < 42$ GW m^{-2}. The existence of kinetic waves in 2D nanomaterials and their contribution to the overall thermal transport of the materials are of great interest to researchers. Furthermore, the interfacial thermal conductance value of hybrid GE/SE monolayers is much lower than those of CNTs and GE, and thus the low-frequency kinetic wave mechanism of these monolayers that provides an additional channel for efficient non-Fourier heat conduction may be more critical for thermal transport.

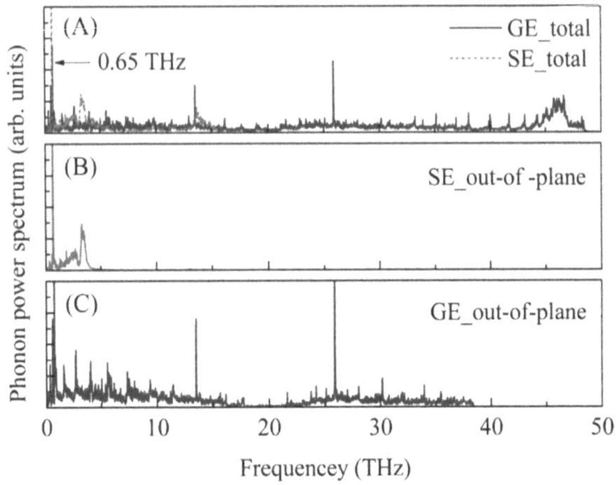

FIGURE 6.15 (A) Total phonon spectra of the GE and SE monolayers for $J = 52.5$ GW m^{-2} and (B–C) their out-of-plane components. Source: *Reprinted with permission from a work by Liu et al. (2014a).*

6.3.3 Interfacial thermal transport of a silicene/graphene hybrid bilayer

Besides the aforementioned hybrid monolayer, an SE/GE bilayer heterostructure can be formed by stacking SE and GE monolayers on top of each other with weak van der Waals interactions between the layers. This bilayer heterostructure fits well with the current hot trend toward developing advanced van der Waals heterostructures that are constructed by stacking different 2D crystals vertically on top of each other. By carefully planning the stacking sequence, heterostructures with novel functionalities can be obtained. In this section, an MD simulation work on the interfacial thermal transport of an SE/GE hybrid bilayer is presented. The effects of temperature, interface coupling strength, thermal transport direction, and hydrogenation of GE on the interfacial thermal transport behavior of the system are investigated by analyzing the phonon spectra. Different GE hydrogenation patterns and coverages are explored.

6.3.3.1 Simulation model

To model the SE/GE hybrid bilayer, an SE layer is deposited onto a GE layer, as illustrated in Fig. 6.16. Since the lattice constant of SE ($a_{SE} = 3.98$ Å) is ~60% larger than that of GE ($a_{GE} = 2.46$ Å), a supercell of the SE/GE heterostructure that consists of a 2 × 2 supercell of SE and a 3 × 3 supercell of GE is created, as delineated by the red dotted line. Accordingly, the lattice constant of the heterostructure is taken to be $a_{GE/SE} = (3a_{GE} + 2a_{SE})/2 = 7.67$ Å with a small lattice mismatch of ~3% between GE and SE. The packing patterns of GE and SE, which describe the relative in-plane position between them, have negligible effects on the structural stability and physical properties of the SE/GE heterostructure and thus are not considered. The interlayer distance D, defined as the average Si–C distance along the out-of-plane z-direction, is set at 3.60 Å. The in-plane dimensions of the heterostructure are set at 10.5 × 10 nm^2. PBCs are applied along the in-plane x- and y-directions to remove the edge effect, and a free-boundary condition is applied along the out-of-plane z-direction. Both the interactions of the C–C atoms and C–H atoms are modeled by the AIREBO potential (see Chapter 2: Potential Energy Functions), while the SW potential is used to describe the interactions among Si atoms. To model the weak van der Waals interactions between the SE and GE monolayers, the Lennard–Jones (LJ) potential is adopted as follows:

$$V(r_{ij}) = 4\chi\varepsilon\left[\left(\frac{\sigma}{r_{ij}}\right)^{12} - \left(\frac{\sigma}{r_{ij}}\right)^{6}\right]. \tag{6.36}$$

Here, χ is an artificial introduced scaling factor to tune the interaction strength. The parameters ε and σ take on the following values:

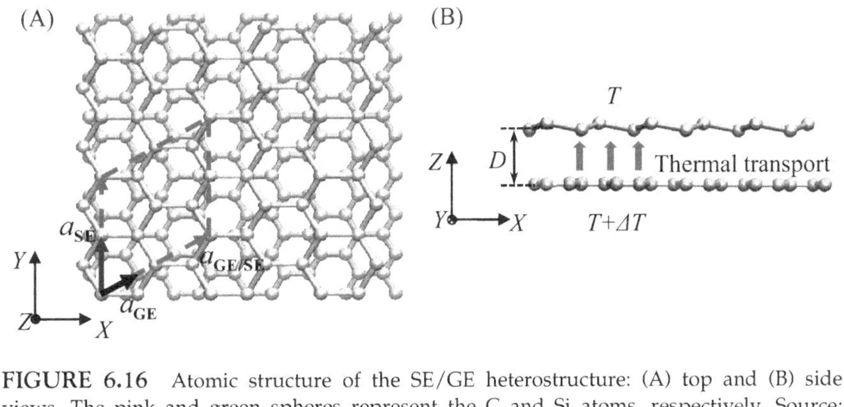

FIGURE 6.16 Atomic structure of the SE/GE heterostructure: (A) top and (B) side views. The pink and green spheres represent the C and Si atoms, respectively. Source: *Modified and reprinted with permission from a work by Liu et al. (2014b).*

$\varepsilon_{Si-C} = 8.909$ meV, $\sigma_{Si-C} = 3.326$ Å, $\varepsilon_{Si-H} = 5.767$ meV, and $\sigma_{Si-H} = 2.932$ Å. The cutoff distance of the LJ potential is set at 10 Å, which is approximately three times the value of σ_{Si-C}.

The thermal relaxation simulation, which mimics the experimental pump-probe approach, is performed to calculate the interfacial thermal conductance G of the heterostructure. In the beginning, the initial configuration of the SE/GE bilayer is equilibrated at $T = 300$ K using the NVT (constant number of species, volume, and temperature) ensemble for 50 ps with a time step of 0.1 fs. Upon realization of the equilibrium state, a temperature difference ΔT is implemented between the GE and SE monolayers by rapidly increasing the temperature of the GE monolayer to $T + \Delta T$ through the velocity rescaling method while keeping the temperature of the SE monolayer at T (Fig. 6.16B). To stabilize ΔT, the simulation is conducted for another 50 ps during which the GE and SE monolayers are coupled with two separate thermostats to remain at $T + \Delta T$ and T, respectively.

Afterward, the NVT ensemble is switched to the NVE ensemble with the thermostats removed. In the absence of temperature control, the system is allowed to relax thermally, and the temperature difference between the GE and SE monolayers decays in time as heat is transferred between the monolayers due to weak van der Waals interactions between them. The rate of decay of ΔT reflects how rapidly heat is being transferred between the two monolayers and corresponds to the interfacial thermal conductance of the system. The temperature of the GE monolayer is defined as T_G with its initial value being T_G^0, while the temperature of the SE monolayer is defined as T_S with its initial value T_S^0. Under the condition of energy conservation imposed by the NVE ensemble, the following equations can be obtained:

$$\frac{d}{dt}\left[C_{GE}\left(T_G - T_G^0\right)\right] = -GA_{In}(T_G - T_S), \tag{6.37}$$

$$\frac{d}{dt}\left[C_{SE}\left(T_S - T_S^0\right)\right] = -GA_{In}(T_S - T_G), \tag{6.38}$$

where A_{In} denotes the area of the interface between the GE and SE monolayers, and C_{GE} and C_{SE} denote their respective constant volume heat capacities, which can be calculated by

$$C = 3N_a k_B \int_0^{\infty} d\omega P(\omega)\left(\frac{\hbar\omega}{k_B T}\right)^2 \exp\left(\frac{\hbar\omega}{k_B T}\right) \Big/ \left(\exp\left(\frac{\hbar\omega}{k_B T}\right) - 1\right)^2. \tag{6.39}$$

Here, N_a denotes the number of atoms, k_B is the Boltzmann constant, $P(\omega)$ is its phonon spectrum power at frequency ω and can be calculated by Eq. (6.35), and T is the temperature of the monolayer.

It is assumed that the heat capacities of the GE and SE monolayers do not change significantly during the thermal relaxation process and can thus be regarded as constants. Eq. (6.36) can be rewritten as

$$\frac{dT_G}{dt} = -\frac{GA_{In}}{C_{GE}}(T_G - T_S), \tag{6.40}$$

$$\frac{dT_S}{dt} = -\frac{GA_{In}}{C_{SE}}(T_S - T_G). \tag{6.41}$$

The substitution of T_S in Eq. (6.41) with its expression in terms of T_G obtained from Eq. (6.40) gives

$$\frac{d^2T_G}{d\,t^2} + (\alpha + \beta)\frac{dT_G}{dt} = 0, \tag{6.42}$$

with $\alpha = GA/C_{GE}$ and $\beta = GA/C_{SE}$. Similarly, the substitution of T_G in Eq. (6.40) with its expression in terms of T_S obtained from Eq. (6.41) gives

$$\frac{d^2T_S}{dt^2} + (\alpha + \beta)\frac{dT_S}{dt} = 0. \tag{6.43}$$

Subtracting Eq. (6.43) from Eq. (6.42) gives

$$\frac{d^2(T_G - T_S)}{dt^2} + (\alpha + \beta)\frac{d(T_G - T_S)}{dt} = 0. \tag{6.44}$$

The solution of this second-order homogeneous linear differential equation with the initial conditions of $\Delta T(t) = T_G - T_S = T_G^0 - T_S^0$ at $t = 0$ and $\Delta T(t) = T_G - T_S = 0$ at $t \to \infty$ is given as

$$\Delta T(t) = \Delta T(0)\exp\left(-\frac{t}{\zeta}\right). \tag{6.45}$$

Here, $\zeta = (\alpha+\beta)^{-1} = \left[GA_{In}\left(C_{GE}^{-1} + C_{SE}^{-1}\right)\right]^{-1}$ is known as the thermal relaxation time and is determined by the simulation. Once the value of ζ is obtained, the interfacial thermal conductance G can be calculated by

$$G = \frac{C_V}{A_{In}\zeta}, \tag{6.46}$$

where C_V is the effective constant volume heat capacity of the bilayer system and equals $C_{GE}C_{SE}/(C_{GE} + C_{SE})$.

In the simulation, 200 K is determined to be an appropriate value for the temperature difference ΔT. If ΔT is too small, the thermal relaxation process will be extremely vulnerable to external noise, thereby rendering the value of the thermal relaxation time ζ obtained by applying Eq. (6.46) inaccurate. However, if ΔT is too large, it will be difficult to determine the temperature dependence of the interfacial thermal conductance G because of the difficulties in determining the average system temperature of the hybrid bilayer.

6.3.3.2 Interface thermal conductance

Fig. 6.17A shows the temperature evolution for both the GE and the SE monolayer during the thermal relaxation process at $T=300$ K. As time passes, the temperature of the GE monolayer decreases, while that of the SE monolayer increases, indicating a net heat transfer from the GE monolayer to the SE monolayer. Since the GE and SE monolayers have different heat capacities per unit area, their bilayer heterostructure does not equilibrate at the average temperature of $T+\Delta T$ and T. The final equilibrium temperature of the SE/GE bilayer is closer to the initial temperature of the GE monolayer than that of the SE monolayer because GE possesses a greater heat capacity per unit area than SE. The simulation shows that when the temperatures of the GE and SE monolayers are set at 500 K and 300 K, respectively, their bilayer heterostructure eventually equilibrates at a temperature of ~440 K. In contrast, if the temperatures of the GE and SE monolayers are set at 300 K and 500 K, respectively, the heterostructure will reach an equilibrium temperature of ~360 K.

As shown in Fig. 6.17B, the temperature difference ΔT between the GE and SE monolayers decays exponentially, which is in good agreement with Eq. (6.45). An exponential fitting of the curve determines the relaxation time ζ to be 31.95 ps. Therefore, the dependence of ΔT

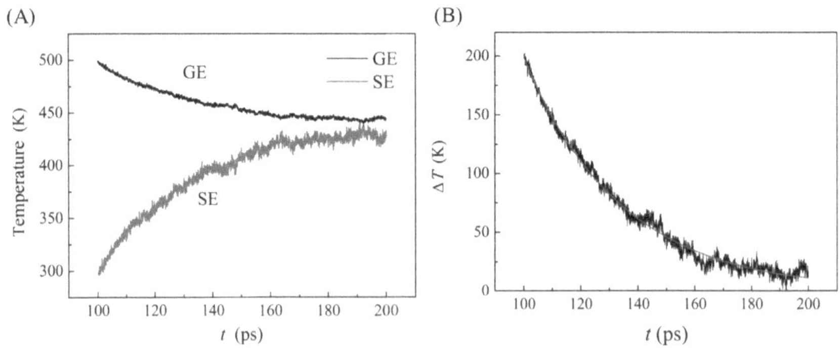

FIGURE 6.17 (A) Evolution of the temperatures of SE and GE, and (B) their difference during thermal relaxation. The exponential fitting in (B) is performed to obtain the relaxation time ζ. Source: *Modified and reprinted with permission from a work by Liu et al. (2014b).*

on t can be modeled by the relationship $\Delta T(t) = \Delta T(0)\exp(-t/31.95)$. Using Eq. (6.46), the interfacial thermal conductance G of the SE/GE heterostructure at 300 K is calculated to be 11.74 MW m^{-2} K^{-1}. A simulation is also conducted for the initial conditions of $\Delta T(0) = 100$ K and $\Delta T(0) = 150$ K. The results indicate the presence of large fluctuations in the $\Delta T(t)$ curve when $\Delta T(0) = 100$ K, and these fluctuations are reduced when ΔT increases to 150 K. For $\Delta T(0) = 150$ K, G is calculated to be 11.28 MW m^{-2} K^{-1}, which is close to the value obtained for $\Delta T(0) = 200$ K. The values of G for the SE/GE heterostructure are of the same order as those obtained for several other GE-based interfaces, e.g., 21 MW m^{-2} K^{-1} for the GE/resin interface (Hu et al., 2011) and 28 MW m^{-2} K^{-1} for the GE/Si interface (Zhang et al., 2013).

6.3.3.3 Temperature and interface coupling strength effects

The dependence of the interfacial thermal conductance G on temperature T is investigated for various scaling factors χ. The temperature difference is fixed at $\Delta T(0) = 200$ K. The simulation results indicate that the interlayer distance D is nearly independent of χ. Fig. 6.18A indicates that G increases monotonically with both T and χ. As T increases from 200 K to 700 K, G shows a twofold to threefold increase. This temperature dependence of G is in good consistency with those observed at the interfaces of GE/Cu (Chen et al., 2013) and CNT/SiO_2 (Ong and Pop, 2010) as well as that of the aforementioned GE/SE hybrid monolayer. According to Eq. (6.46), the interfacial thermal conductance G of a system depends on its effective heat capacity C_V and the thermal relaxation time ζ for a given interface area A. Fig. 6.18B shows that ζ decreases monotonically with T. As T increases from 200 K to 700 K, the value of

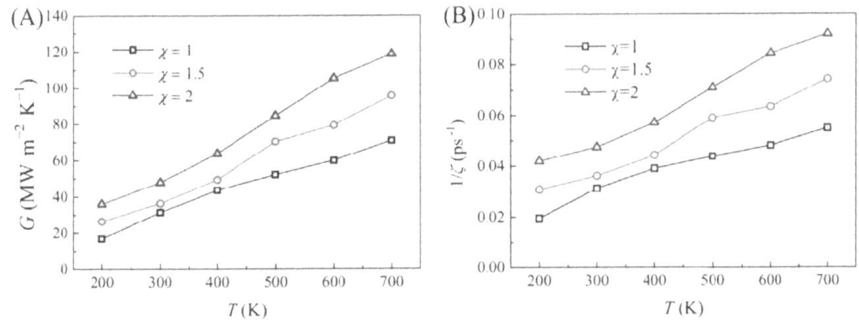

FIGURE 6.18 Dependence of (A) the interfacial thermal conductance G and (B) thermal relaxation time ζ on the temperature T for different values of χ. Source: *Modified and reprinted with permission from a work by Liu et al. (2014b).*

ζ for $\chi = 1$ is reduced by 65% (which corresponds to a 186% increase for G), and C_V increases by 25%. These results indicate that a change in the initial temperature T of the SE/GE heterostructure will affect its interfacial thermal conductance mainly by changing the thermal relaxation time ζ. The temperature dependence of the interfacial thermal conductance of the heterostructure has the same underlying mechanism as the GE/SE hybrid monolayer discussed in the previous section.

The increase in the interfacial thermal conductance G of the system with the interface coupling strength χ can be explained in terms of two aspects: (1) the coupling between the phonons of the GE and SE monolayers is enhanced, which directly boosts the efficiency of interfacial thermal transport and (2) the coupling between the in-plane and out-of-plane phonons in the GE monolayer is strengthened and indirectly facilitates interfacial thermal transport.

In a free-standing GE monolayer, its in-plane and out-of-plane phonons are well decoupled. Most of the in-plane phonons have frequencies beyond the overlapping range of 1–17 THz (Fig. 6.11) and cannot contribute to heat transport. When an SE monolayer is located on top of a GE monolayer, various symmetry types (e.g., reflection, translation, and rotation) are broken, and vibrations of the C atoms in the GE monolayer are altered by GE-SE interactions, especially at sites where the C and Si atoms share the same in-plane coordinates. As a result, the C atoms at these sites serve as scattering centers for the in-plane phonons of the GE monolayer, which leads to the coupling of its in-plane and out-of-plane phonons. Part of the thermal energy stored in the in-plane phonons of the GE monolayer can be transferred to its out-of-plane phonons and then transmitted to the SE monolayer. This process facilitates heat transfer between the GE and SE monolayers, thereby enhancing the interfacial thermal conductance of the hybrid bilayer.

6.3.3.4 GE hydrogenation effect

To enhance the interfacial thermal conductance of a system, surface engineering approaches, including chemical functionalization and atomic intercalation, are usually adopted. In this section, the effect of the hydrogenation of GE on the interfacial thermal conductance G of the heterostructure is discussed. There are two typical types of hydrogenated GE (HGE): double-side HGE (DHGE) and single-side HGE (SHGE). Both patterns are utilized to construct SE/DHGE and SE/SHGE heterostructures here.

Fig. 6.19A shows the atomic configuration of the SE/DHGE heterostructure. In practical applications, the hydrogen coverage f of the SE/DHGE heterostructure may not be complete. Usually, a partial hydrogen coverage is obtained by randomly removing an equal number of H atoms on both sides of the DHGE, i.e., a random distribution pattern of the H atoms is obtained. Thus, the coverage ratio of the H atoms is considered instead of the details of their distribution, although the latter also plays a role in determining the interfacial thermal conductance of the system. The hydrogen coverage is defined as $f = N_1/N_0$, where N_0 denotes half the total number of C atoms in GE and N_1 denotes the number of H atoms on one side of it.

By removing all the H atoms on one side of the DHGE, the SHGE can be obtained. As shown in Fig. 6.19B and C, according to the positions of the H atoms on the GE monolayer, the SE/SHGE heterostructure is further classified into two types: SE/THGE (top side) and SE/BHGE (bottom side). Note that after hydrogenation, the interlayer distance D increases slightly with the hydrogen coverage f, especially when the H atoms are located at the SE/GE interface. For $f = 100\%$, D equals 4.12 Å, 4.02 Å, and 3.70 Å for SE/DHGE, SE/THGE, and SE/BHGE, respectively.

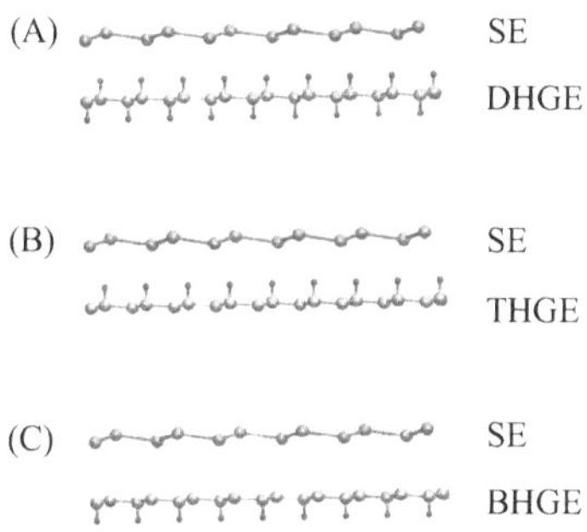

FIGURE 6.19 Atomic configurations of (A) SE/DHGE, (B) SE/THGE, and (C) SE/BHGE. Source: *Reprinted with permission from a work by Liu et al. (2014b).*

Fig. 6.20 shows the dependence of the interfacial thermal conductance G on the hydrogen coverage f for all three types of SE/HGE heterostructures. It is discovered that for both SE/THGE and SE/DHGE that possess H atoms at their interfaces, G dramatically increases as f increases up to 50% and then decreases.

In contrast, for the SE/BHGE heterostructure that does not possess interfacial H atoms, the value of G first decreases as f increases and then increases slightly beyond $f > 10\%$. Overall, the change in G for the SE/BHGE heterostructure is almost negligible compared with those for the SE/THGE and SE/DHGE heterostructures. This difference indicates that hydrogenation is more efficient in tuning G when the H atoms are located between SE and GE. The maximum values of G for the SE/THGE and SE/DHGE heterostructures are 61.75 $\mathrm{MW\,m^{-2}\,K^{-1}}$ and 39.14 $\mathrm{MW\,m^{-2}\,K^{-1}}$, respectively, which are three to five times higher than that of the pristine SE/GE heterostructure.

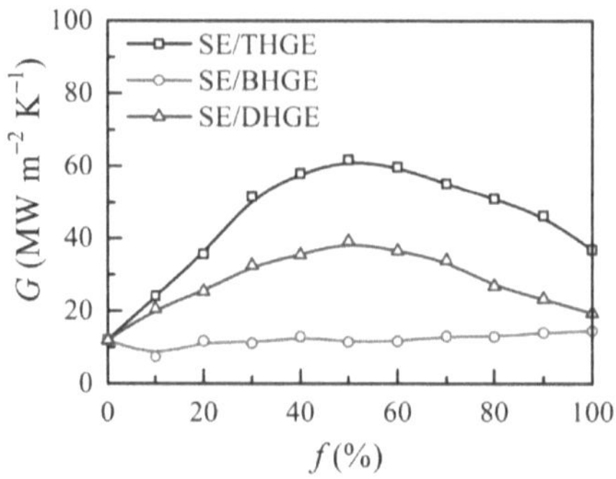

FIGURE 6.20 Dependence of the interfacial thermal conductance G on the hydrogen coverage f for the three types of SE/HGE heterostructures. Source: *Modified and reprinted with permission from a work by Liu et al. (2014b).*

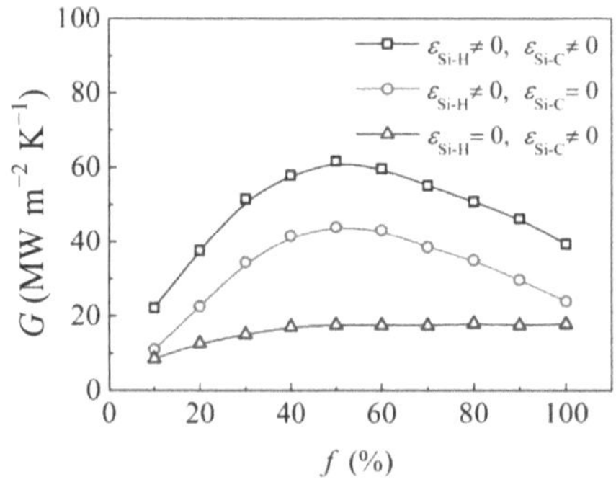

FIGURE 6.21 Interfacial thermal conductance G of the SE/THGE heterostructure vs. its hydrogen coverage f for different atomic interaction conditions. Source: *Modified and reprinted with permission from a work by Liu et al. (2014b).*

The interfacial thermal transport of the SE/HGE heterostructures is dependent on both Si–H and Si–C interactions. To illustrate the contribution of the Si–H interaction to the interfacial thermal transport of the system, the van der Waals interactions between the Si atoms and C atoms are turned off by setting $\varepsilon_{Si-H} \neq 0$ and $\varepsilon_{Si-C} = 0$. Alternatively, to explore the Si–C interaction exclusively, the van der Waals interactions between the Si atoms and H atoms are turned off by setting $\varepsilon_{Si-H} = 0$ and $\varepsilon_{Si-C} \neq 0$. For both cases, the interfacial thermal conductance G for the SE/THGE heterostructure with $\chi = 1$ is calculated for various values of hydrogen coverage f, as shown in Fig. 6.21. For comparison, the results for the case with full interactions ($\varepsilon_{Si-H} \neq 0$ and $\varepsilon_{Si-C} \neq 0$) are presented as well.

When $\varepsilon_{Si-H} \neq 0$ and $\varepsilon_{Si-C} = 0$ (red curve in Fig. 6.21), the value of G increases with f until $f \approx 50\%$ and decreases afterward. Such a nonmonotonic relationship is related to two mechanisms. On the one hand, as f increases, more H atoms participate in interactions with Si atoms and thus enhance the interfacial thermal transport of the heterostructure. On the other hand, H atoms possess a small atomic mass and tend to vibrate at high frequencies. Fig. 6.22 presents the phonon spectra of the H atoms in the low-frequency range for three different values of f. It indicates that, as f increases, the phonon spectrum shifts toward a higher frequency, and the spectrum power decreases dramatically, particularly in the range of 15–20 THz. As a result, the coupling between the phonons of the H atoms and those of the Si atoms is weakened, thereby resulting in a reduction in the interfacial thermal conductance of the system. When f is small, the former mechanism dominates the interfacial thermal transport within the system; when f is large, the latter becomes the dominant mechanism, and thus there exists a nonmonotonic dependence of the interfacial thermal conductance G on the hydrogen coverage f.

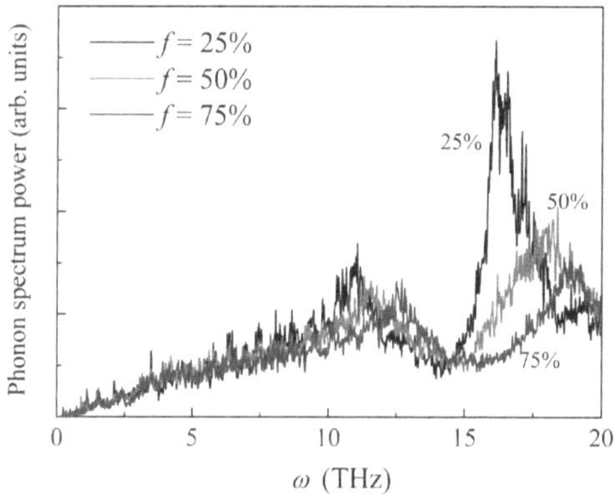

FIGURE 6.22 Phonon spectra of H atoms in the low-frequency range for three different values of hydrogen coverage f. Source: *Modified and reprinted with permission from a work by Liu et al. (2014b).*

When $\varepsilon_{Si-H} = 0$ and $\varepsilon_{Si-C} \neq 0$ (blue curve in Fig. 6.21), the value of G increases with f until $f \approx 50\%$ and then stabilizes. This increase in G is due to the enhanced coupling between the in-plane and out-of-plane phonons of the C atoms. Fig. 6.23 presents the in-plane and out-of-plane components of the phonon spectra for the C atoms in the THGE heterostructure. It shows that, as f increases from 25% to 50%, the overlap between the in-plane and out-of-plane spectrum components increases dramatically, particularly in the frequency range of 15–30 THz. This corresponds to an enhanced coupling between the in-plane and out-of-plane phonons, which facilitates heat transfer from the THGE monolayer to the SE monolayer. As f increases beyond 50%, the overlap does not undergo any significant changes, and thus the value of G remains virtually constant for $f > 50\%$.

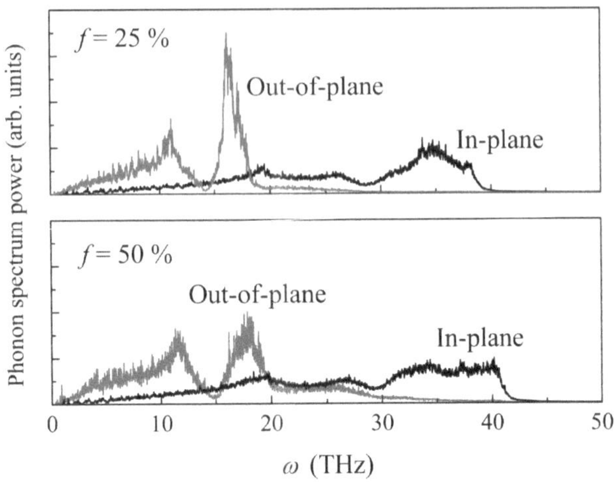

FIGURE 6.23 In-plane and out-of-plane components of the phonon spectra for the C atoms in the THGE heterostructure for two different values of hydrogen coverage f. Source: *Modified and reprinted with permission from a work by Liu et al. (2014b).*

When $\varepsilon_{Si-H} \neq 0$ and $\varepsilon_{Si-C} \neq 0$ (black curve in Fig. 6.21), the interfacial thermal conductance G of the SE/THGE heterostructure changes with its hydrogen coverage f in the same manner as the case in which $\varepsilon_{Si-H} \neq 0$ and $\varepsilon_{Si-C} = 0$. A similar phenomenon is observed for the SE/DHGE heterostructure, which indicates that hydrogenation affects the interfacial thermal conductance of the SE/HGE heterostructure mainly through the interactions between its constituent Si atoms and H atoms.

Fig. 6.21 also shows that the overall interfacial thermal conductance G of the SE/THGE heterostructure with all interactions taken into account ($\varepsilon_{Si-H} \neq 0$ and $\varepsilon_{Si-C} \neq 0$) is approximately the sum of the value of G with only Si–C interactions ($\varepsilon_{Si-H} = 0$ and $\varepsilon_{Si-C} \neq 0$) and the value of G with only Si–H interactions ($\varepsilon_{Si-H} \neq 0$ and $\varepsilon_{Si-C} = 0$). This relationship implies that the effect of Si–H interactions on the thermal transport across the SE/HGE interface is decoupled from that of the Si–C interactions.

Besides the hydrogen coverage f, the hydrogenation pattern can significantly affect the interfacial thermal conductance of the SE/HGE heterostructure. The value of G for the SE/DHGE heterostructure always lies between those of the SE/THGE and SE/BHGE heterostructures, regardless of the value of f (Fig. 6.20). This phenomenon is mainly due to changes in the average vertical distance d_{Si-H} of the Si–H atoms in the SE/GE heterostructure along its out-of-plane direction because of the different hydrogenation patterns.

When H atoms are bonded to C atoms in a GE monolayer, the latter are pulled away from their initial positions with an average out-of-plane displacement d_{out} of ~ 0.25 Å. Such displacements are labeled positive when H atoms are located on the top side of the GE monolayer and negative when H atoms are located on the bottom side. As a result, for the SE/THGE heterostructure, the Si and H atoms have an average vertical distance $d_{Si-H} = D - d_{C-H} - d_{out} = 4.02 - 1.10 - 0.25 = 2.67$ Å, which is much smaller than the value of $d_{Si-H} = D + d_{C-H} + d_{out} = 3.70 + 1.10 + 0.25 = 5.05$ Å for the SE/BHGE heterostructure. Here, $d_{C-H} = 1.1$ Å is the averaged C–H bond length. With a larger value of d_{Si-H}, the van der Waals interactions among the Si and H atoms become weaker. Since the Si–H interaction plays a dominant role in determining the interfacial thermal conductance G of the SE/HGE heterostructure, as demonstrated above, the value of G for the SE/THGE heterostructure is much larger than that of the SE/BHGE heterostructure. In addition, due to the large value of d_{Si-H} for the SE/BHGE heterostructure, the interactions among its Si atoms and H atoms are very weak and contribute minimally to its interfacial thermal transport. As a result, the interfacial thermal conductance of the SE/BHGE heterostructure is almost independent of its hydrogen coverage f (Fig. 6.20).

For the SE/DHGE heterostructure, the H atoms on the top side of its GE monolayer are much closer to the Si atoms than those located on the bottom side, thereby contributing more to its interfacial thermal transport. These topside H atoms possess an average Si–H distance of $d_{Si-H} = 4.12 - 1.10 - 0.25 = 2.77$ Å, a slightly larger value than the value of 2.67 Å for the SE/THGE heterostructure. Moreover, the interlayer distance $D = 4.12$ Å for the SE/DHGE heterostructure is larger than the interlayer distance $D = 4.02$ Å for the SE/THGE heterostructure. Therefore, both the Si–H and Si–C interactions are weaker in the case of the SE/DHGE heterostructure, which leads to its lower interfacial thermal conductance than that of the SE/THGE heterostructure.

More importantly, due to the double-sided hydrogenation, the DHGE has twice as many H atoms as the THGE for the same hydrogen coverage. Consequently, more phonons have high frequencies in the DHGE, and the coupling between the phonons of the H atoms and those of the Si atoms in the SE/DHGE heterostructure is therefore weaker than the coupling in the SE/THGE heterostructure, as indicated by Fig. 6.22. Such weakening is also responsible for the

SE/DHGE heterostructure possessing a smaller value of G than the SE/THGE heterostructure.

6.4 Thermal rectification effects

An interesting phenomenon that can be observed from Figs. 6.10 and 6.12 is that for a preset heat flux, the interfacial thermal conductance for the GE/SE hybrid monolayer changes when the heat flux direction is reversed. Specifically, the interfacial thermal conductance is larger when the temperature of the SE side is higher, which indicates that heat prefers to flow from the SE side to the GE side. Such an asymmetrical heat flux gives rise to the thermal rectification effect, which is analogous to the electrical rectification effect observed in an electric diode.

Generally, thermal rectification is a phenomenon in which thermal transport along a specific axis is dependent on the sign of the temperature gradient. Thermal rectification has been a topic of interest dating back to its initial experimental observation by Starr (1935). Since then, several experimental and theoretical studies have been performed to understand the underlying mechanisms of thermal rectification. With an improved understanding of how thermal rectification is achieved, devices such as thermal transistors, thermal logic circuits, and thermal diodes can be developed and utilized in nanoelectronics (Roberts and Walker, 2011).

Currently, several factors accounting for thermal rectification have been proposed, including surface roughness or flatness at material interfaces, thermal potential barriers between material contact areas, differences in the temperature dependence of thermal conductivity between dissimilar materials in contact, structural asymmetry, and anharmonic lattices [typically one-dimensional (1D)]. Among these factors, the surface roughness/flatness and the thermal potential barrier are the major driving forces behind thermal rectification in bulk materials, whereas the introduction of structural asymmetry is the main strategy for designing thermal rectifiers with nanomaterials and nanostructures. Several methods have been proposed to realize such structural asymmetries, including the addition of asymmetric mass loadings, the design of asymmetric geometries, imposing asymmetric strains, and the creation of interfaces with dissimilar materials.

An experimental study conducted by Chang et al. (2006) is the first demonstration of thermal rectification on the nanoscale. In their experiments, trimethyl-cyclopentadienyl platinum ($C_9H_{16}Pt$) is loaded onto both CNTs and boron nitride nanotubes along the length of the tubes in a nonuniform manner, and the thermal conductivity of the nanotubes is measured in both directions. The measurements indicate that for both types of nanotubes with asymmetric mass loading, heat flows more easily from the high-mass end to the low-mass end than in the opposite direction. By defining the rectification ratio $\varepsilon = (\lambda^+ - \lambda^-)/\lambda^-$ with $\lambda^+ > \lambda^-$, where λ^+ denotes thermal conductivity in the forward direction and λ^- denotes thermal conductivity in the opposite direction, the rectification ratio is ~2% for the CNTs and 7% for the boron nitride nanotubes. Although these rectification ratios are small, this pioneering experimental work has greatly inspired the design of nanostructured thermal rectifiers with improved rectification efficiency.

Due to the difficulties encountered in exerting precise control over the atomic structures of nanomaterials and nanostructures in experiments, most of the research work on the thermal rectification effect on the nanoscale is performed based on MD simulations. For instance, Alaghemandi et al. (2010) have performed nonequilibrium MD simulations on CNTs with asymmetric mass loadings to investigate their thermal rectification mechanisms. In their simulation model, the mass loading is imposed by artificially varying the atomic masses of the C atoms so that a mass or density gradient is created in the axial direction of the nanotube. The thermal conductivity of the asymmetric CNT is then calculated by the direct method with the model shown in Fig. 6.1. In contrast to the experimental observations made by Chang et al. (2006), an MD simulation work by Alaghemandi et al. (2010) indicates that in an asymmetrically mass-loaded nanotube, heat flows preferentially in the direction from its low-mass end to its high-mass end with a rectification ratio in the range of 0.015–0.074, depending on its temperature and mass gradient. The preference for the opposite direction of the heat flux is attributed to the anharmonicities caused by transverse motions (motions perpendicular to the direction of the longitudinal axis) that are stronger at the low-mass end.

By introducing cone cavities in a silicon nanofilm, Ju and Liang (2012) have successfully rendered it geometrically asymmetric, as shown in Fig. 6.24. The direct method is used to calculate its thermal conductivity by employing the simulation model shown in Fig. 6.1. The velocity exchange approach is then adopted to generate a heat flux. The simulation results indicate that the effective thermal resistance as a result of the cone cavities is ~11% larger for a heat flux in the direction from the base of the cone to its vertex than for a heat flux in the opposite direction, which corresponds to a significant thermal rectification effect. The authors have attributed the source of such a thermal rectification effect to asymmetric phonon transmission rates along the two opposite directions. Since the base of the cone is perpendicular to the direction of the heat flux, when heat flows in the direction from the base of the cone to its vertex, phonons incident on the base are reflected, and those passing through the narrow paths beside the cone cavities can reach the low-temperature side. When the direction of the heat flow is reversed, these narrow paths are still available for phonons to pass through. However, in this case, phonons scattered at the sloped edges of the

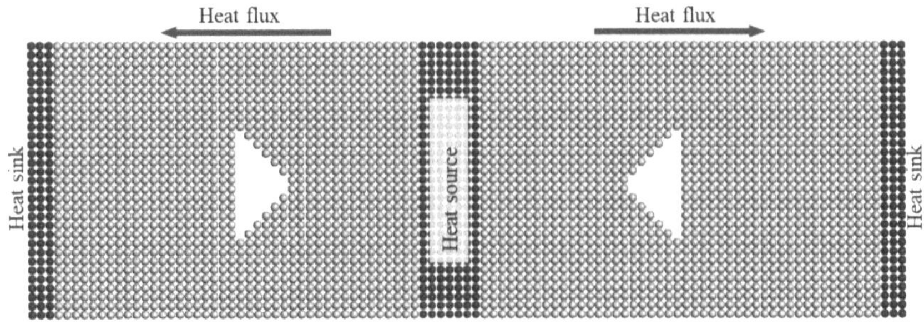

FIGURE 6.24 Schematic of the MD simulation model for the thermal conductivity calculation of a silicon nanofilm with cone cavities. The simulation box is divided into N layers, and the cold layer and hot layer are located at the first and $(N/2)$th layers, respectively. The cold layer enclosed with the dashed line at the right end is the image of the cold layer at the left end under the PBC along the heat flow direction (Ju and Liang, 2012).

cone cavities will have a much greater probability of being transmitted through these narrow paths after being scattered and reaching the lower temperature region. As a result, more phonons can circumvent the cavities when the heat flux is directed from the vertex of the cone to its base, thereby resulting in a higher transmission coefficient and lower thermal resistance of the system.

Geometrical asymmetry has also been implemented in GE nanoribbons to induce the thermal rectification effect. Yang et al. (2009) have reported that a high thermal rectification ratio of 0.64 can be achieved for asymmetric GE nanoribbons based on MD simulations. In their simulations, two types of GE nanoribbons are considered. The first is triangular, while the second consists of two connected rectangular GE nanoribbons with the same length but significantly different widths. For both types of GE nanoribbons, heat prefers to flow from the wider side to the narrower side. In addition, the triangular graphene nanoribbon structure has a larger thermal rectification ratio than the structure composed of rectangular GE nanoribbons.

Hu et al. (2009b) have also performed MD simulations to investigate single-layer and bilayer GE nanoribbons with triangular and trapezoidal shapes. Their results indicate that the thermal rectification effect is maximized for a single-layer triangular GE nanoribbon with smooth edges. With a decrease in the length of the nanoribbon and an increase in the roughness of the nanoribbon edges, its thermal rectification ratio reduces. The thermal rectification effect is also observed for the trapezoidal single-layer GE nanoribbon, but its rectification ratio is much lower than that of the triangular nanoribbon. For both the asymmetric GE nanoribbons, the change in the phonon spectrum along their lengths is considered the source of the thermal rectification effect.

Besides GE nanoribbons, geometrical asymmetry has been introduced in CNTs to generate thermal rectification. For instance, MD simulations have been conducted to investigate the thermal rectification effect in a system composed of two CNTs of different diameters joined at their ends (Wu and Li, 2007). This structure is analogous to the GE nanoribbons with asymmetric rectangular morphologies studied by Yang et al. (2009). When the temperature gradient imposed on the nanotube structure is reversed, a significant change can be observed in the overlap between the phonon spectra of each side of the junction. Such a change in the spectra overlap gives rise to the thermal rectification effect. The thermal rectification effect in carbon nanohorns has also been investigated by MD simulation (Wu and Li, 2008). Carbon nanohorns are CNTs with varying diameters along their tube axes (Fig. 6.25) and thus possess a linear geometrical irregularity similar to those of the triangular graphene nanoribbons discussed above. In the simulation, the

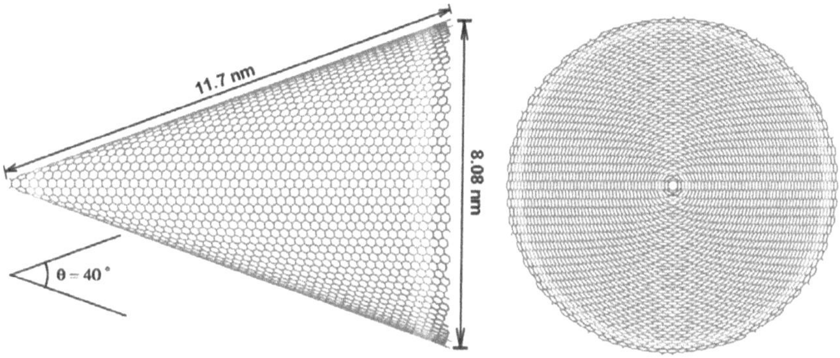

FIGURE 6.25 Atomic structure of a carbon nanohorn. Source: *Reprinted with permission from a work by Wu and Li (2008).*

dependence of the thermal rectification effect on the flare angle of the nanohorn, the tensile load applied to the nanohorn along its axial direction, and the magnitude of the temperature gradient are considered. The simulation results indicate that heat flows preferentially in the direction of decreasing tube radius. In addition, for a given length of the nanohorn, a large flare angle and/or a mild tensile strain can result in an enhanced thermal rectification effect. The underlying mechanisms of the thermal rectification effect and its dependence on the tensile strain applied to the system are also attributed to the mismatch in the phonon spectra of the two ends of the nanohorn.

In addition to geometric asymmetry, the interface between two dissimilar materials is frequently utilized to construct thermal rectifiers.

Taking the GE/SE hybrid monolayer discussed in Section 6.3.2 as an example, the thermal rectification effect within the structure can be clearly observed from Figs. 6.10 and 6.12. The interfacial thermal conductance G of the system is greater when heat flows from the SE monolayer to the GE monolayer than that for the reverse direction. The existence of the interfacial thermal rectification effect can be explained by the asymmetry in the phonon transmission caused by the mismatch of the phonon spectra across the interface.

Fig. 6.11 shows that the GE monolayer has a much wider phonon frequency range than the SE monolayer and can accommodate phonons of higher frequencies. Consequently, the GE monolayer can accommodate most of the phonons that originate from the SE monolayer, while the high-frequency phonons from the GE monolayer are unable to be contained by the SE monolayer. The only way for high-frequency phonons in the GE monolayer to enter the SE monolayer is for them to undergo inelastic scattering beforehand and form phonons of lower frequencies. When the heat flux is in the direction from the GE monolayer to the SE monolayer, the average temperature of the GE monolayer is higher than that of the SE monolayer. Hence, more high-frequency phonons are excited in the GE monolayer but cannot be transmitted across the interface without undergoing inelastic scattering first. When the direction of the heat flux is reversed, the average temperature of the SE monolayer is now higher than that of the GE monolayer, and it contains more high-frequency phonons that can be easily transmitted across the interface and reach the GE monolayer. Therefore, an interfacial thermal rectification effect can be observed for the GE/SE hybrid monolayer, and its interfacial thermal conductance G becomes different for the two heat fluxes in opposite directions, as discussed above.

When one of the two materials that form an interface is in a liquid state, the underlying mechanisms for the thermal rectification effect can be more complex. Hu et al. (2009a) conducted MD simulations to investigate the thermal rectification effect occurring at an interface between water and functionalized silica. The simulation results indicate a rectification ratio of 0.54, and heat flows preferentially from the silica region to the water region. By calculating the radial distribution function of water, it is established that such a thermal rectification effect is a result of the characteristics of hydrogen bonds in water being highly dependent on temperature. In a situation where the temperature of the water is higher than that of silica, the network of hydrogen bonds in water is weakened due to an enhanced kinetic energy fluctuation of the water molecules. Consequently, the transfer of heat across the interface is impeded compared with the case in which the temperature of water is lower, thereby leading to the thermal rectification effect.

References

Alaghemandi, M., Leroy, F., Algaer, E., Böhm, M.C., Müller-Plathe, F., 2010. Thermal rectification in mass-graded nanotubes: a model approach in the framework of reverse nonequilibrium molecular dynamics simulations. Nanotechnology 21, 1–6.

Cahangirov, S., Topsakal, M., Aktürk, E., Şahin, H., Ciraci, S., 2009. Two- and one-dimensional honeycomb structures of silicon and germanium. Phys. Rev. Lett. 102, 236804-1–236804-4.

Chang, C.W., Okawa, D., Majumdar, A., Zettl, A., 2006. Solid-state thermal rectifier. Science 314, 1121–1124.

Che, J., Cagin, T., Deng, W., Goddard III, W.A., 2000. Thermal conductivity of diamond and related materials from molecular dynamics simulations. J. Chem. Phys. 113, 6888–6900.

Chen, J., Zhang, G., Li, B., 2012. Thermal contact resistance across nanoscale silicon dioxide and silicon interface. J. Appl. Phys. 112, 064319-1–064319-7.

Chen, L., Huang, Z., Kumar, S., 2013. Phonon transmission and thermal conductance across graphene/Cu interface. Appl. Phys. Lett. 103, 123110-1–123110-4.

De Padova, P., Kubo, O., Oivieri, B., Quaresima, C., Nakayama, T., Aono, M., et al., 2012. Multilayer silicene nanoribbons. Nano Lett. 12, 5500–5503.

Diao, J., Srivastava, D., Menon, M., 2008. Molecular dynamics simulations of carbon nanotube/silicon interfacial thermal conductance. J. Chem. Phys. 128, 164708-1–164708-5.

Geim, A.K., Novoselov, K.S., 2007. The rise of graphene. Nat. Mater. 6, 183–191.

Ghosh, S., Calizo, I., Teweldebrhan, D., Pokatilov, E.P., Nika, D.L., Balandin, A.A., et al., 2008. Extremely high thermal conductivity of graphene: prospects for thermal management applications in nanoelectronic circuits. Appl. Phys. Lett. 92, 151911-1–151911-3.

Howell, P.C., 2011. Thermal conductivity calculation with the molecular dynamics direct method I: more robust simulations of solid materials. J. Computat. Theor. Nanosci. 8, 2129–2143.

Howell, P.C., 2012. Comparison of molecular dynamics methods and interatomic potentials for calculating the thermal conductivity of silicon. J. Chem. Phys. 137, 224111-1–224111-14.

Hu, L., Desai, T., Keblinski, P., 2011. Determination of interfacial thermal resistance at the nanoscale. Phys. Rev. B 83, 195423-1–195423-5.

Hu, M., Goicochea, J.V., Michel, B., Poulikakos, D., 2009a. Thermal rectification at water/functionalized silica interfaces. Appl. Phys. Lett. 95, 151903-1–151903-3.

Hu, J., Ruan, X., Chen, Y., 2009b. Thermal conductivity and thermal rectification in graphene nanoribbons: a molecular dynamics study. Nano Lett. 9, 2730–2735.

Hu, W., Li, Z., Yang, J., 2013. Structural, electronic, and optical properties of hybrid silicene and graphene nanocomposite. J. Chem. Phys. 139, 154704-1–154704-5.

Hu, B., Yang, L., Zhang, Y., 2006. Asymmetric heat conduction in nonlinear lattices. Phys. Rev. Lett. 97, 124302-1–124302-4.

Ju, S., Liang, X., 2012. Thermal rectification and phonon scattering in silicon nanofilm with cone cavity. J. Appl. Phys. 112, 054312-1–054312-5.

Jund, P., Jullien, R., 1999. Molecular-dynamics calculation of the thermal conductivity of vitreous silica. Phys. Rev. B 59, 13707–13711.

Kapitza, P.L., 1941. The study of heat transfer in helium II. J. Phys. USSR 4, 181–210.

Lee, Y.H., Biswas, R., Soukoulis, C.M., Wang, C., Ho, K.M., 1991. Molecular-dynamics simulation of thermal conductivity in amorphous silicon. Phys. Rev. B: Condens. Matter 43, 6573–6580.

Lee, C., Wei, X., Kysar, J.W., Hone, J., 2008. Measurement of the elastic properties and intrinsic strength of monolayer graphene. Science 321, 385–388.

Li, X., Maute, K., Dunn, M.L., Yang, R., 2010. Strain effects on the thermal conductivity of nanostructures. Phys. Rev. B 81, 245318-1–245318-11.

Lindsay, L., Broido, D.A., 2010. Optimized Tersoff and Brenner empirical potential parameters for lattice dynamics and phonon thermal transport in carbon nanotubes and graphene. Phys. Rev. B 81, 205441-1–205441-6.
Liu, B., Baimova, J.A., Reddy, C.D., Dmitriev, S.V., Law, W.K., Feng, X.Q., et al., 2014a. Interface thermal conductance and rectification in hybrid graphene/silicene monolayer. Carbon 79, 236–244.
Liu, B., Baimova, J.A., Reddy, C.D., Law, A.W.K., Dmitriev, S.V., Wu, H., et al., 2014b. Interfacial thermal conductance of a silicene/graphene bilayer heterostructure and the effect of hydrogenation. ACS Appl. Mater. Interfaces 6, 18180–18188.
Lo, W.C., Wang, L., Li, B., 2008. Thermal transistor: heat flux switching and modulating. J. Phys. Soc. Jpn. 77, 1–5.
Luo, T., Lloyd, J.R., 2012. Enhancement of thermal energy transport across graphene/graphite and polymer interfaces: a molecular dynamics study. Adv. Funct. Mater. 22, 2495–2502.
Matthes, L., Hannewald, K., Bechstedt, F., 2012. *Ab initio* investigation of graphene-based one-dimensional superlattices and their interfaces. Phys. Rev. B 86, 205409-1–205409-9.
Müller-Plathe, F., 1997. A simple nonequilibrium molecular dynamics method for calculating the thermal conductivity. J. Chem. Phys. 106, 6082–6085.
Novoselov, K.S., Geim, A.K., Morozov, S.V., Jiang, D., Zhang, Y., Dubonos, S.V., et al., 2004. Electric field effect in atomically thin carbon films. Science 306, 666–669.
Ong, Z.-Y., Pop, E., 2010. Molecular dynamics simulation of thermal boundary conductance between carbon nanotubes and SiO_2. Phys. Rev. B 81, 155408-1–155408-7.
Rajabpour, A., Vaez Allaei, S.M., 2012. Tuning thermal conductivity of bilayer graphene by inter-layer sp^3 bonding: a molecular dynamics study. Appl. Phys. Lett. 101, 053115-1–053115-4.
Roberts, N.A., Walker, D.G., 2011. A review of thermal rectification observations and models in solid materials. Int. J. Therm. Sci. 50, 648–662.
Samvedi, V., Tomar, V., 2009. The role of interface thermal boundary resistance in the overall thermal conductivity of Si-Ge multilayered structures. Nanotechnology 20, 1–11.
Schelling, P.K., Phillpot, S.R., Keblinski, P., 2002. Comparison of atomic-level simulation methods for computing thermal conductivity. Phys. Rev. B 65, 144306-1–144306-12.
Sellan, D.P., Landry, E.S., Turney, J.E., Mcgaughey, A.J.H., Amon, C.H., 2010. Size effects in molecular dynamics thermal conductivity predictions. Phys. Rev. B 81, 214305-1–214305-10.
Starr, C., 1935. The copper oxide rectifier. J. Appl. Phys. 7, 15–19.
Vogt, P., De Padova, P., Quaresima, C., Avila, J., Frantzeskakis, E., Asensio, M.C., et al., 2012. Silicene: compelling experimental evidence for graphenelike two-dimensional silicon. Phys. Rev. Lett. 108, 155501-1–155501-5.
Wang, L., Li, B., 2007. Thermal logic gates: computation with phonons. Phys. Rev. Lett. 99, 177208-1–177208-4.
Wang, L., Li, B., 2008a. Phononics gets hot. Phys. World 21, 27–29.
Wang, L., Li, B., 2008b. Thermal memory: a storage of phononic information. Phys. Rev. Lett. 101, 267203-1–267203-4.
Wu, G., Li, B., 2007. Thermal rectification in carbon nanotube intramolecular junctions: molecular dynamics calculations. Phys. Rev. B 76, 085424-1–085424-8.
Wu, G., Li, B., 2008. Thermal rectifiers from deformed carbon nanohorns. J. Phys.: Condens. Matter 20, 27–29.
Xie, Z., Li, K., Tang, L., Pan, C., Chen, K., 2012. Nonlinear phonon transport and ballistic thermal rectification in asymmetric graphene-based three terminal junctions. Appl. Phys. Lett. 100, 183110-1–183110-4.
Xu, Z., Buehler, M.J., 2012. Heat dissipation at a graphene-substrate interface. J. Physics: Condens. Matter 24, 1–7.

Yang, N., Zhang, G., Li, B., 2009. Thermal rectification in asymmetric graphene ribbons. Appl. Phys. Lett. 95, 033107-1–033107-3.

Zhang, X., Hu, M., Poulikakos, D., 2012. A low-frequency wave motion mechanism enables efficient energy transport in carbon nanotubes at high heat fluxes. Nano Lett. 12, 3410–3416.

Zhang, J., Wang, Y., Wang, X., 2013. Rough contact is not always bad for interfacial energy coupling. Nanoscale 5, 11598–11603.

Zhang, X., Xie, H., Hu, M., Bao, H., Yue, S., Qin, G., et al., 2014. Thermal conductivity of silicene calculated using an optimized Stillinger-Weber potential. Phys. Rev. B 89, 054310-1–054310-7.

Zheng, K., Wang, L., Bai, S., Yu, J., Tang, Z., Huang, Z., 2014. An anomalous wave-like kinetic energy transport in graphene nanoribbons at high heat flux. Phys. B: Condens. Matter 434, 64–68.

Zhou, Y., Anglin, B., Strachan, A., 2007. Phonon thermal conductivity in nanolaminated composite metals via molecular dynamics. J. Chem. Phys. 127, 184702-1–184702-11.

CHAPTER

7

Application of molecular dynamics simulation in mass transport problems

This chapter presents the application of molecular dynamics (MD) simulation in solving mass transport problems. It begins with the discussion of various driven methods and models for fluids in nanoconfinement, followed by the unique phenomena observed for water flow in carbon nanotubes (CNTs), porous monolayer graphene, and multilayer graphene and graphene oxide membranes. Afterward, the MD simulation models for the nanofiltration with porous thin films using reverse osmosis (RO), forward osmosis (FO), and capacitive deionization (CDI) processes are illustrated. Finally, MD simulation models for liquid–vapor phase transitions, including the droplet evaporation in a gaseous environment and the liquid evaporation on a substrate, are demonstrated.

7.1 Fluids in nanoconfinement

Fluids in nanoconfinement are usually defined as fluids contained within channels or structures with a size of 0–100 nm in one or more dimensions. Studies of nanofluids are mainly driven by the development of novel nanotechnologies for energy, biological, medical, and environmental applications. Some of these studies are devoted to discovering new transport phenomena in which the fluidic dimensions are close to or smaller than the characteristic dimensions that determine the mechanical and electrochemical behaviors of macroscopic systems.

The computational modeling of transport phenomena in nanofluidic systems is significantly different from modeling transport phenomena in microfluidic systems. This difference is because on the nanoscale, the

DOI: https://doi.org/10.1016/B978-0-12-816419-8.00012-X

close contact or interactions between the fluid and the channel walls leads to unique fluidic behaviors and mass transport phenomena. Consequently, models developed for macrofluidic or microfluidic systems fail on the nanoscale due to several reasons. First, for nanofluidic systems, the fluids cannot be regarded as a continuum, and the discrete nature of molecules should be considered. For instance, a spherical water droplet with a diameter of one nanometer contains fewer than 40 molecules; modeling the water droplet by applying continuum equations would lead to errors, and individual water molecules and their mutual interactions must be considered instead. Second, equations for macro- and microscale liquid transport through convection, diffusion, or migration are no longer applicable on the nanoscale due to the breakdown of the basic assumptions behind them. Third, the effects imposed by the channel walls are significant in nanofluidic systems. The specific wall structure, surface charge states, roughness, interactions between the walls and the fluid particles, and the adsorption and desorption behaviors of these molecules on the wall surface are all important factors that should be considered in the investigation of nanofluidic behaviors. Unfortunately, these factors are seldom included in macro- and microscale liquid transport models.

To compensate for the limitations of classical equations in the study of nanofluids, MD simulation is being increasingly utilized for the investigation of fluid transport on the nanoscale. In MD simulation, the structure, dynamics, and transport properties of many-body systems can be efficiently explored by calculating the motion of the atoms in a molecular assembly using Newtonian dynamics with atomic interactions being described by appropriate empirical potential functions. In this section, the application of MD simulation in nanofluid modeling is introduced with examples of water transport in nanochannels composed of advanced carbon-based nanomaterials, including carbon nanotubes (CNTs) and graphene. The main modeling and analyzing techniques of these systems are introduced, and their unique mass transport behaviors are highlighted with the underlying mechanisms illustrated.

7.1.1 Fluid-driven methods

Fluid transport in nanochannels can be driven by adopting various approaches that are distinguished by their driving forces. The most frequently adopted approach is the pressure-driven method, and other approaches include the temperature gradient-driven, electric field-driven, and mechanical wave-driven methods. The MD simulation models for these approaches are illustrated as follows.

7.1.1.1 Pressure-driven method

The most widely applied pressure-driven approach is the gravity-fed method in which a uniform external pseudo-gravitational field is imposed on the fluidic system to drive its particles or molecules. Fig. 7.1 shows the schematic of an MD simulation model for the pressure-driven methane transport inside a single-wall CNT using the gravity-fed method. A periodic boundary condition (PBC) is applied along the axis of the nanotube. In such a simulation, the pseudo-gravitational field is realized by applying a constant force to each fluid particle. When the simulation system reaches a steady state, the external force acting on each fluid particle is balanced by viscous forces arising from interactions between the walls and the fluid particles. To prevent the temperature of the system from rising because of viscous heating, a thermostat is coupled with the system to control its temperature. The gravity-fed method is usually adopted to model incompressible fluids, and its advantages mainly lie in its simplicity and low computational cost.

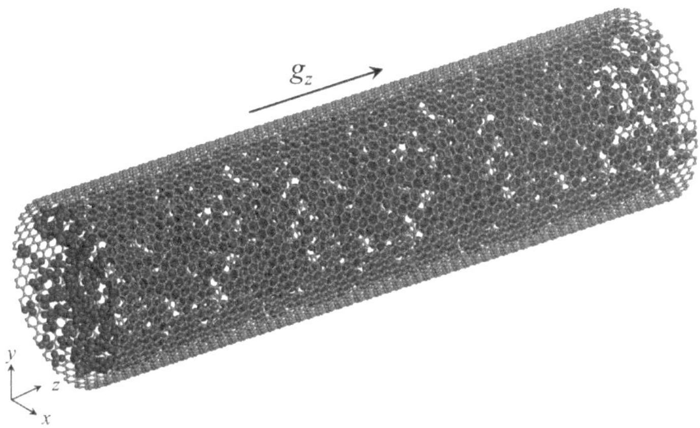

FIGURE 7.1 Schematics of the MD simulation model for pressure-driven methane transport inside a single-wall CNT. An external gravitational field is applied along the axis of the tube to drive the fluid (Sokhan et al., 2002).

However, this method possesses certain limitations when modeling compressible fluids, which are caused by applying a PBC (in the fluid flow direction) that renders the fluid parameters (e.g., density, temperature, and pressure) at the inlet and outlet of the fluid channel identical. For incompressible fluids, variations in their properties along the flow direction are negligible. Hence, the pressure distribution of such a fluid can be regarded as uniform in the flow direction, and the fluid can be modeled by the gravity-fed method. On the contrary, for compressible fluids, including low-density gases, a tiny irregularity in their pressure distribution may lead to large nonuniformity in their properties. For a nanochannel with a length of over tens of nanometers, accumulations of such deviations across the entire channel can amount to significant discrepancies between the fluid properties at both ends of the channel, thereby compromising the validity of the gravity-fed method.

Several other methods have been developed to overcome the limitations of the gravity-fed method in modeling the transport of fluids with nonuniform properties along the flow direction. One of these methods is the dual control volume grand canonical MD (DCV−GCMD) method established by Heffelfinger and van Swol (1994). In the DCV−GCMD method, Monte Carlo (MC) simulation is combined with MD simulation method to generate a pressure gradient by inserting or deleting particles from two controlled volumes located at the inlet and outlet of the nanofluidic system. However, the DCV−GCMD method incurs a substantial computational cost to model fluidic systems with high densities due to the extremely low acceptance rates in the MC simulation for the insertion and deletion of the molecules. Hence, the DCV−GCMD method is usually employed to simulate gaseous fluids. Meanwhile, it is difficult to choose appropriate starting velocities of the inserted particles, and the ratio of the MC simulation steps to the MD simulation steps is a crucial parameter and should be selected with caution.

Takaba et al. (2007) have proposed a pressure control method in which two fluctuating walls are introduced, as shown in Fig. 7.2. The simulation cell is a rectangular box, and PBCs are applied in the

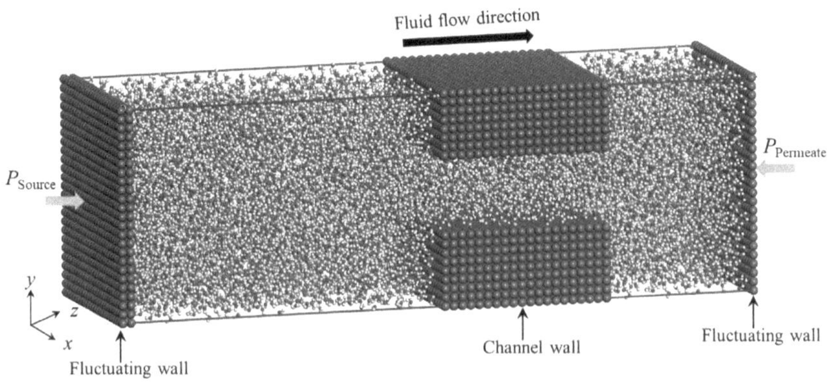

FIGURE 7.2 Schematic for modeling fluid transport through a channel with nanoconfinement driven by an external pressure induced by two fluctuating walls (Takaba et al., 2007).

directions perpendicular to the fluid flow direction. A constant pressure condition is realized by placing the two fluctuating walls at both sides of the simulation cell perpendicular to the direction of fluid flow. Each of these walls has a fictitious mass and is allowed to interact with the fluid particles. Constant pressure is applied to the walls to counteract the forces imposed on them by the fluid particles and further drive the fluid to penetrate the pore region, making the walls behave similarly to frictionless pistons.

The component of the force exerted on a wall by the fluid particles along the flow direction, f_w, can be expressed as

$$f_w = -\sum_i^N f_{iw}(|z_i - z_w|) = \sum_i^N \left[\frac{d}{dz_{iw}} V_w(|z_i - z_w|)\right], \tag{7.1}$$

where f_{iw} is the force imposed on the ith fluid particle due to its interaction with the wall, Z_i and Z_w are the z-coordinates of the ith fluid particle and the fluctuating wall, respectively, V_w is the potential energy due to the interaction between the ith particle and the wall at a separation distance of z_{iw}, which equals $z_i - z_w$. The summation spans over all the N fluid particles. The equation of motion of the fluctuating walls along the flow direction is given by

$$m_w \ddot{z}_w = f_w + P_{ext} A, \tag{7.2}$$

where m_w is the fictitious mass of the wall, P_{ext} is the external pressure, and A is the surface area of the wall. The pressure exerted on the fluctuating wall P_f by the fluid particles is $P_f = f_w / A$.

This pressure control method has been widely adopted to investigate fluid transport through nanochannels or nanopores in which the PBC along the flow direction is not required, for example, water transport through a porous membrane during the desalination of seawater. The main disadvantage of this method is that most of the fluid particles are outside the pores or channels, and the transport behavior of the system is characterized by only a small portion of the fluid particles. In addition, the simulation time is determined by the number of fluid particles in the source region. When all the particles in the source region have passed through the pores, the simulation is forced to end.

Liang and Tsai (2012) have proposed another method to generate a pressure gradient to simulate pressure-driven fluids in nanochannels. This method has proven to be applicable for investigating both compressible and incompressible nanofluids. In addition, it can be used to simulate fluid flow in nanochannels where sudden geometric changes occur. Fig. 7.3 presents the schematic of the simulation model of the method.

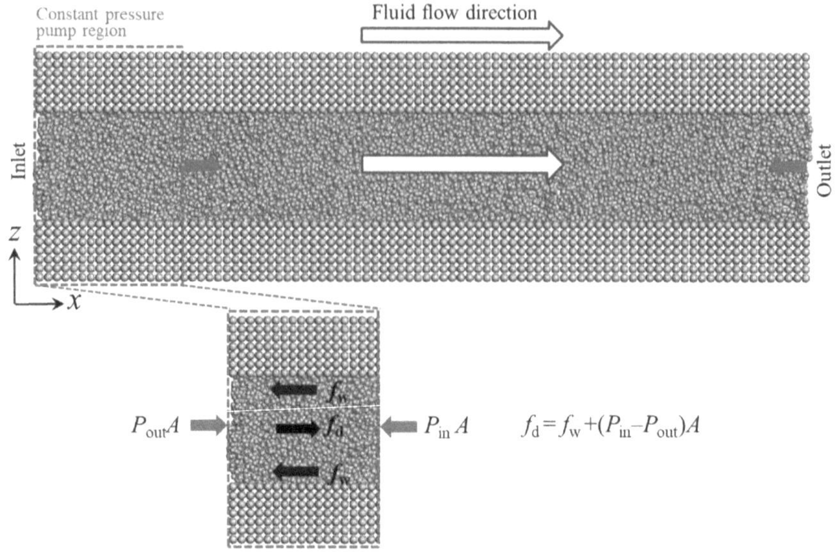

FIGURE 7.3 Schematic of the simulation model for investigating pressure-driven fluids in nanochannels with PBCs applied (Liang and Tsai, 2012).

The simulation system is divided into two regions. The region on the left with a relatively small size is assigned to be the "pump" region where a driving force f_d is applied to all the atoms within it to generate a constant pressure difference between the "pump" region and the other region. The latter region is denoted as the pressure-driven region where the fluid flows under the influence of the aforementioned pressure difference. PBCs are applied for all directions except the z-direction. In the "pump" region, the pressure increases in the flow direction due to the accumulative effects of the driving force applied to the fluid atoms. In the pressure-driven region, the pressure gradually decreases along the flow direction because of the friction imposed by the solid walls on the fluid. For simplicity, this method is called the constant-pressure-pump method hereafter in this chapter.

To determine an appropriate driving force f_d for generating the desired pressure difference, a control volume analysis is conducted on the volume of the fluid in the "pump region." The forces acting on the fluid particles in the "pump" region include the driving force f_d, the force f_w exerted by the channel walls on the fluid particles, and the force f_p that arises from the pressure difference at the boundary of the "pump" region. Here, the force f_p can be expressed in terms of the inlet and outlet pressure P_{in} and P_{out} of the pressure-driven flow as $f_P = (P_{in} - P_{out})A$ with A denoting the cross-sectional area of the nanochannel. Hence, at a steady state, the linear momentum equation can be established to describe the pump region as

$$f_{\mathrm{d}} - f_{\mathrm{w}} - f_{\mathrm{P}} = \left[\int \rho v_x^2 \mathrm{d}A\right]_{\mathrm{out}} - \left[\int \rho v_x^2 \mathrm{d}A\right]_{\mathrm{in}}, \tag{7.3}$$

where the subscripts "in" and "out" represent the inlet and outlet of the "pump," respectively. Due to the PBC applied in the flow direction, the outlet and inlet of the "pump" region are equivalent to the inlet and outlet of the pressure-driven region, respectively. In Eq. (7.3), ρ is the local density of the fluid at the inlet or outlet of the "pump" region, v_x is the corresponding x-component of the local average stream velocity (i.e., velocity of the center of mass of this local region), and A is the cross-sectional area of the nanochannel. If the fluid properties are homogeneous across the flow direction, the two terms on the right-hand side of Eq. (7.3) will cancel each other out. In this case, the "pumping" force f_{d} is $f_{\mathrm{d}} = f_{\mathrm{w}} + f_{\mathrm{P}}$.

At a steady state, the force $f_{\mathrm{P}} = (P_{\mathrm{in}} - P_{\mathrm{out}})A$ that arises because of the pressure difference at the boundary of the "pump" region is determined by $(P_{\mathrm{in}} - P_{\mathrm{out}})$, which equals the preset desired pressure difference. In the simulation, only the force f_{w} exerted by the constraint wall on the fluid particles is calculated at each time step by

$$f_{\mathrm{w}} = \sum_i^{N_{\mathrm{f}}} \sum_j^{N_{\mathrm{w}}} f_{ij}, \tag{7.4}$$

where N_{f} and N_{w} are the numbers of atoms in the "pump" region and the constraint wall, respectively, and f_{ij} is the force imposed by the jth atom in the wall on the ith atom in the "pump" region. Based on the calculation of f_{w}, the "pump" force f_{d} can be updated accordingly.

If the fluid properties are not homogeneous across the nanochannel, the local density ρ and local stream velocity component v_x will not be constant along the flow direction. Consequently, the two terms on the right-hand side of Eq. (7.3) will not cancel each other out and should be considered in the calculation of the driving force f_{d}. However, this is usually not an issue as these two terms are only significant under the conditions that the velocity component v_x surpasses $\sim 100\ \mathrm{m\ s^{-1}}$, and the maximum variation in the fluid density along the flow direction exceeds $\sim 100\ \mathrm{kg\ m^{-3}}$, which are usually not met if the "pump" region is appropriately selected. Hence, the two terms on the right-hand side of Eq. (7.3) are rendered negligible and usually ignored.

Fig. 7.4 shows the MD simulation results for the velocity, pressure, and density distributions of the flow of liquid Ar inside a nanochannel formed by two thin Ag plates based on the constant-pressure-pump method as discussed above and the gravity-fed method. Ar in the liquid state is considered incompressible as a pressure variation of 5 MPa only leads to a change of less than 1% in its density. The channel width (i.e., the channel opening along the z-direction in Fig. 7.3) is set at 4 nm, and the Ar–Ag and Ar–Ar interactions are described by the Lennard–Jones (LJ) potential with the interaction strength (i.e., the potential energy well in the LJ potential) of ε_{Ag-Ag} and ε_{Ag-Ar}, respectively.

It is evident from Fig. 7.4A that the velocity profiles obtained from both methods are consistent, and stronger solid–fluid interactions lead to a higher fluid viscosity and thus a lower flow velocity. In addition, at $\varepsilon_{Ag-Ar} = 204$ K, the strong solid–fluid interactions result in a parabolic Poiseuille-type velocity profile with almost zero velocity near the Ag surface, which fulfills the no-slip boundary condition.

Fig. 7.4B shows that for the constant-pressure-pump method, the pressure of the fluid increases in the "pump" region and decreases linearly in the pressure-driven region, while the overall pressure variance of the fluid in the pressure-driven region is ~5 MPa. Such variations in pressure lead to a slight decrease in the fluid density with a maximum variance of 0.5% at the outlet of the pressure-driven region. For the results obtained from the gravity-fed method, both the pressure and density of the fluid remain constant in the pressure-driven region. The variations of the pressure and density for the constant-pressure-pump method can be attributed to two factors. First, the driving force applied to the fluid particles in the "pump" region pushes them toward the pressure-driven region and thus increases the density and pressure of the fluid at its inlet. Second, this force also pulls the fluid particles away from the pressure-driven region, which leads to a reduction in the density and pressure of the fluid at its outlet.

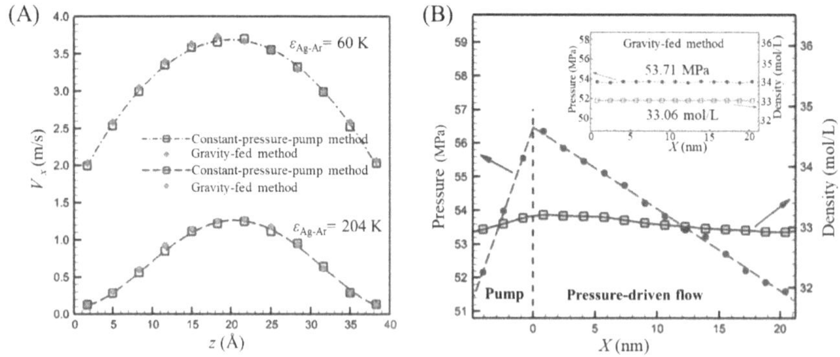

FIGURE 7.4 Comparison of simulation results obtained from the constant-pressure-pump method with those from the gravity-fed method for liquid Ar flowing inside the nanochannel formed by two thin planar Ag plates: (A) velocity distribution along the channel width direction at an interaction strength of $\varepsilon_{Ag-Ar} = 60$ K and $\varepsilon_{Ag-Ar} = 204$ K (corresponding to $\varepsilon_{Ag-Ar} = 0.119$ kcal mol^{-1} and $\varepsilon_{Ag-Ar} = 0.405$ kcal mol^{-1}) and (B) pressure and density distributions along the flow direction. The results obtained from the gravity-fed method are shown in the inset of panel (B). Source: *Modified and reprinted with permission from a work by Liang and Tsai (2012).*

For compressible gaseous Ar inside a Ag nanochannel, the simulation results obtained from the constant-pressure-pump method and the gravity-fed method diverge significantly. Fig. 7.5 presents the steady-state distributions of pressure and density for the gaseous Ar in its flow direction. It is observed that the pressure of the fluid also exhibits a linear drop in the pressure-driven flow region, which is similar to the case of liquid Ar. The density of the gaseous Ar exhibits a much more significant drop along the flow direction, but this drop is nonlinear with an overall value corresponding to 19% of the original. Such variations are absent in the results obtained from the gravity-fed method, as shown in the inset.

According to the principle of conservation of mass, a decrease in the density of a fluid with respect to its flow direction leads to an increase in its velocity. For both models, a Poiseuille-type velocity profile can be observed for the gaseous Ar. However, the simulation results indicated that the velocity profile obtained from the constant-pressure-pump method gradually rises in the flow direction, while that obtained from the gravity-fed method remains unchanged. From these comparisons, it can

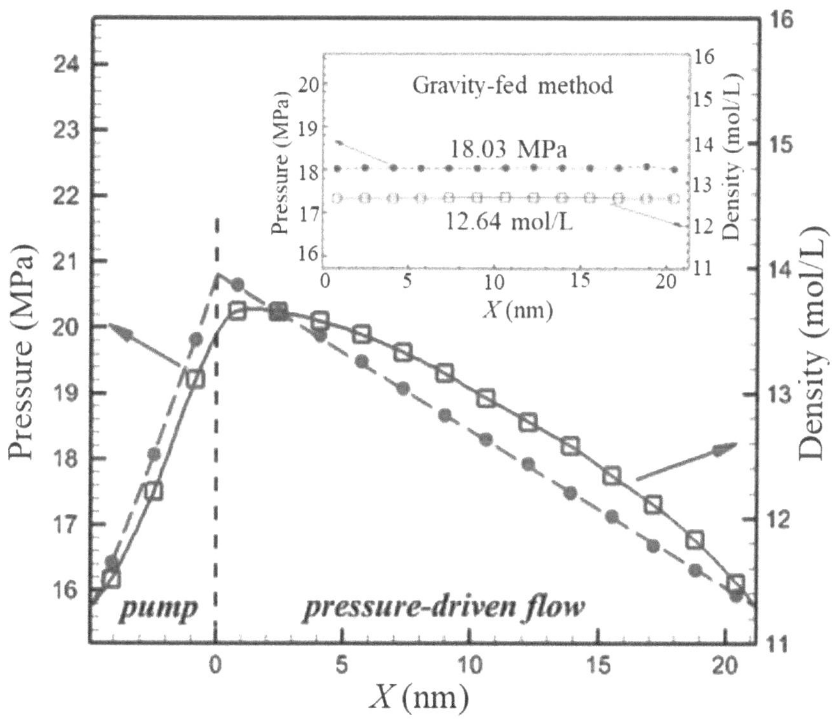

FIGURE 7.5 Comparison of simulation results obtained from the constant-pressure-pump method with those from the gravity-fed method (shown in the inset) for gaseous Ar flowing inside the nanochannel formed by two thin planar Ag plates: pressure and density distributions along the flow direction. Source: *Modified and reprinted with permission from a work by Liang and Tsai (2012).*

be verified that when a flowing fluid undergoes a considerable change in its density along its flow direction, it is inappropriate to replace an actual pressure gradient with a gravity-like field, and the constant-pressure pump method would be more applicable.

7.1.1.2 Temperature-gradient driven method

It has been recently demonstrated that a temperature gradient coupled with specific surface effects can also be utilized to drive fluids in nanoconfinement (Liu and Li, 2010; Liu et al., 2012). Fig. 7.6 presents a typical MD simulation system for a temperature gradient-driven fluid. In this system, liquid Ar flows across a nanochannel formed by two parallel solid walls. A fluid reservoir is connected to both ends of the nanochannel. To eliminate any lateral movements of the solid walls, the atoms in the outermost layers of the walls are fixed in place. The other solid-wall atoms are allowed to vibrate and interact with the fluid atoms. The atomic interactions within the system are described by the LJ potential; PBCs are applied in all directions to ensure that the two fluid reservoirs exhibit the same pressure, and infinitely wide walls are modeled. A temperature gradient is generated across each of the two walls along the direction of the fluid flow by setting the two ends of each wall at different constant temperatures. The temperature T_H of the right end of the channel is higher than that (T_L) of its left end.

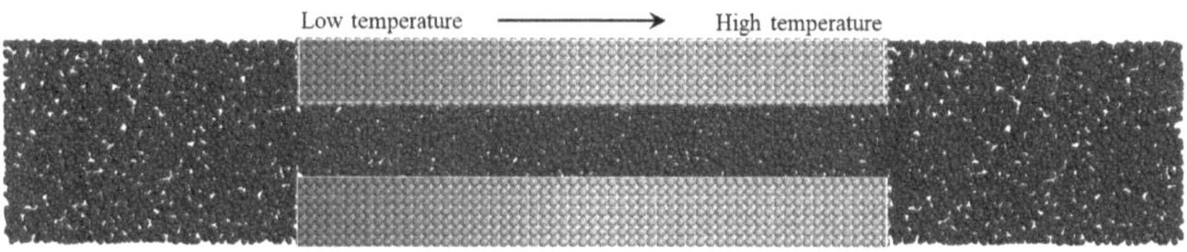

FIGURE 7.6 Schematic of the MD simulation system for a fluid in a nanochannel driven by a temperature gradient (Liu et al., 2012).

Note that the fluid flow direction under the temperature gradient is determined not only by the gradient direction but also by the strength of the interactions between the solid walls and the fluid particles. Fig. 7.7 presents the flux Q of the liquid Ar as a function of the fluid-wall interaction strength, which is represented by the potential energy well (or binding energy) ε_{fw} of the LJ potential. The width and length of the nanochannel are set at 2 nm and 8 nm, respectively. It is observed that the value of Q changes from negative to positive as ε_{fw} increases. With a weak interaction strength, the fluid flows from the high-temperature region to the low-temperature region, and this direction is reversed as the interaction strength increases. In addition, the volumetric flux Q reaches its maximum

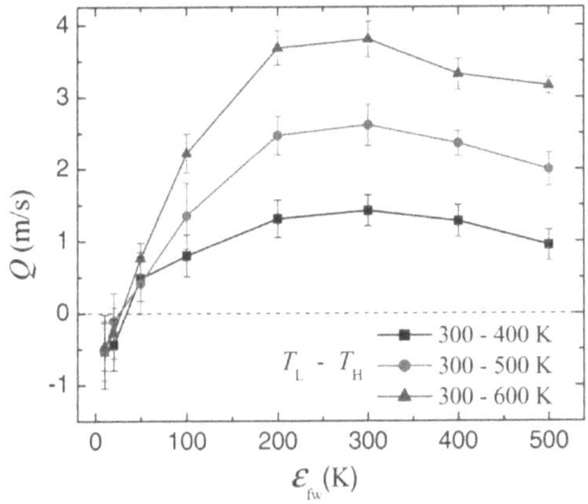

FIGURE 7.7 Dependence of the volumetric flux Q of the fluid on the fluid-wall interaction strength represented by ε_{fw} under different temperature gradients. Source: *Reprinted with permission from a work by Liu et al. (2012).*

value at an interaction strength of $\varepsilon_{fw} = 300$ K (corresponding to $\varepsilon_{fw} = 0.596$ kcal mol^{-1}) and decreases slightly when ε_{fw} further increases.

The flow direction depends on the fluid-wall interaction because the wall effects are significant if the fluid channels are of nanoscale sizes. In these nanochannels, the density distribution of the fluid becomes non-uniform, which leads to pressure variations. For the fluid confined in the nanochannel shown in Fig. 7.6, the internal pressure P of the fluid can be calculated via the virial equation of state as

$$P = \rho_N k_B T + \frac{\sum_i \sum_{j>i} r_{ij} f_{ij}}{3V}, \tag{7.5}$$

where ρ_N is the number density of atoms, k_B is the Boltzmann constant (1.380649×10^{-23} J K^{-1}), f_{ij} is the force of a pair of atoms, and V is the volume. The index i spans over all the atoms in the fluid, while the index j spans over all the atoms (including the wall atoms) interacting with atom i. The condition $j > i$ prevents double counting in the summation. The first term in the equation corresponds to the kinetic component of the pressure, and the second term represents the potential component. For a given molecular interaction, the kinetic component is determined by the number density ρ_N and temperature T, whereas the potential term only depends on the number density.

Fig. 7.8 shows the two-dimensional (2D) contour plots of the fluid density and temperature of the liquid Ar in the nanochannel, with a liquid–wall interaction strength $\varepsilon_{fw} = 300$ K. The density of the fluid fluctuates significantly along the channel width direction, and the fluid molecules form apparent layered structures near the walls because of

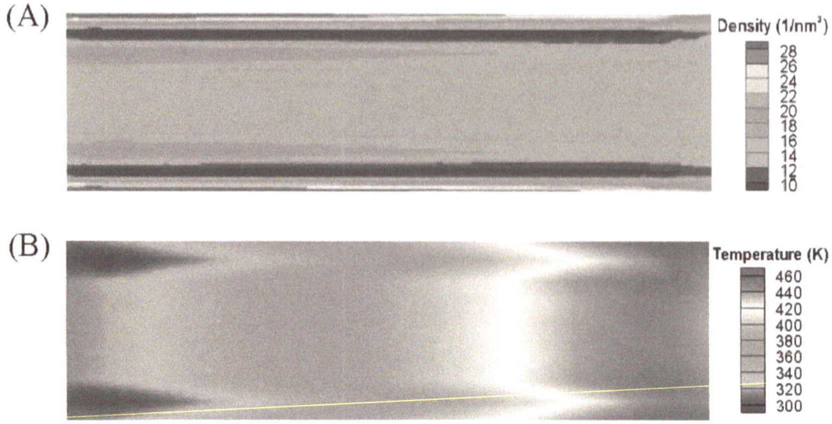

FIGURE 7.8 2D contour plots of the (A) density and (B) temperature of liquid Ar in the nanochannel with strong liquid–wall interactions. Source: *Modified and reprinted with permission from a work by Liu et al. (2012).*

the strong liquid–wall interactions. In addition, the neighboring molecular layers that exhibit high densities are separated by regions with low densities where the potential energy is sufficiently high for the Ar atoms to stay. Furthermore, the temperature gradient of the channel wall creates density variation in the flow direction, and the highest density is located near the wall at the low-temperature end. This phenomenon occurs because a low temperature favors the adsorption of fluid molecules due to their low kinetic energy, which is insufficient to overcome the attraction effects of the channel wall. The nonuniform density distribution leads to nonuniform intermolecular interactions and pressure variation, which becomes the driving force for fluid flow.

The difference between the adsorption processes at the low-temperature and high-temperature ends becomes small for weak liquid–wall interactions. Thus, the density gradient along the fluid flow direction also becomes small and is mainly caused by the thermal expansion of the fluid. The thermal expansion of the incompressible liquid Ar is small, and thus its kinetic component in Eq. (7.5) becomes the dominant factor for the pressure. Hence, at a higher temperature, the fluid molecules at the high-temperature end are subjected to a higher pressure than those at the low-temperature end, and therefore the fluid flows from the high-temperature end to the low-temperature end. Theoretically, fluid adsorption on solid surfaces is generally determined by the ratio of the fluid-wall binding energy to the kinetic energy, i.e., $\varepsilon_{fw}/(k_B T)$. Physically, ε_{fw} represents the attraction effect of the wall, whereas $k_B T$ represents the kinetic energy of the fluid molecules and reflects the possibility of them escaping from the wall surface by overcoming the attractive effect. Consequently, if the ratio meets the condition $\varepsilon_{fw}/k_BT > 1$, fluid adsorption tends to occur. Based on this relation, the fluid flow direction can be tuned by controlling the temperature gradient or the liquid–solid interaction strength.

7.1.1.3 Electric-field driven method

An electric field provides a contactless method to drive particles with net charges or dipoles. For molecules or ions with net charges, the electric field functions similarly to the gravitational field shown in Fig. 7.1. However, in most cases, each fluid molecule possesses only a net dipole and no net charge. For such cases, a static electric field is unable to drive the molecules translationally. Hence, a gradient electric field (Wang et al., 2011), an electric field generated by an alternating current (AC) (Rinne et al., 2012), or a rotating electric field (De Luca et al., 2014) is applied to generate a resultant force on molecules with zero net charges.

Consider the water flow in a single-wall CNT as an example. A linear gradient electric field can be applied along the axial direction of the CNT, with the electric field strength increasing linearly along this direction. Since water molecules are polar, they will orientate themselves toward the direction of the electric field. Fig. 7.9A shows the probability distribution of the time-averaged dipole orientation angle $\langle\theta\rangle$ of the water molecules inside a CNT with a diameter of ~8 Å (which only allows a single file of water molecules to be transmitted through) during an MD simulation lasting for 150 ns. When no electric field is applied, two peaks

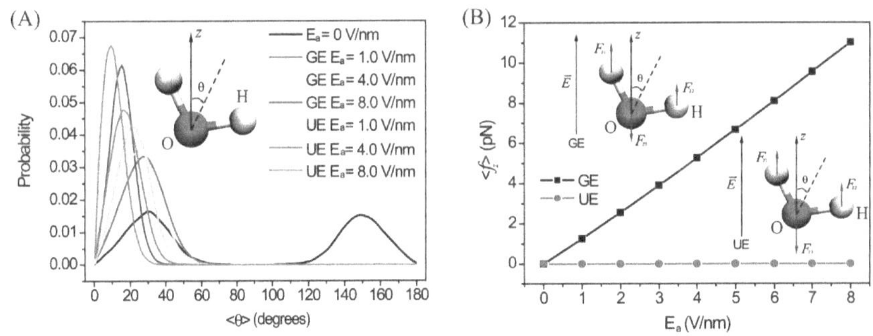

FIGURE 7.9 (A) Probability distributions of the dipole orientation of water molecules inside a single-wall CNT with a diameter of ~8 Å under different electric field conditions. $\langle\theta\rangle$ denotes the time-averaged angle between the dipole vector and the direction of the electric field; E_a denotes the intensity of the electric field; GE denotes the gradient electric field; UE denotes the uniform electric field. (B) Magnitude of the net force $\langle f_z\rangle$ acting on a water molecule within the CNT as a function of E_a. $\langle f_z\rangle$ is averaged over all the water molecules inside the CNT. Source: *Modified and reprinted with permission from a work by Wang et al. (2011).*

symmetric about $\langle\theta\rangle = 90^{\circ}$ can be observed. This observation indicates that in the absence of an external electric field, the water molecules can exhibit the "+ dipole" $(0 < \langle\theta\rangle < 90^{\circ})$ and "− dipole" $(90^{\circ} < \langle\theta\rangle < 180^{\circ})$ states with equal probability. When a linear gradient or a uniform electric field is applied, the probability distribution of the dipole orientation appears to concentrate at a value of $\langle\theta\rangle$ smaller than 40°. In other words, only the "+ dipole" state remains, and the "− dipole" state ceases to exist because each water molecule exhibits a propensity to reorientate its dipole parallel to the electric field direction to achieve a state of minimum electric energy, which is given by $-p \cdot E$. Here, P denotes the dipole moment of a water molecule, and E is the electric field strength vector.

In addition to the external electric field, the water molecules inside the CNT are affected by their neighbors and thermal fluctuations. Hence, the probability distribution peak will always correspond to a value of $\langle\theta\rangle$ that is greater than 0°. In addition, the probability distribution peak shifts toward smaller values of $\langle\theta\rangle$ as the electric field intensity increases.

Fig. 7.9B shows the magnitude of the net electrostatic force $\langle f_z\rangle$ averaged over all the water molecules inside the CNT as a function of the electric field intensity E_a for both the linear gradient and uniform electric fields. As the water molecule is dipolar, if it is placed in a linear gradient electric field, its H atoms that are positively charged (denoted by the white spheres) will always correspond to a position of higher electric field density than its negatively charged O atom (denoted by the red sphere).

Therefore, the net electrostatic force acting on the water molecule is nonzero, and it drives the water molecule to move in the direction of the electric field. This behavior is in contrast to that of a water molecule in a uniform electric field, where the electrostatic forces acting on its H atoms are balanced by the electrostatic force acting on its O atom. Thus, the net electrostatic force acting on it is zero, and no net fluid flow is induced.

In practical applications, the generation of a gradient electric field is usually challenging. Alternatively, a rotating electric field may be more feasible. It has been established that continuous fluid flow can be generated when the fluid is confined between planar surfaces with different levels of hydrophilicity and driven by an external rotating electric field (De Luca et al., 2014). The intrinsic dipole moment of the water molecules causes them to continuously reorientate themselves according to the rotating electric field. Consequently, the torque exerted on the water molecules by the rotating electric field is continuously supplied to the fluid, thereby enabling a steady conversion of spin angular momentum to linear momentum.

Water flow in a nanochannel formed by two planar plates has been explored as an example of the rotating electric-field driven method. The atomic configuration of the MD simulation model is shown in Fig. 7.10. The upper and lower plates are composed of hydrophobic and hydrophilic materials, respectively. To induce fluid flow in the channel, an electric field rotating within the $x-y$ plane is applied. The rotation of the electric field is described by its field vector $\boldsymbol{E} = (E\cos(2\pi v t), E\sin(2\pi v t), 0)$ with E, v, and t denoting the electric field strength amplitude, rotational frequency, and time, respectively.

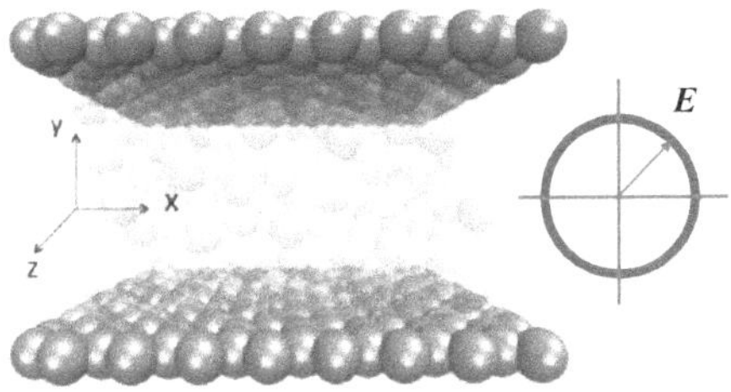

FIGURE 7.10 Atomic schematic of an MD simulation model of water flow inside a nanochannel driven by a rotating electric field. Source: *Modified and reprinted with permission from a work by De Luca et al. (2014).*

The torque exerted on a single water molecule is given by

$$\boldsymbol{\Gamma} = \sum_i (\boldsymbol{r}_i - \boldsymbol{r}_{\mathrm{cm}}) \times q_i \boldsymbol{E}_i, \tag{7.6}$$

where $\boldsymbol{r}_i$ and q_i denote the position and partial charge of the ith atom ($i = 1, 2,$ or 3) that constitutes the water molecule, respectively; $\boldsymbol{r}_{\mathrm{cm}}$ is the center of mass of the water molecule; $\boldsymbol{E}_i$ is the electric field vector at the position $\boldsymbol{r}_i$. The torque reorientates the intrinsic dipole to make it parallel to the electric field, which causes the water molecule to possess a nonzero average angular velocity in the z-direction. However, this alignment process is impeded by several factors, including intermolecular interactions among the water molecules, system thermal fluctuations that revert the dipoles to randomized orientations, and the channel wall surface that hinders the dipole from aligning itself with the electric field. The flow velocity $v_x(y)$, of which x denotes the flow direction and y denotes the channel width direction (Fig. 7.6), can be obtained by dividing the momentum of the flux density $J_x(y,t)$ by the mass density $\rho(y,t)$. Here, $J_x(y,t)$ and $\rho(y,t)$ are expressed as

$$J_x(y,t) = \frac{1}{L_x L_z} \sum_i m_i v_{x,i} \delta(y - y_i), \tag{7.7}$$

$$\rho(y,t) = \frac{1}{L_x L_z} \sum_i m_i \delta(y - y_i), \tag{7.8}$$

where L_x and L_z are the channel lengths along the x- and z-directions, respectively; i is the index of the water molecule; m_i is the mass of the water molecule; $v_{x,i}$ is the x-component of the velocity; y_i is the y-component of the position of the center of mass of the water molecule; $\delta(y-y_i)$ is the Dirac delta function. At the end of the simulation, the streaming velocity is obtained as

$$v_x(y) = \frac{\langle \sum_i m_i v_{x,i} \delta(y-y_i) \rangle}{\langle \sum_i m_i \delta(y-y_i) \rangle}. \tag{7.9}$$

The angular brackets represent the time average upon the fluid flow reaching a steady state.

7.1.1.4 Surface-wave driven method

In addition to the pressure-driven and electric-field-driven methods, it has been reported that surface waves can be utilized to drive fluids across nanochannels (Insepov et al., 2006; Xie and Cao, 2017). There are several approaches for introducing traveling waves on the surface of a channel wall. For example, thermoacoustic waves can be generated on a single-wall CNT by utilizing short laser pulses. Alternatively, ultrasound waves can be transmitted to the nanotube through a liquid or a dense gaseous medium. Rayleigh surface waves are activated when a longitudinal wave traveling in a liquid/gas impinges on a solid surface at an incidence angle corresponding to the Rayleigh angle θ, with $\theta = \arcsin(v_P/v_S)$. Here, v_P is the velocity of the incident wave, and v_S is the velocity of the surface wave on the solid.

Based on MD simulation, Insepov et al. (2006) have introduced surface waves to a CNT surface by periodically displacing the C atoms along the radial direction of the nanotube at frequencies ranging from 1 THz to over 30 THz. Upon the activation of Rayleigh surface waves, gas molecules residing in the CNT experience frequent reflections from the nanotube wall. For a specific combination of the molecular mass of the gas, wave frequency, and phase velocity of the surface waves, the gas inside the nanotube begins flowing with a high velocity in the direction of the traveling surface waves. The driving force for the fluid flow is attributed to the friction between the gas particles and the nanotube wall.

It has been reported that for a fluid confined in a nanochannel formed by two planar nanoplates, generating a surface acoustic wave (propagating along the length of the channel) in the plates can drive the fluid to flow in the same direction as the wave (Xie and Cao, 2017). In addition, for a pressure-driven fluid inside a nanochannel, the existence of surface acoustic waves significantly changes the density and distribution pattern of the fluid molecules near the channel walls and dramatically reduces the friction between the fluid and the walls, particularly for strong fluid–wall interactions. Consequently, the fluid flow rate can be increased with the introduction of surface acoustic waves. The nanochannel fluid dynamics under the effect of traveling acoustic surface waves largely depend on the wave amplitude, wavelength, and phase velocity.

7.1.2 Water flow in carbon nanotubes

CNTs have attracted intensive research interest since their discovery. The unique mechanical and physical properties of CNTs render them promising candidates for many applications, including nanoelectronics, biomedical devices, composite structures, energy storage, and water treatment. Single-, double-, and multiwall CNTs of various sizes have been synthesized, and novel fabrication techniques have enabled the arrangement of aligned CNTs with controllable sizes within membranes (Miller et al., 2001; Hata et al., 2004; Hinds et al., 2004; Talapatra et al., 2006; Yamada et al., 2006), which has opened the door to applications involving nanofiltration. Previous experimental (Holt et al., 2006) and simulation works (Hummer et al., 2001; Kalra et al., 2003) have demonstrated that water can pass through narrow CNTs with unexpectedly high flow rates that are independent of the lengths of the CNTs, which is in contrast with predictions based on macroscopic hydrodynamics (Majumder et al., 2005). These results appear counter-intuitive as it is commonly recognized that it is energetically unfavorable for polar water molecules to reside in narrow, nonpolar nanotubes. For single-wall CNTs, there is a critical diameter of ~7 Å at which only water molecules can pass through them, while ions cannot. This characteristic has established CNTs and CNT-based membranes as promising materials for fabricating semipermeable membranes for nanofiltration applications, such as seawater desalination. Hence, numerous studies have been conducted to investigate water flow through CNTs.

The ultrahigh water flux within CNTs is believed to be due to the higher ordering of the water molecules and their stronger hydrogen bonds than those of bulk water, which causes the water molecules to move or diffuse rapidly along the axial direction of the nanotube in a coordinated manner. Moreover, the hydrophobic nature of the nanotube wall provides an almost frictionless surface for fast water transport.

The strong and highly ordered hydrogen bonding between the water molecules inside a CNT can be illustrated by the MD simulation conducted by Hummer et al. (2001). In the simulation, a short uncapped single-wall CNT with a length of 13.4 Å and a diameter of 8.1 Å is solvated in a water reservoir. Afterward, the dynamics of the water molecules are monitored for a simulation time of 66 ns. Despite the strong hydrophobic nature of the inner wall of the CNT, water molecules from the surrounding reservoir can readily enter the initially empty inner nanotube and occupy it during the entire simulation process. Upon entering the nanotube, the water molecules form a single-file chain structure via hydrogen bonding (Fig. 7.11A), and the average binding energy among them is 2 kcal mol^{-1} lower than that of bulk water (Fig. 7.11B). The distribution of the binding energy of the water

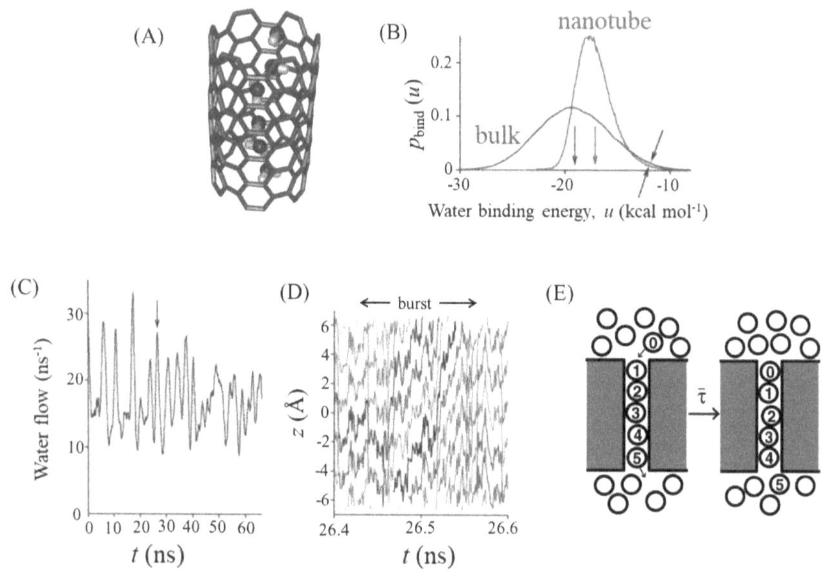

FIGURE 7.11 (A) Structure of the hydrogen-bonded single water chain inside a CNT. (B) Probability distributions of the binding energy of water molecule pairs for bulk water and water inside the CNT with the vertical arrows indicating the average binding energies and the tilted black arrows indicating the cross-over region in which weakly bound states are more populated for bulk water. (C) Number of water molecules transported through the CNT during the simulation with the vertical arrow indicating a typical pulse in the water flow. (D) Motions of individual water molecules illustrated by colored lines inside the CNT along the axial direction during the conduction "burst" at 26.5 ns. (E) Schematics showing the coordinated back-and-forth motion of the entire chain of water molecules along the channel axis. Particles labeled 0 and 5 enter and exit the channel, respectively. Source: *For (A)–(D) Reprinted with permission from a work by Hummer et al. (2001). (E) Reprinted with permission from a work by Berezhkovskii and Hummer (2002).*

molecules inside the nanotube is narrower than that of bulk water. The high-energy states dominating the free energy are less frequently occupied (as indicated by the cross-over region delineated by the tilted arrows). Therefore, although the average binding energy of water molecules at the center of the CNT (within 0.8 Å of the cylinder axis) is lower than that of bulk water, their average free energy is also reduced, indicating stronger hydrogen bonding among them.

In addition, the hydrogen-bonding structure inside the CNT is protected by the nanotube wall from thermal fluctuations in the water reservoir. Inside the nanotube, only 0.02% of the water molecule pairs in contact (with the distance between the two O atoms smaller than 3.5 Å) are unbound during the entire simulation duration of 66 ns. This ratio is insignificant compared with the corresponding value of 15% for bulk water. Here, a water molecule pair is considered to be unbound if its pair energy is positive. Furthermore, the hydrogen bonds inside the CNT are highly oriented with more than 85% of their angles being smaller than 30°, which is much higher than a value of 63% for bulk water. Here, the angle of a hydrogen bond refers to the angle of an OH bond that forms the hydrogen bond with respect to the reference line connecting the two O atoms of the donor and acceptor water molecules that form the same hydrogen bond.

Such strong and highly ordered hydrogen bonds lead to a distinctive conduction behavior of the water molecules inside the CNT compared with that of bulk water. During the simulation duration of 66 ns, around 1100 water molecules enter the CNT from one end and exit from the other, which corresponds to an average of ~17 water molecules passing through the CNT per nanosecond. In addition, water conduction occurs in a burst-like manner through the nanotube with the highest flow rate exceeding 30 water molecules per nanosecond (Fig. 7.11C). The burst-like water conduction is mainly caused by the strong hydrogen bonds between the water molecules inside the nanotube. Breaking the single-file chain structure of water molecules is energetically unfavorable and thus seldom occurs. Thermal fluctuations outside the nanotube result in a highly coordinated, rapid, and drift-like motion of the water molecule chain along the nanotube axis, thereby leading to short bursts in the water flow (Fig. 7.11D). During the bursts, the water molecule chain moves along the nanotube wall with negligible resistance due to its hydrophobic nature.

The bursts in the water flow are attributed to the collective transport of a single chain of water molecules inside the CNT (Berezhkovskii and Hummer, 2002). During the collective transport, the motion of a water molecule chain requires all the molecules constituting it to move concomitantly. Consequently, the water transport through the CNT occurs in the form of coordinated back-and-forth motion of the entire water molecule chain along the nanotube axial direction, as shown in Fig. 7.11E. Motions of sufficiently large amplitudes can lead to the entry and exit of water molecules at the ends of the CNT.

For a CNT with a small diameter and a length much greater than 13.4 Å, as shown in the above-mentioned simulation, the coordinated motion of water molecules can be observed inside the nanotube. Previous MD simulation results indicate that for an infinitely long narrow CNT, water transport through it occurs in the form of coordinated motion, regardless of the degree of occupation of the nanotube (Striolo, 2006), as shown in Fig. 7.12A. Such coordinated motion results from the long lifetime of hydrogen bonds among the water molecules and the weak interactions between the water and nanotube wall.

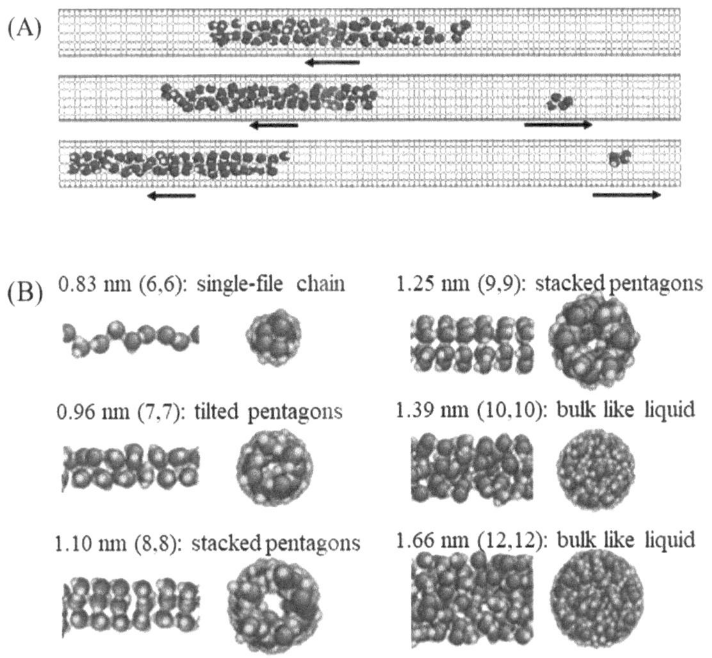

FIGURE 7.12 (A) Sequence of simulation snapshots obtained for 68 water molecules confined within a single-wall CNT with a diameter of 1.10 nm at 298K. The sequence illustrates the formation of a small cluster of water molecules that escapes from a large hydrogen-bonded cluster, flows through the nanotube, and is adsorbed again by the large cluster. The black arrows indicate the direction of the coordinated motion of the clusters of water molecules. The lengths of the arrows are indicative of the speeds of the clusters. (B) Water structure inside the CNTs with different diameters. Source: *For (A) reprinted with permission from a work by Striolo (2006); (B) reprinted with permission from a work by Thomas and McGaughey (2009).*

A small CNT diameter would be preferable to achieve a high seawater desalination rate because it facilitates both high water permeation and high salt rejection. However, the experimental synthesis of CNTs with subnanometer diameters and uniform size distributions is exceptionally challenging. Therefore, water transport in CNTs with relatively large diameters are explored as an alternative. With the increase in the diameter of CNTs, the structure of water molecules within them is significantly altered from an ordered single-file to a bulk-like liquid, as shown in Fig. 7.12B. These structural changes affect the water transport behaviors in a CNT. For a narrow CNT with a radius of ~0.4 nm, the high water flow rate results from the single-file chain structure of the water molecules and their tight hydrogen-bonding network. A high water flow rate can also be observed in a wide CNT with a diameter of several nanometers. However, such a high flow rate does not correspond to a Newtonian fluid governed by the Hagen–Poiseuille (HP) equation but results from special distributions of the density, dipole orientation, and velocity of the water molecules within the CNT, which are caused by the weak interactions between them and the nanotube wall. These factors are discussed in detail below.

Using MD simulation, Joseph and Aluru (2008) thoroughly explored the underlying mechanisms of rapid water transport in CNTs with large diameters. During the simulation, water is considered incompressible, and the gravity-fed method shown in Fig. 7.1 is applied. A constant external force (equivalent to an effective pressure of 13.86 MPa) is applied to each water molecule to drive it into the nanotube and enable its flow along the axial direction.

Fig. 7.13A shows the MD simulation results of the water density and velocity distributions inside a CNT with a diameter of ~2.20 nm. These two quantities are computed and averaged over both the axial length of the nanotube and the entire simulation duration during which a steady-state velocity profile is retained. For comparison, results for simulations involving a CNT with the same geometry but enhanced water–CNT interactions are also presented.

It is observed that the water molecules form a layered structure near the nanotube wall due to nanoconfinement imposed by the CNT, as indicated by the peaks of their density distributions. In addition, there exists a "depletion" region near the nanotube wall where the water density drops drastically to only ~5% of the value in the central region. Moreover, in the depletion region, a sharp velocity "jump" can be observed, as depicted by the orange dashed line. However, the height of the velocity "jump" decreases as the density of the first water layer increases with the enhancement in water–CNT interactions.

The continuum theory for mass flow in a pipe states that the velocity components of the mass along the radial direction of a pipe should be zero. However, for the water flow inside the CNT, it is observed from the radial velocity distribution in Fig. 7.13B that the radial velocities of the water molecules are nonzero, and the greatest value corresponds to the region between the first density peak and the tube walls. These nonzero velocities result from the layered structure of the water molecules near the nanotube wall as a quasi-steady-state density gradient can result in a steady-state flow. By enhancing the interaction strength between the water molecules and the nanotube wall, the height of the density peak increases, which leads to an increase in the radial velocities of the water molecules.

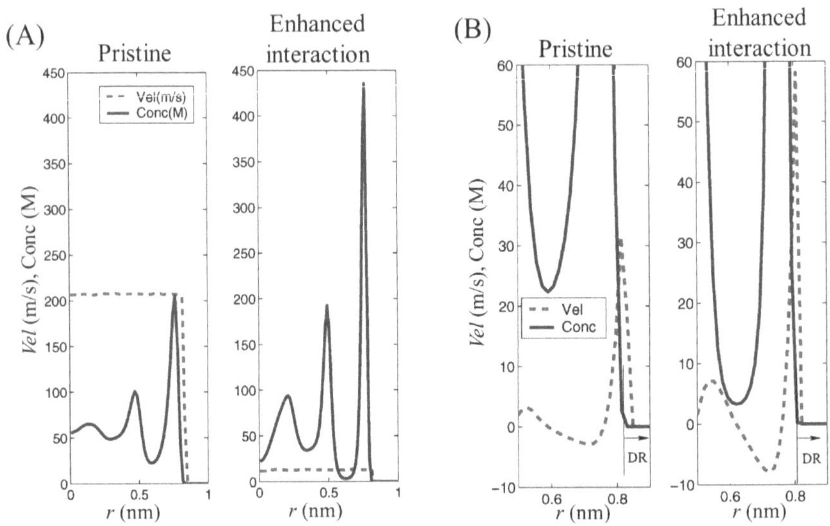

FIGURE 7.13 (A) Concentration and axial velocity profiles of water molecules as functions of the radial distance from the center of the CNT toward its wall. (B) Concentration and radial velocity profiles as functions of the radial distance from the center of the CNT toward its wall. The radial velocity is defined to be positive in the direction from the center toward the wall. Source: *Reprinted with permission from a work by Joseph and Aluru (2008).*

In the depletion region, the water structure exhibits significant differences from that of bulk water. Fig. 7.14A presents orientations of the water dipoles and O–H bonds of the water molecules within and near the depletion region. The dipole orientation is defined as the angle between the water dipole vector and the nanotube axis, and the O–H bond orientation angle is similarly defined. The top panels show the distributions of the average water dipole orientations with respect to the radial direction (0° corresponds to the dipole being perpendicular to the nanotube wall) at different radial locations for both the pristine CNT and the CNT with enhanced interactions. The bottom panels show the orientation distributions of the O–H bonds. Three radial locations are chosen and identified by their distance r from the nanotube axis: (1) the depletion region (with $r = 0.807$ nm for the CNT with enhanced interactions, and $r = 0.821$ nm for the pristine CNT), (2) the location of the first water density peak ($r = 0.764$ nm), and (3) the valley near the first water density peak ($r = 0.636$ nm).

It is observed that in the depletion region, the dipole orientation mainly remains in the range of $55^\circ < \theta < 80^\circ$. The orientation of O–H bonds shows peaks in the ranges of $0^\circ < \theta < 15^\circ$ and $95^\circ < \theta < 120^\circ$. This orientation indicates that water molecules in the depletion region tend to have one of their O–H bonds directed toward the nanotube wall and the other toward the water molecules in the peak region, as illustrated in Fig. 7.14B. Such an orientation of the O–H bonds is similar to the experimental observation in which water molecules near a hydrophobic surface show "free" O–H bonds or "dangling" bonds at the interface. Near the first water density peak, both the water dipoles and O–H bonds are mainly perpendicular to the radial direction so that the water molecules form a planar structure, which can facilitate the formation of hydrogen bonds among them. In the valley near the first peak, the water molecules tend to direct one O–H bond to the nanotube center with the bond orientation angle lying in the range of $160 - 180^\circ$ but the other O–H bond with the bond orientation angle in the range of $60 - 80^\circ$.

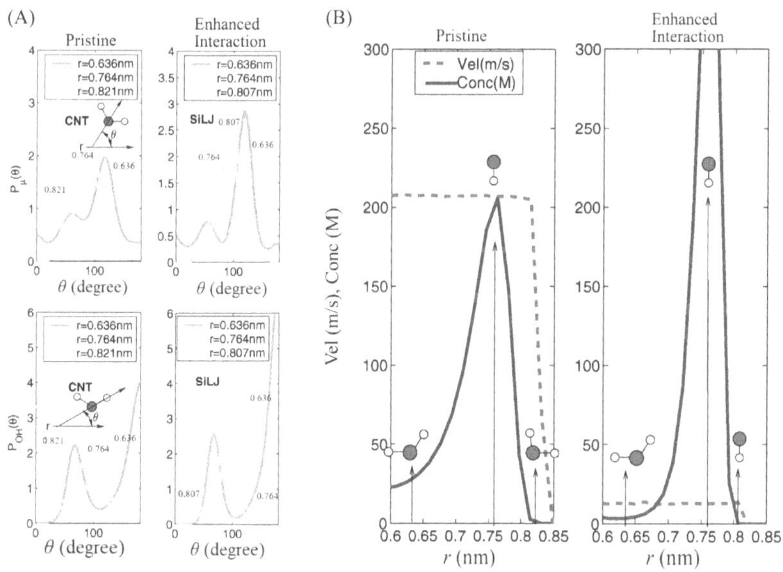

FIGURE 7.14 (A) Dipole orientation and O–H bond orientation distributions at various radial locations. (B) Schematic of water dipole orientations near the nanotube wall at different water concentrations for both types of CNTs. Source: *Modified and reprinted with permission from a work by Joseph and Aluru (2008).*

In the case of the CNT with enhanced water–CNT interactions, most water dipoles exhibit a perpendicular orientation to the wall and form planar layered structures in the depletion region and near the first water layer. There are only a few "free" or "dangling" O–H bonds compared with the case of the pristine CNT. In the valley near the first water peak, the orientations of the O–H bonds and dipoles are not significantly different, except that the distributions of their orientation angles become more concentrated. Due to the different O–H bond orientations in the depletion regions of the pristine CNT and the CNT with enhanced interactions, the degrees of hydrogen bonding in this region also differ significantly for the two types of CNTs.

Due to an abundance of "free" O–H bonds, hydrogen bonds among the water molecules decrease in number in the water-depletion region of the pristine CNT. The combined effect of the hexagonal lattice structure of a CNT, the specific orientations of water molecules with sufficient "free" O–H bonds directed at the nanotube wall, and the reduction of hydrogen bonds in the depletion region leads to ultrafast water flow within CNTs with large diameters.

Besides the diameter, many other factors can significantly affect the water transport inside a CNT, including surface topology, surface roughness, and charges. For instance, using MD simulation, Liu et al. (2008) have investigated the diffusion behavior of confined water in CNTs with a zigzag or armchair helicity and reported that the diffusivity of water in a zigzag CNT is much lower than that in an armchair CNT when both CNTs have a similar diameter, particularly when the diameter is small. To understand the underlying mechanisms of water transport through the CNTs, water diffusion in a (14, 14) armchair CNT and a (24, 0) zigzag CNT with similar pore diameters (1.90 nm and 1.88 nm, respectively) have been analyzed, and potential energy map (PEM) calculations have been performed. Due to the different arrangement of the C atoms, water molecules exhibit distinctive diffusive behaviors inside CNTs with different helicities. The diffusion dynamics of the water molecules highly depend on the distance between them and the nanotube wall. For both types of CNTs, water molecules near the nanotube wall (denoted as the outer water shell) diffuse much slower than those near the central axis (denoted as the inner water shell). This phenomenon is because despite its hydrophobic nature, the nanotube wall imposes a much stronger attraction on the water molecules in the outer water shell than on those in the inner water shell. In addition, the diffusion of water molecules along the axial direction is always faster than that along the radial direction, regardless of the helicity of the CNT and the position of the water molecules. The water diffusivity within the armchair CNT is more rapid than that in the zigzag CNT.

The overall averaged diffusivity of water confined in CNTs can be expressed as a weighted linear combination of the diffusivities of water molecules in the inner and outer water shells based on their respective ratios. In the case of a small nanotube diameter, more water molecules are located in the outer water shell where the diffusivity is very low because of the stronger interactions between the water molecules and the nanotube wall. Thus, the overall diffusivity is low because of the predominance of water molecules in the outer shell. By increasing the diameter of the CNT, there will be a greater proportion of water molecules in the inner water shell, which will in turn dominate the overall diffusivity of the water molecules. Hence, the overall averaged diffusivity increases with the diameter of the CNT. These results indicate that the helicity of a CNT plays a critical role in affecting the dynamic behaviors of water molecules passing through it, particularly for nanotubes with small diameters where water–wall interactions play a dominant role.

To quantitatively investigate the diffusion dynamics of water molecules near the surface of a CNT, the activation energy of water diffusion inside it can be calculated using the activation transition theory, which states that the diffusion coefficient D_c is correlated with the activation energy E_a via the relationship $\ln D_C = -\left[E_a/(R_I T)\right] + C$, where R_I is the ideal gas constant, T denotes the temperature, and C is a constant. By calculating the diffusion coefficients from a series of MD simulations conducted at different temperatures, the logarithmic values of the axial diffusion coefficients can be plotted as a function of the reciprocal of temperature for both the zigzag and armchair CNTs. A least-squares linear fitting of the data points provides the relations $\ln D_C = -1216.884/T + 3.794$ and $\ln D_C = -1939.77/T + 4.26$ for the armchair CNT and zigzag CNT, respectively. Hence, the activation energy of water diffusion in these two types of CNTs can be obtained as $E_a = 2.42$ kcal mol^{-1} for the (14, 14) armchair CNT and $E_a = 3.86$ kcal mol^{-1} for the zigzag (24, 0) CNT. The higher activation energy for water diffusion within the zigzag CNT directly corresponds to a much lower water diffusion rate, particularly for water molecules near the nanotube wall.

A PEM describes the interaction energy of a system with different configurations and is commonly utilized to study its dynamics on the atomic scale. Hence, during the investigation of water diffusion within a CNT, its PEM profile is also a subject of interest. During the PEM calculation, a water molecule is selected as a probe to scan the inner wall surface of the CNT. The cylindrical coordinates (r, θ, z) are used to describe the location of the water probe, where r, θ, and z denote the coordinates corresponding to the radial, tangential, and axial directions, respectively, as shown in Fig. 7.15. For the PEM computation for both CNTs, the scanning scope of θ ranges from 0° to 359° with a resolution of 1°, and the scanning

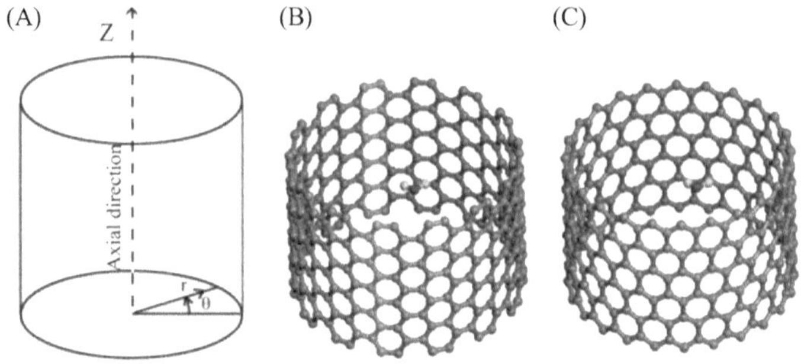

FIGURE 7.15 (A) Cylindrical coordinates used to describe the position of the water molecule probe on the interior surface of a CNT, (B) atomic configuration of a water molecule probe in a (14, 14) armchair CNT, and (C) in a zigzag (24, 0) CNT. Source: *Reprinted with permission from a work by Liu et al. (2008).*

scope of z ranges from 1.0 Å to 5.5 Å with a resolution of 0.1 Å. The radial location r is selected based on the positions where water molecules have the highest local density.

Fig. 7.16A and B shows three-dimensional (3D) plots of the PEMs $E(\theta, z)$ of the armchair and zigzag CNTs. The lowest potential energy is

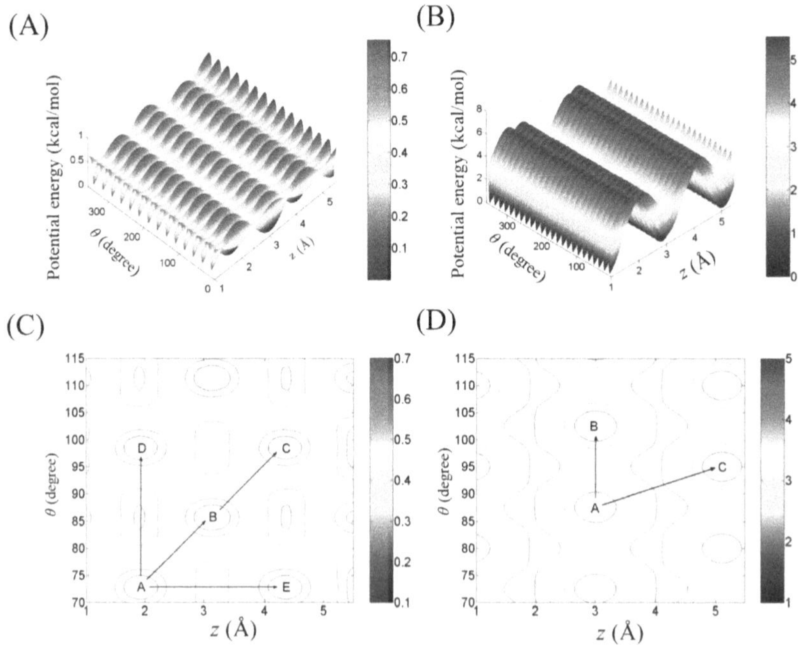

FIGURE 7.16 (A), (B) 3D and (C), (D) 2D contour plots of the PEMs for the CNTs. (A), (C) The armchair (14, 14) CNT. (B), (D) The zigzag (24, 0) CNT. Source: *Reprinted with permission from a work by Liu et al. (2008).*

taken to be zero. The values of $E(\theta, z)$ for the two CNTs periodically change along both the axial (z) and angular (θ) directions. However, the PEM profiles for the two CNTs exhibit significant differences in terms of the number of energy minima along the axial and angular directions. Such differences strongly indicate that the helicity of a CNT is a critical factor that influences water diffusion within it. For a more detailed exploration of the relationship between the PEM profiles of the CNTs and the water diffusion within them, 2D contour plots for the polar angle θ in the range of 70–115° are extracted from the 3D plots, as shown in Fig. 7.16C and D. The letters in the contour plots denote the positions of the local energy minima.

There are three possible pathways for water diffusion on the inner surface of the armchair (14, 14) CNT, as indicated by the arrows AD, AB, and AE. Arrow AD indicates that the water molecules diffuse in the angular (θ) direction, specifically, around the central axis. Arrow AE illustrates that the water molecules diffuse along the nanotube axial (z) direction. Water molecules diffusing along these two directions are associated with an energy penalty of $\sim 0.7\ \mathrm{kcal\ mol^{-1}}$. The diffusion route of the water molecules from site A to site B possesses components along both the z and θ directions. The energy barrier along this route is $\sim 0.5\ \mathrm{kcal\ mol^{-1}}$. Hence, it is more energetically favorable for water molecules to diffuse along the AB pathway. After a water molecule has diffused to site B, it will further diffuse to site C because the potential energy profile of this site is similar to that of site A. Following this pattern, water molecules will diffuse along the cylindrical surface in a spiral pattern. For the zigzag (24, 0) CNT, both AB and AC are potential diffusion pathways. However, the energy barrier for the AB pathway is $\sim 3.3\ \mathrm{kcal\ mol^{-1}}$, which is $1.8\ \mathrm{kcal\ mol^{-1}}$ lower than that for the AC pathway. Consequently, the water molecules in the zigzag (24, 0) CNT will exhibit a propensity to circle around its central axis instead of moving in a spiral manner as in the case of the armchair (14, 14) CNT. Overall, although the diameters of the armchair (14, 14) CNT and the zigzag (24, 0) CNT are almost the same, their different helicities result in their distinctive PEM profiles and significantly alter the water diffusion dynamics within them.

7.1.3 Water flow in porous monolayer graphene

Graphene is a 2D nanomaterial composed of a single layer of C atoms arranged in a honeycomb lattice structure. Recently, graphene has attracted intensive interest for its potential applications in seawater desalination due to its ultrahigh mechanical strength, chemical inertness, thermal stability, ultrathin thickness, and frictionless surface for ultrafast water conduction. Pristine graphene is impermeable to even the smallest gas molecules. Hence, pores with suitable diameters and functionalized edges are constructed within graphene nanosheets to fabricate nanoporous membranes. The removal of salt ions from seawater via such nanoporous membranes is achieved through size exclusion or electrostatic repulsion between the pore rims and the targeted ions.

Using MD simulation, Suk and Aluru (2010) have investigated water transport through a porous graphene monolayer under the effect of external pressure and compared it with that through a CNT. Fig. 7.17A and B presents an atomic illustration of the simulation model, which corresponds to the pressure-driven method shown in Fig. 7.3. Various pore diameters in the range from 0.75 nm to 2.75 nm are considered. A water reservoir that is 6 nm in length is placed on each side of the monolayer. A constant pressure of 100 MPa is applied to the left water reservoir by imposing a constant force on each water molecule in the region shed in yellow on the left side of the simulation box. The simulation model for water transport through a CNT is the same as the model in Fig. 7.17B, except that the graphene monolayer is replaced with a CNT.

Fig. 7.17C and D shows the water flux through the porous graphene monolayer and CNT for both small (0.75 nm) and large (2.75 nm) pore or nanotube diameters. The length of the CNT varies from 2 nm to 10 nm with an incremental step of 2 nm. For the small pore (or nanotube) diameter of ~0.75 nm, the water flux through the porous graphene monolayer is lower than that through the CNT, while for the larger pore (or nanotube) diameter of 2.75 nm, the opposite is true.

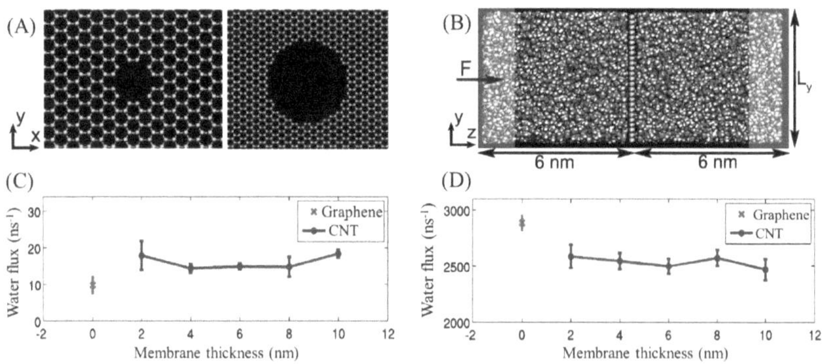

FIGURE 7.17 (A) Graphene monolayer with a nanopore of different diameters: 0.75 nm (left) and 2.75 nm (right). (B) MD simulation model for water transport through a porous graphene monolayer. (C) Water flux through the graphene monolayer containing a nanopore with a diameter of 0.75 nm and water flux through a (10, 0) CNT with a diameter of 0.78 nm. (D) Water flux through the graphene monolayer containing a nanopore with a diameter of 2.75 nm and water flux through a (10, 0) CNT with a diameter of 2.71 nm. Source: *Reprinted with permission from a work by Suk and Aluru (2010).*

For the case of the small nanotube diameter, a single-file chain of water molecules is formed inside the CNT, and its transport proceeds through the collective transport mechanism as previously described in Fig. 7.11. For the porous graphene monolayer, the pore with the diameter of 0.75 nm also only allows for the formation of a single-file water structure composed of a pair of water molecules connected by a hydrogen bond, as shown in Fig. 7.18A. However, due to the different dipole orientations and hydrogen bonding networks of water molecules near the nanopores, the transport mechanism of single-file water molecules through the nanoporous graphene monolayer is distinct from that of single-file water molecules through a CNT.

Fig. 7.18B shows the averaged dipole orientation of a pair of water molecules in a single-file structure across the nanopore as a function of time. It is observed that the averaged dipole orientation experiences a huge number of inversions, which indicates that the hydrogen bonds break frequently. The hydrogen bonding dynamics of two water molecules can be characterized by the autocorrelation function $c(t)$ that describes the possibility of a hydrogen bond lasting for a duration of t. Thus, $c(t)$ measures the relaxation speed of the hydrogen bond and can serve as a representation of its lifetime.

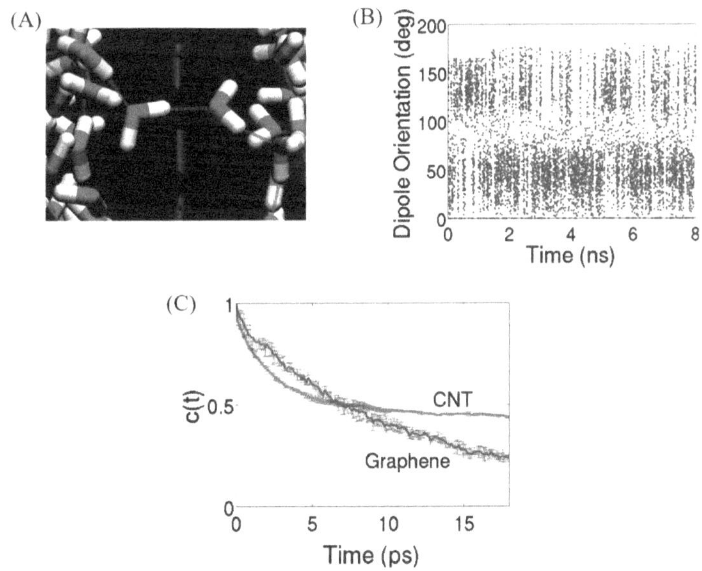

FIGURE 7.18 (A) Snapshot of the single-file water structure across the nanopore in graphene with a small diameter of 0.75 nm. (B) Averaged dipole orientation as a function of time for water molecules in the single-file water structure in the pore of the nanoporous graphene monolayer with a pore diameter of 0.75 nm. (C) Hydrogen bonding autocorrelation function of single-file water molecules in the CNT and nanoporous graphene monolayer. Source: *Modified and reprinted with permission from a work by Suk and Aluru (2010).*

Fig. 7.18C shows the autocorrelation functions *c*(*t*) of the single-file water molecule pairs in the nanoporous graphene monolayer with a small pore and the CNT with a small tube diameter. The autocorrelation function *c*(*t*) of the water molecules in the porous graphene monolayer decays much faster than that of water molecules in the CNT, indicating that the hydrogen bonds of the water molecule pair for the graphene case are relatively weak. Both the frequent inversions of the dipole orientation and the frequent breaking of hydrogen bonds between water molecules forming a single-file structure in the pores of the nanoporous graphene monolayer cause the relatively low water flux through it.

The transport of single-file water molecules through the narrow pores of a porous graphene monolayer and a CNT with a small diameter is highly dependent on the water dipole orientations and hydrogen bonds among the molecules. In contrast, water transport through nanopores with large diameters is primarily determined by the resistance at the entrance of the pores. Fig. 7.19A shows the pressure decline along the direction of water flow for the nanoporous graphene monolayer and a CNT with a large pore diameter (or tube diameter) of ~2.75 nm. From the left, the first vertical dashed line indicates the position of the porous graphene monolayer and the inlet of the CNT. The second and third dashed lines indicate the outlets of the CNTs with lengths of 4 nm and 10 nm, respectively. It is observed that the overall declines in pressure of both the nanoporous graphene monolayer and CNT are nearly identical, and dramatic pressure declines occur near the entrances. Inside the CNT, the pressure decline is almost negligible because of its smooth inner wall surface, while the pressure decline at the outlet is much less significant than that at the inlet. This difference indicates that the entrance resistance is a dominant factor in determining the rate of water flow through CNTs with large diameters.

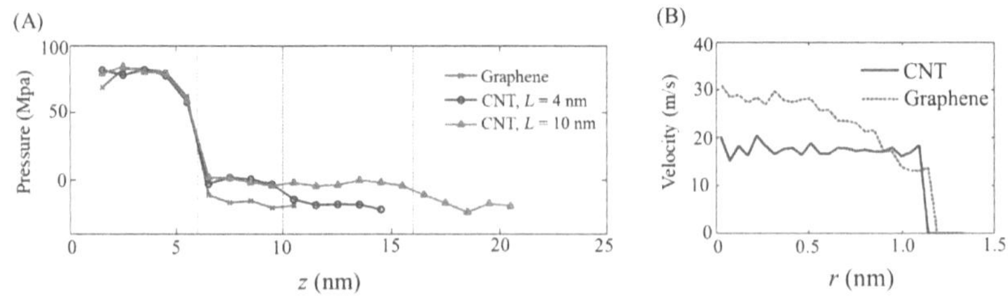

FIGURE 7.19 (A) Pressure distributions along the water flow direction for the nanoporous graphene monolayer and CNTs with lengths of 4 nm and 10 nm. (B) Axial velocity distributions of water molecules along the radial direction of the pore and nanotube. Source: *Modified and reprinted with permission from a work by Suk and Aluru (2010).*

The energetics of water permeation through the porous graphene monolayer and CNT with a pore or tube diameter of ~2.75 nm shows that the energy barrier for a water molecule permeating the porous graphene monolayer is 0.32 k_BT, which is lower than 0.43 k_BT for the CNT (k_B is the Boltzmann constant and T is the temperature). This phenomenon occurs because the entry of a water molecule into the pore of the graphene monolayer or CNT is associated with breaking hydrogen bonds between that water molecule and its neighboring water molecules, which results in an energy penalty. For the porous graphene monolayer, the direct connection of the two water reservoirs on both sides reduces the breakage of hydrogen bonds and thus results in a lower energy barrier for water molecules permeating the monolayer. Therefore, it is more energetically favorable for water molecules to enter the pores of the graphene monolayer. The axial velocities of water molecules through the pores in the graphene monolayer are higher than those inside the CNT, as shown in Fig. 7.19B. Although the axial velocities of water molecules near the pore edge of the nanoporous graphene monolayer are lower than those near the CNT wall, the axial velocities of water molecules at the center of the pore are much higher than those at the center of the CNT. The combination of a lower energy barrier to entry and a higher average axial velocity of the water molecules leads to a higher water flow rate across the porous graphene monolayer than across CNTs when their tube diameters are large.

By functionalizing the pore edge of the nanoporous graphene membrane, its pore size and chemical properties can be effectively tuned, and the interactions between the water molecules and the pore can also be significantly modified, thereby leading to distinct water transport behaviors across the membrane. Based on MD simulation, Cohen-Tanugi and Grossman (2012) have reported that for a given pore size, the water permeability through the pores in a graphene monolayer can be significantly enhanced by the hydroxylation of its pore edges. Specifically, the water permeability across hydroxylated pores with open areas of ~25 Å^2 and ~50 Å^2 are 69% and 115% greater than those of hydrogenated pores of the same sizes, respectively. This behavior is attributed to the fact that the hydroxylation of the pore edge causes the hydrophobic pores to become hydrophilic, which allows for a larger number of hydrogen bonds and thus provides a smoother energy landscape for water molecules to pass through at high flow rates.

In short, for a porous graphene monolayer with a small pore diameter, the water transport is slower than that for a CNT with a similar tube diameter because of the lower stability of the hydrogen bonding structure inside the pore of the graphene monolayer. The porous graphene monolayer with a large pore diameter surpasses the CNT in water transport because of the lower energy barrier presented at the entrance of its pore than that at the CNT entrance. Moreover, the hydroxylation of the pore edge enhances the hydrophilicity of the pore and can serve to further enhance the water transport rate. For practical water desalination applications, a proper filtration membrane should be chosen based on the hydration radii of the target ions. For ions with smaller radii, a CNT-based membrane may be a better choice, while for ions with larger radii, a nanoporous graphene monolayer may be more feasible.

7.1.4 Water flow in multilayer graphene and graphene oxide membranes

Despite their excellent seawater desalination performance in water permeation and ion filtration, the practical application of nanoporous graphene monolayers is mostly limited by difficulties in fabricating and transferring graphene nanosheets on a large scale. Furthermore, creating nanopores in monolayer graphene with well-defined pore sizes and distributions presents a massive challenge. Compared with graphene, its derivatives, which mainly include graphene oxide (GO) and reduced GO, possess intrinsic nanopores within a multilayer structure and are typically obtained by dip or spray coating, vacuum filtration, or layer-by-layer self-assembly. In such multilayer structures, the interlayer channel, interedge space, and nanopores in the individual layers provide channels for rapid water transport, as shown in Fig. 7.20. Both theoretical and experimental studies have demonstrated that graphene-based ultrathin membranes can potentially exhibit high salt rejection rates and facilitate rapid transport of water molecules (Cohen-Tanugi and Grossman, 2012; Nair et al., 2012; Huang et al., 2013). The ion filtration process involving these membranes comprises several underlying mechanisms and thus is more complicated than that involving the

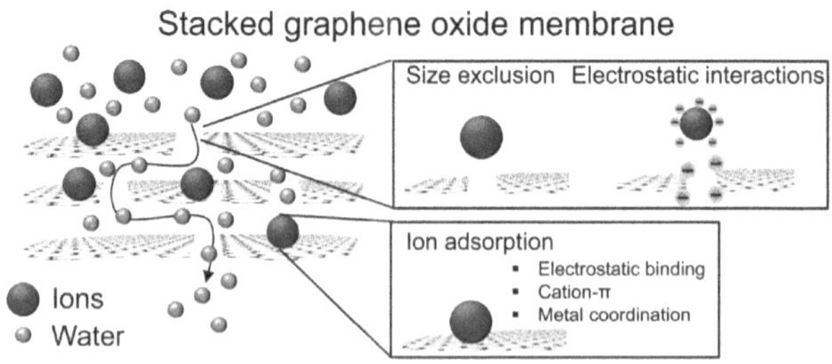

FIGURE 7.20 GO membrane composed of stacked GO nanosheets. The interlayer channel, interedge space, and nanopores in individual layers of such membranes provide channels for rapid and selective mass transport. Source: *Reprinted with permission from a work by Perreault et al. (2015).*

aforementioned porous monolayer graphene. Size exclusion and electrostatic interactions still play critical roles in molecular separation, while the physical and chemical adsorption of ions on the surface of individual graphene nanosheets that constitute the multilayer structure also contributes immensely to molecular separation.

The desalination performance of a GO-based membrane is characterized by both its salt rejection and water flux. To achieve high salt rejection, the pores within the membrane must be small, which will usually reduce water permeation. Hence, a delicate balance must be struck regarding the pore size of a GO-based membrane to achieve optimal desalination performance. Understanding the behaviors of water transport through a GO-based membrane is helpful in designing advanced membrane structures with improved seawater desalination performance. Such an understanding begins by investigating water transport inside the nanochannel of a graphene bilayer.

Fig. 7.21 shows the MD simulation model for pressure-driven water transport through the nanochannel of a pristine graphene bilayer. In this model, the pressure-driven method shown in Fig. 7.2 is adopted, except that no piston is involved. The graphene bilayer is located in the middle of the simulation system and connects the two water reservoirs on both sides. Two extra graphene nanosheets are attached to each layer of the graphene bilayer to serve as constraint walls and restrict the movements of the water molecules. Consequently, the water reservoirs on the left and right can only be connected by the nanochannel or by invoking a PBC along the x-direction.

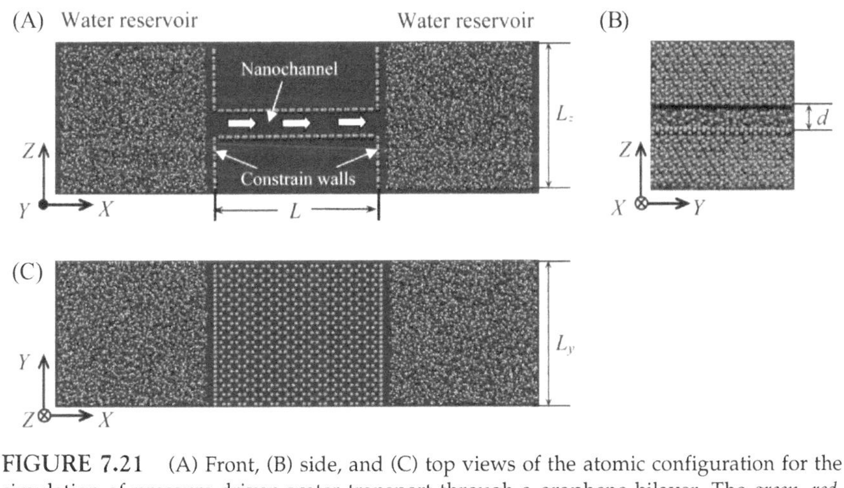

FIGURE 7.21 (A) Front, (B) side, and (C) top views of the atomic configuration for the simulation of pressure-driven water transport through a graphene bilayer. The *green*, *red*, and *white* spheres represent C, O, and H atoms, respectively. Source: *Reprinted with permission from a work by Liu et al. (2016a).*

The graphene bilayer surfaces are parallel to the x–y plane, and its center plane is located at $z = 19$ Å. The interlayer distance d of the graphene bilayer is set to vary from 7 Å to 13 Å, and the length L of the graphene bilayer is set to vary from 1 nm to 10 nm. The interactions between the water molecules and graphene bilayer are described by a modified 12–6 LJ potential as

$$V(r) = 4\chi\varepsilon\left[\left(\frac{\sigma}{r_{ij}}\right)^{12} - \left(\frac{\sigma}{r_{ij}}\right)^{6}\right]. \tag{7.10}$$

A scaling factor χ is introduced to the original LJ potential (discussed in Chapter 2: Potential Energy Functions) to tune the interaction strength. During the simulation, a driving pressure ΔP is applied along the channel length direction by imposing a constant force on all the O atoms in the left water reservoir. After a specific simulation time, the system reaches a steady state in which the average number N of water molecules in the nanochannel and their average velocity $\overline{v}$ along the x-direction are constant. The water flow rate Q is defined as the number of water molecules that are transmitted through the right end of the nanochannel per unit time per unit area as

$$Q = \frac{\overline{v}N}{dLL_y}. \tag{7.11}$$

Fig. 7.22 shows the dependence of the water flow rate Q on the applied pressure difference ΔP for a case of $L = 5$ nm and $d = 10$ Å. It is observed that Q increases linearly with the driving pressure ΔP. By linearly extrapolating Q with respect to ΔP, a value of 41.89 $\text{nm}^{-2}\,\text{ns}^{-1}$ is obtained for Q at a low driving pressure of 5 MPa (a driving pressure typically used for industrial nanofiltration systems). For a Poiseuille

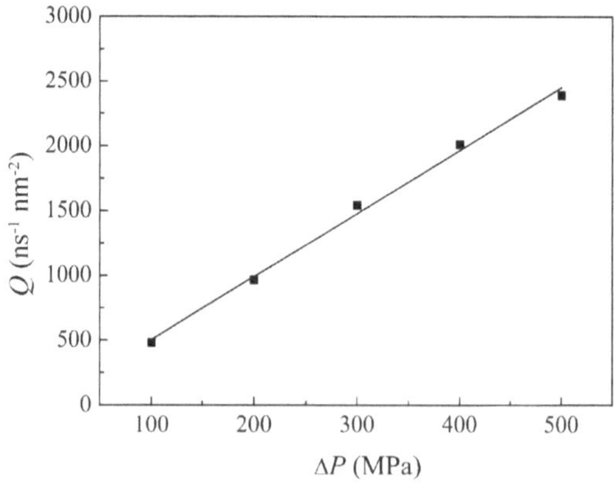

FIGURE 7.22 MD simulation results of the water flow inside a graphene bilayer with an interlayer distance of 1 nm and a length of 5 nm at different driving pressures. Source: *Reprinted with permission from a work by Liu et al. (2016a).*

flow confined between two flat plates with the no-slip boundary condition applied, the flow rate Q_{HP} estimated based on the HP equation is

$$Q_{HP} = \frac{d^2 \Delta P \rho N_A}{12\eta L}, \tag{7.12}$$

where d is the separation distance of the two plates; ρ and η are the molar density and viscosity of bulk water, respectively; N_A is the Avogadro constant. By defining the enhancement factor $s = Q/Q_{HP}$ and taking $\eta = 10^{-3}$ Pa s and $\rho = 0.0555$ mol cm^{-3}, s is calculated to be 15.34 at $\Delta P = 5$ MPa, which indicates that the water flow rate in the graphene bilayer is much higher than that predicted based on classical continuum mechanics.

The ultrafast water transport inside the graphene bilayer is closely related to both the frictionless surfaces of the graphene layers and the unique water structure inside the nanochannel. Fig. 7.23A shows the PEM of a water molecule moving on a graphene surface with a vertical distance of 3.20 Å between them. The minimum potential energy of the molecule corresponds to a position above the hollow center of the honeycomb lattice structure of graphene, while the maximum energy of the molecule corresponds to points on top of the C atoms. The difference between the minimum and maximum energy is about 0.08 kcal mol^{-1}, which is one order of magnitude lower than the corresponding energy difference of 0.5 kcal mol^{-1} for a water molecule moving on the inner wall of a CNT (Liu et al., 2008). This small energy corrugation indicates a smooth potential energy landscape for water molecules moving laterally on the surface of graphene.

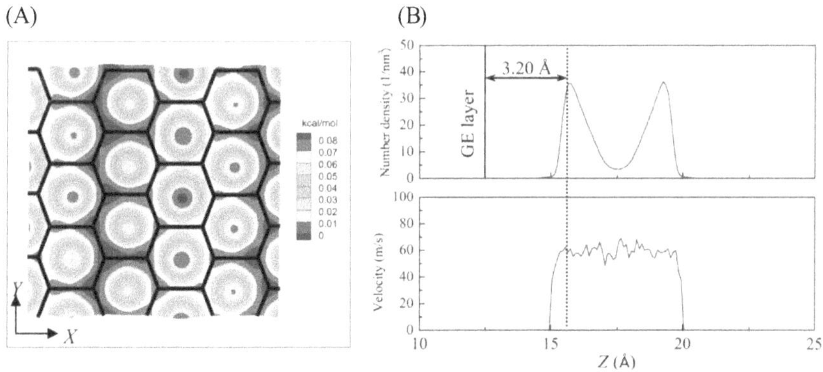

FIGURE 7.23 (A) PEM of a water molecule moving on the surface of a graphene layer with a vertical distance of 3.20 Å. (B) Number density (upper) and velocity distribution (lower) along the direction of the channel thickness for an interlayer distance d = 10 Å and a driving pressure of 300 MPa. Source: *Reprinted with permission from a work by Liu et al. (2016a).*

The smoothness of the graphene surface results in a plug-like flow of the water molecules within the graphene bilayer. Fig. 7.23B shows the number density and velocity profiles of water molecules in the channel thickness direction. Two peaks appear in the density profile of the water molecules in the direction of the channel thickness, each corresponding to a layer of water molecules near the graphene surface. A depletion region with a thickness of ~3.20 Å is formed between the water layer and graphene surface. Furthermore, at the inner boundary of this depletion region, the flow velocity undergoes a sharp change. These results demonstrate that no boundary water layer is attached to the graphene layer, and thus a plug-like flow occurs inside the nanochannel.

An important factor influencing the transport behavior of water molecules inside the graphene bilayer is the interlayer distance (or the thickness of the nanochannel). For different interlayer distances, the water molecules within the graphene bilayer form different structures, as shown in Fig. 7.24. As the channel thickness d increases from 7 Å to 10 Å (Fig. 7.24A), the single density peak splits into two new peaks that shift toward the channel walls. As d further increases from 10 Å to 13 Å

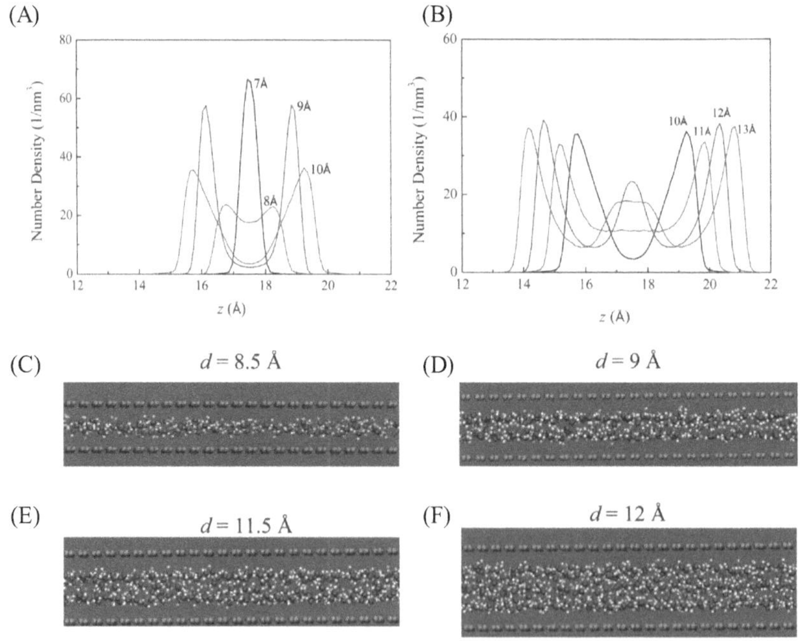

FIGURE 7.24 (A), (B) Density profile evolution of water molecules inside the nanochannel of the graphene bilayer as the channel thickness d increases from 7 Å to 13 Å. (C)–(F) Snapshots of the arrangement of water molecules for channel thicknesses of 8.5 Å, 9 Å, 11.5 Å, and 12 Å, respectively. Source: *Reprinted with permission from a work by Liu et al. (2016a).*

(Fig. 7.24B), a third density peak appears in the middle of the channel. The appearance of these new peaks corresponds to the formation of new water layers inside the nanochannel, as shown in Fig. 7.24C–F.

The change in the distribution of water molecules in the channel leads to significant transport behavior changes. Fig. 7.25A shows the variance of the number of water molecules inside the channel and their average velocity in the channel length direction for different channel thicknesses. It is observed that both the number of water molecules and their average velocity increase with the channel thickness. The formation of new water layers within the nanochannel leads to significant changes in both the number of water molecules and their average velocity for channel thicknesses of 8.5 Å and 12 Å. While the formation of a new water layer inside the channel causes the number of water molecules to experience a sharp increase, the average velocity of the water molecules decreases drastically because of the increased friction between the water molecules and the two graphene layers. Overall, the water flow rate increases monotonically with the channel thickness, as shown in Fig. 7.25B.

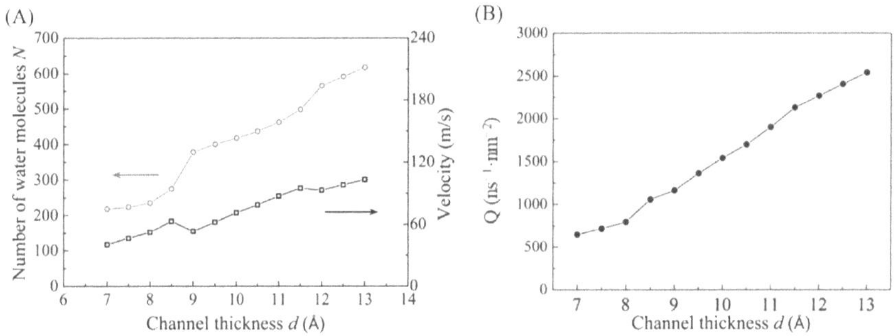

FIGURE 7.25 (A) Dependence of the number of water molecules N inside the channel and their average velocity $\overline{v}$ on the channel thickness d. (B) Dependence of the water flow rate Q on the channel thickness d. Source: *Reprinted with permission from a work by Liu et al. (2016a).*

A later MD simulation work by Liu et al. (2016b) has indicated that the corrugation of a graphene bilayer also plays a significant role in affecting the water transport inside its channel. The simulation results show that the corrugation significantly changes the distribution, dipole orientations, and interactions of the water molecules inside the channel. Meanwhile, the energy barrier for the transport of water molecules through the channel also significantly increases compared with that for the case of flat graphene layers. Under the same driving force, the water flow rate decreases as the corrugation of the graphene layers increases, particularly when there is a mismatch in the corrugations of the two graphene layers.

7.2 Nanofiltration with porous thin films

Nanofiltration is a filtration method utilizing ultrathin membranes containing pores of nanoscale sizes. As an important mass separation technique, nanofiltration was initially used for water treatment, in particular, water softening by removing scale-forming hydrated divalent ions (e.g., Ca^{2+} and Mg^{2+}) while allowing water molecules and monovalent ions with small hydration diameters to pass through a membrane. RO, FO, and CDI techniques are commonly applied to realize nanofiltration.

Conventional filtration membranes are mostly fabricated through modifying polymer thin films. The pore dimensions of such thin films are controlled by factors involving pH, temperature, and time during the film development, and their pore densities range from 1 cm^{-2} to 100 cm^{-2}. With the rapid development of synthesis techniques of novel nanomaterials including graphene and CNTs, the porous structures and properties of nanofiltration membranes have been greatly improved, and the range of their practical applications has also been significantly broadened. Recently, the promise of high water permeation through graphene- and CNT-based membranes as well as their fine and tunable porous structures has aroused intense research interest for their applications in various water desalination techniques. In this section, the mass transport behaviors regarding several nanofiltration processes involving porous thin films are discussed by highlighting the ion filtration mechanisms of RO, FO, and CDI desalination processes utilizing graphene-based membranes.

7.2.1 Reverse osmosis process

In an RO desalination process, seawater is driven by an external pressure to permeate a filtration membrane with an appropriate porous structure that rejects the passage of ions and molecules larger than the pores. Fig. 7.26A presents the MD simulation model for an RO system utilizing a

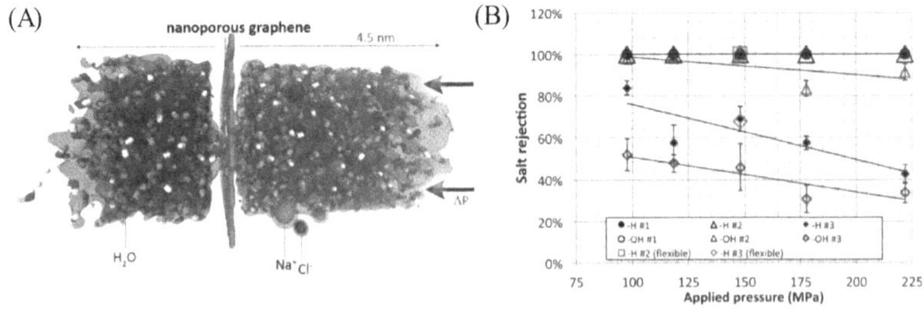

FIGURE 7.26 (A) Atomic configuration of an RO system using a porous graphene monolayer as the filtration membrane. (B) Salt rejection as a function of the pore type and driving pressure. The indices #1, #2, and #3 indicate pore diameters of ~4.5 Å, ~6 Å, and ~9 Å, respectively. Source: *Reprinted with permission from a work by Cohen-Tanugi and Grossman (2012).*

porous graphene monolayer as the filtration membrane (Cohen-Tanugi and Grossman, 2012). The pore edge is functionalized with −H or −OH groups. The saline water is modeled as 16 NaCl ion pairs dissolved within 825 water molecules, which corresponds to a salt concentration of 72 g L^{-1}. This concentration is approximately twice the salinity of seawater and is chosen to increase the frequency of ions interacting with the pore, thus achieving a sufficiently large sample to obtain more precise statistical results for a given simulation time and system size.

The simulation results indicate that the desalination performance of the RO system depends significantly on the pore size and the type of functional groups present at the pore edge. The nanopore functionalized with H-groups (i.e., hydrogenated) exhibits a lower water permeability and a higher salt rejection rate than the nanopore functionalized with −OH groups (i.e., hydroxylated), as shown in Fig. 7.26B. This behavior is due to the hydrophobic nature of the hydrogenated pore and the need for both water molecules and hydrated NaCl ions to break more hydrogen bonds when penetrating through the pore. For a sufficiently small nanopore (with a diameter smaller than 4.5 Å), a complete salt rejection can be attained, but the rejection drops to 84% for the hydrogenated pore and 52% for the hydroxylated pore when the diameter increases to ~9 Å.

For the nanoporous graphene monolayer, there are five main mechanisms facilitating salt rejection: (1) size exclusion, (2) dehydration effects, (3) charge repulsion, (4) entropic differences, and (5) interactions between solutes with specific chemical structures at the edge of the nanopore. The first two mechanisms involve the size exclusion of ions, with or without hydration shells, and are the most direct salt rejection mechanisms. To achieve a complete rejection of NaCl from seawater while enabling rapid water permeation, the critical nanopore diameter should lie between 0.6 nm and 0.8 nm. The third mechanism is associated with the electrostatic repulsion between the target ions and the charges at the nanopore edge.

The fourth mechanism is related to the entropic differences of the target ions across the nanopore. Since the filtration of salt ions is governed by a free energy barrier through each nanopore, a nanoporous graphene monolayer can reject salt ions by restricting their physical configurations (e.g., geometric orientation) when they pass through the nanopore. For instance, Fig. 7.26B shows that the hydrogenated nanopore rejects salt ions more effectively than the hydroxylated nanopore. This outperformance is because the hydrogenated nanopore imposes a more radical restriction on the permitted geometric and hydrogen-bonding configurations of the hydrated ions passing through it.

The final mechanism refers to the interactions between specific chemical functional groups at the nanopore edge and the salt ions. Such interactions can potentially alter the energy barrier for the permeation of water molecules and ions through the nanopore, which can be characterized by the change of the potential of mean forces (PMFs) of both the water molecules and ions. During the PMF calculation, a water molecule or an ion is selected and forced to move along a reaction coordinate (e.g., a geometrical path), and the energy of this molecule or ion along this coordinate is sampled over a certain duration (typically several nanoseconds). An umbrella sampling technique is commonly used to enhance sampling efficacy. The weighted histogram analysis method is often applied to achieve a one-dimensional (1D) free energy profile. Fig. 7.27 shows the PMF calculation results of a single water molecule passing through a nanopore in the nanoporous graphene monolayer with different diameters and functional groups.

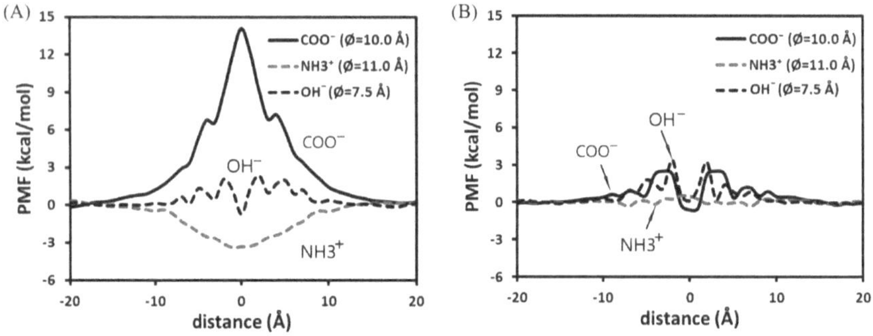

FIGURE 7.27 PMF of a water molecule as a function of its axial distance from the center of a functionalized nanopore of the graphene monolayer. The results are shown for NaCl solutions with a concentration of (A) 0.025 M and (B) 0.25 M under ambient conditions. Three chemical groups, COO^- (*black* continuous line), NH_3^+ (*red* dashed line), and OH^- (*black* dashed line) are considered for the functionalization of the nanopore. Φ denotes the diameter of the accessible area of the nanopore for water molecule transport. Source: *Modified and reprinted with permission from a work by Konatham et al. (2013).*

It is observed that the PMF profile of the water molecule depends on the functional groups associated with the nanopore, and both PMF peaks and valleys can appear at the center of the pore. In addition, PMF peaks can appear on both sides of the pore. For a dilute NaCl solution with a low concentration of 0.025 mol L^{-1} (Fig. 7.27A) and a nanopore functionalized with COO^- groups, the PMF profile of the water molecule in the nanopore exhibits a sharp peak (with an amplitude of ~14 kcal mol^{-1}) at the center of the pore, which indicates that the pore strongly repels the water molecule. On the contrary, for a pore functionalized with NH_3^+ groups, a valley in the pore center is observed for the PMF profile of the water molecule, indicating that the water molecule is attracted to the pore. For a pore functionalized with OH^- groups, the PMF profile of the water molecule fluctuates between 0 kcal mol^{-1} and 3 kcal mol^{-1}. When the solution concentration increases to 0.25 mol L^{-1} (Fig. 7.27B), the attractive and repulsive forces imposed on the water molecule by the nanopores functionalized with COO^- groups and NH_3^+ groups, respectively, diminish. These observations indicate that the PMF profile of a water molecule is primarily determined by electrostatic effects, which are shielded at high salt concentrations.

Similarly, pore functionalization also significantly impacts the PMF of ions passing through the nanopore of the porous graphene monolayer. Fig. 7.28 shows the PMF profiles of Na^+ and Cl^- ions with respect to their axial distance from the center of the functionalized nanopore at salt concentrations of 0.025 mol L^{-1} (red) and 0.25 mol L^{-1} (black).

It is observed that for the dilute NaCl solution, the COO^- functionalized nanopore attracts Na^+ ions but repels Cl^- ions. On the contrary, the NH_3^+ functionalized pore attracts Cl^- ions but repels Na^+ ions. For the pore functionalized with OH^- groups, the PMF profile for Na^+ ions fluctuates between 4 kcal mol^{-1} and −8 kcal mol^{-1}, and the PMF profile for the Cl^- ions exhibits a peak at the pore center. As the salt concentration increases, the attractive and repulsive effects of the COO^- and NH_3^+ groups on the salt ions are significantly suppressed. However, for the OH-functionalized pore, the PMF profile for the Na^+ ions remains consistent as the salt concentration increases, but the peak in the PMF profile for the Cl^- ions becomes drastically higher. These results indicate that functionalizing the pores of a nanoporous graphene monolayer membrane with OH^- groups may be an efficient approach for boosting its salt rejection because these groups can lead to high free-energy barriers for Cl^- ions permeating the membrane at both low and moderate salt concentrations.

For a multilayer graphene-based membrane, the filtration mechanism is more complicated. Besides the mechanisms mentioned above, the geometry of the multilayer graphene membrane plays a significant role in determining its RO desalination efficacy in terms of water permeation and salt rejection rate. Fig. 7.29A shows the atomic configuration of an RO system utilizing a multilayer graphene-based membrane as the filtration membrane (Cohen-Tanugi et al., 2016). The pore radius, layer separation, and pore offset are denoted by R, H, and O, respectively. The MD simulation results indicate that the water flow rate is highly dependent on the separation distance H when it is below a critical value of 8 Å (Fig. 7.29B). On the other hand, salt rejection is almost entirely independent of the layer separation H, except for cases in which the nanopores are fully aligned (i.e., $O = 0$) and possess a large radius (Fig. 7.29C). At a separation distance $H = 8$ Å, the water flow rate does not undergo any significant change with the nanopore offset (Fig. 7.29D). However, the

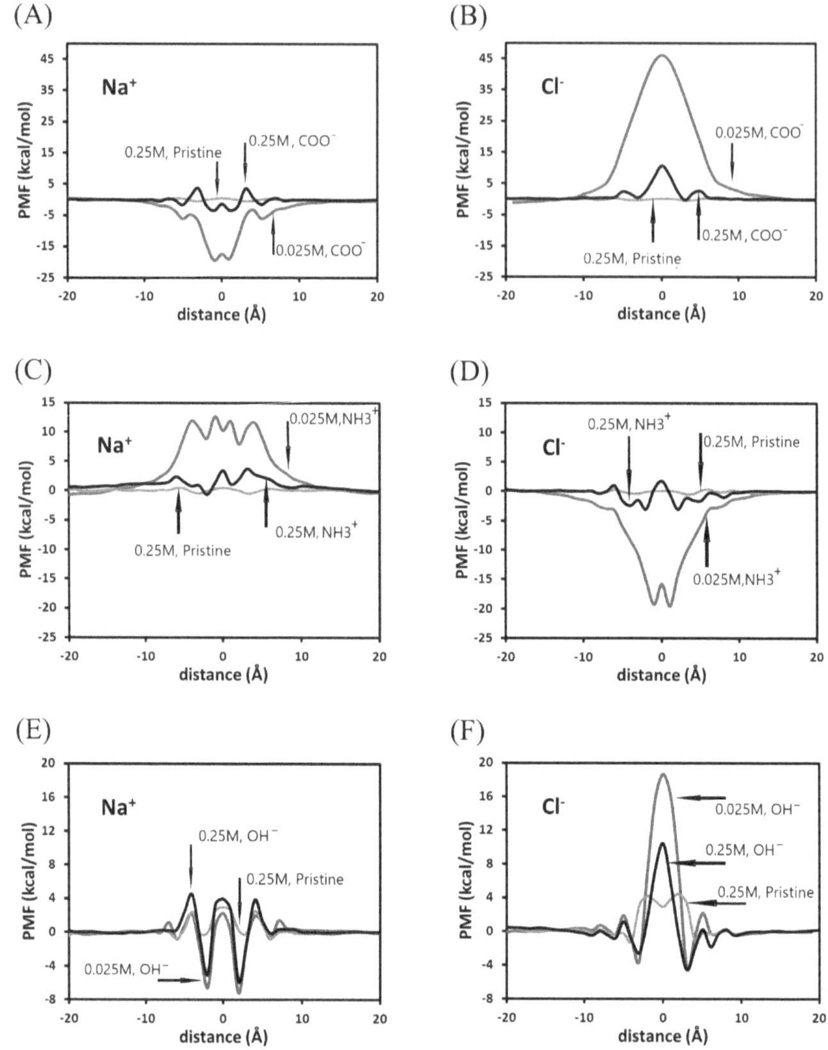

FIGURE 7.28 PMF as a function of the axial distance from the center of the functionalized nanopore for Na^+ and Cl^- ions at a NaCl concentration of 0.025 M (*red*) and 0.25 M (*black*). Three chemical groups, COO^- (A and B), NH_3^+ (C and D), and OH^- (E and F) are considered for the functionalization of the nanopore. For comparison, the results for the pristine nanopore at a NaCl concentration of 0.25 M are also plotted (*gray*). Source: *Modified and reprinted with permission from a work by Konatham et al. (2013).*

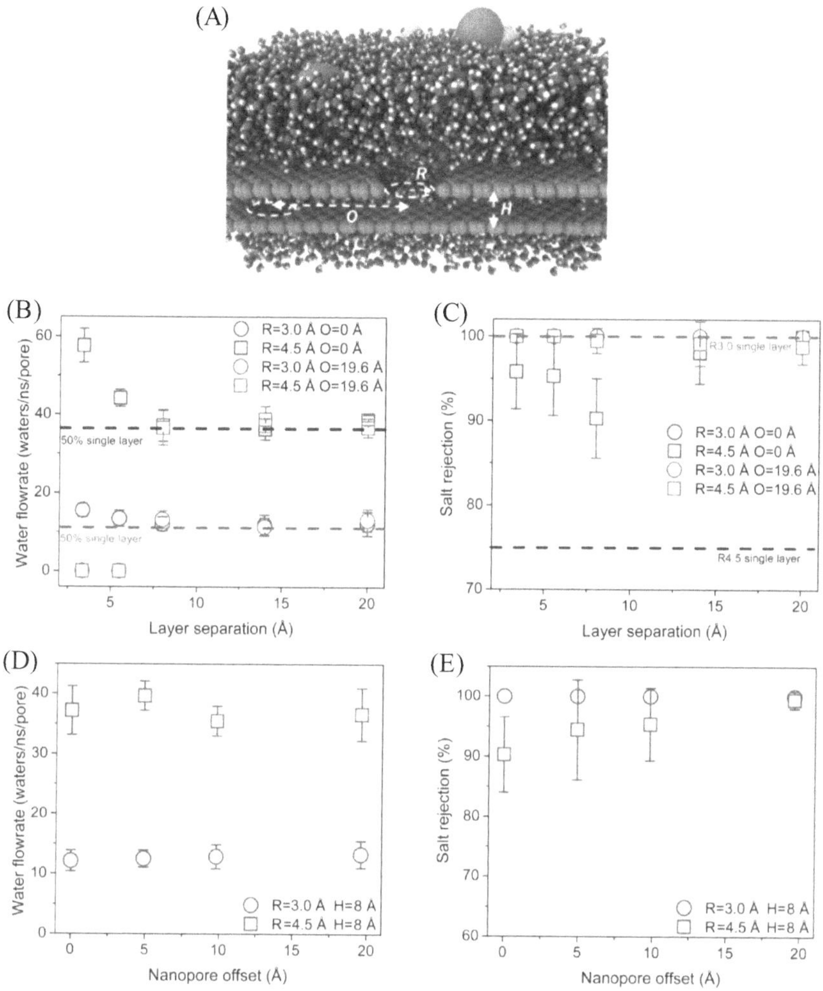

FIGURE 7.29 (A) Atomic illustration for the RO system using a multilayer graphene-based membrane as the filtration membrane. (B) Dependence of the water flow rate and (C) salt rejection rate on the layer separation H for the given pore radii R. (D) Dependence of the water flow rate and (E) salt rejection rate on the pore offset O for the given pore radii R. Source: *Modified and reprinted with permission from a work by Cohen-Tanugi et al. (2016).*

salt rejection rate tends to increase with the pore offset, particularly for cases involving a large pore radius (Fig. 7.29E).

Due to the rapid progress in the design and fabrication of nanomaterials, various types of 2D graphene-analogous nanomaterials have been discovered and synthesized recently (Liu and Zhou, 2019). Some of these newly synthesized materials possess intrinsic periodic nanopores with diameters suitable for ion filtration. For instance, using MD simulation,

Babicheva et al. (2020) have recently investigated the RO seawater desalination performance of semipermeable membranes composed of two new carbon allotropes, cyclicgraphene and graphenylene. It is found that a water flow rate of ~32 L bar^{-1} h^{-1} m^{-2} and complete salt rejection can be obtained at room temperature using graphenylene. Liu et al. (2018) have studied the mechanisms of water and ion permeation across nanoporous membranes that consist of a graphitic carbon nitride (g-C_3N_4) monolayer, which is a novel graphene-analogous 2D nanomaterial with intrinsic nanopores. Their study indicates that the nanoporous g-C_3N_4 membranes can completely reject Ca^{2+} and Cl^- ions and sieve out monovalent cations, such as Na^+ and K^+ at rejection rates exceeding 70%.

Besides graphene-analogous nanomaterials, another type of nanomaterial, metal-organic frameworks (MOFs), is a promising candidate as filtration membranes. MOFs are compounds composed of metal ions being dative-bonded to an organic linker, and they have high porosity (up to 90%) and large surface areas (up to 6000 m^2 g^{-1}) that exceed that of graphene (2630 m^2 g^{-1}). Owing to the huge variety of organic linkers and metal ions, MOFs with varying pore sizes, volumes, and functionalities can be constructed, which render them highly versatile in applications of ion or molecule separation. Fig. 7.30A shows the atomic structure of the unit cell of a typical MOF structure, zeolitic imidazolate frameworks-8 (ZIF-8), which is composed of Zn metal ions and 2-methylimidazole organic linkers (Fig. 7.30B). Two types of pores exist in MOFs (apertures and cages), and they have different sizes as indicated by the gray and yellow circles. Fig. 7.30C shows a ZIF-8 membrane in which the apertures and cages are ordered alternatively.

When MOFs are used as semipermeable membranes for seawater desalination via the RO process, salt rejection is realized through the abovementioned mechanisms of size exclusion and dehydration effects. During RO seawater desalination, it is possible for the target ions to occupy the large cages with the hydration shells of the ions intact. However, it is harder for the ions to pass through the smaller apertures without losing their hydration shells. When an ion passes through an aperture, its hydration shell is peeled off, which is generally associated with a high-energy penalty. In addition, due to the high porosity of MOFs, ions have to pass through several apertures and cages to completely permeate the MOF membranes. Each aperture imposes a high-energy barrier for the ions to enter due to its small size. Therefore, it is common for ions to be trapped within the larger cages near the membrane surface.

An early MD simulation work by Hu et al. (2011) that explores the performance of a semipermeable ZIF-8 membrane for RO seawater desalination has indicated that Na^+ and Cl^- ions cannot be transported across the membrane because of the severe sieving effect arising from the small apertures in ZIF-8, which can lead to a high salt rejection rate.

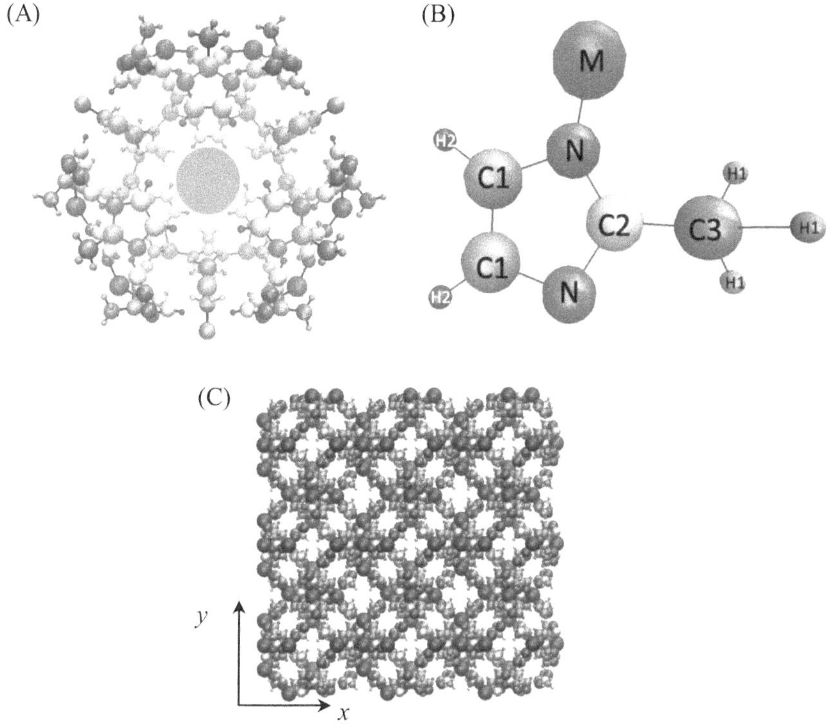

FIGURE 7.30 (A) Atomic structure of the unit cell of a typical MOF structure (ZIF-8), in which the *gray* and *yellow* circles highlight the aperture and cage, respectively. (B) Atomic structure of the 2-methylimidazole organic linker in ZIF-67, in which the atoms are represented by the following color codes: M (zinc) in *red*, N (nitrogen) in *gray*, C1 (carbon) in *orange*, C2 (carbon) in *yellow*, C3 (carbon) in *tan*, H1 (hydrogen) in *green*, and H2 (hydrogen) in *blue*. (C) Atomic structure of a ZIF-8 membrane obtained by duplicating the unit cell in the *x*-, *y*-, and *z*-directions. Source: *Reprinted with permission from a work by Hong et al. (2021).*

However, the rate of water flow across the ZIF-8 membrane is low, and it is much lower than the rates of water flow across membranes comprising of nanoporous graphene and CNTs under the same driving pressure in an RO system. The low water flow rate is attributed to the fact that water molecules in the ZIF-8 membrane exhibit fewer hydrogen bonds than in their bulk-phase. However, these hydrogen bonds possess a considerably longer lifetime due to the interactions between the water molecules and the ZIF membrane as well as the geometrical confinement of the membrane. The stable hydrogen bonds among the water molecules within the ZIF-8 membrane impose a significant energy barrier that impedes their transmission across the membrane.

Hence, MD simulation studies on water desalination involving semipermeable MOF-based membranes tend to focus on the search of suitable combinations of metal atoms and organic linkers to create MOFs with appropriate pore sizes and chemical functionalities, which allow optimal water flow rates and ion rejection performance.

Gupta et al. (2015) have conducted MD simulations to study water desalination across five types of ZIF-based membranes, namely ZIF-25, ZIF-71, ZIF-93, ZIF-96, and ZIF-97. The five ZIFs possess the same rho-topology but different imidazolate linkers with various functional groups. The simulation results reveal that the aperture size and polarity of the functional groups are key factors that govern the rate of water flow across the ZIF membranes. Among the five ZIFs studied, ZIF-25, ZIF-71, and ZIF-93 exhibit larger apertures and correspond to much higher water flow rates than ZIF-96 and ZIF-97. However, a detailed analysis of the interactions between the water molecules and the functional groups in the ZIFs indicates that the rates of water flow across ZIF-25, ZIF-71, and ZIF-93 are more dependent on the polarity of the functional groups than on the aperture size. With the hydrophobic CH_3 group, ZIF-25 displays the highest water flow rate among the five ZIFs, although its aperture size is smaller than those of ZIF-71 and ZIF-93.

Lyu et al. (2019) have discovered that besides chemically modifying their organic linkers, incorporating defects into MOFs can lead to noticeably enhanced water permeation through them. UiO-66, a typical MOF that is stable in water, is modeled. It consists of zirconium (Zr) clusters (e.g., $ZrO_4(OH)_4$), each of which is connected by 12 benzene-1,4-dicarboxylate (BDC) linkers. The atomic structure of a defect-free UiO-66 unit cell is shown in Fig. 7.31A. Two types of defects are created in UiO-66 by removing some BDC linkers and Zr clusters from its defect-free atomic structure, as shown in Fig. 7.31B and C, respectively. After removing the BDC linkers and Zr clusters, the unsaturated metal sites resulting from the defect creation are terminated by one of the following functional groups: chloride (Cl^-), hydroxide (OH^-), and monocarboxylates [including formate ($HCOO^-$), acetate (CH_3COO^-), and trifluoroacetate (CF_3COO^-)], as shown in Fig. 7.31D–H.

The MD simulation results show that both the aperture size and water affinity of the UiO-66 membranes can be optimized to substantially improve water permeability by one order of magnitude while maintaining an excellent salt rejection performance. Most noticeably, the UiO-66 membrane with certain Zr clusters removed and saturated by CF_3COO^- groups at the defect sites exhibits a water flow rate reaching 800 L m^{-2} h^{-1} bar^{-1}, compared with ~50 L m^{-2} h^{-1} bar^{-1} for the defect-free membrane. Meanwhile, this defective UiO-66 membrane can achieve almost complete salt rejection. The outcome of this work suggests that creating structural defects within MOF RO membranes can significantly enhance their desalination efficiency.

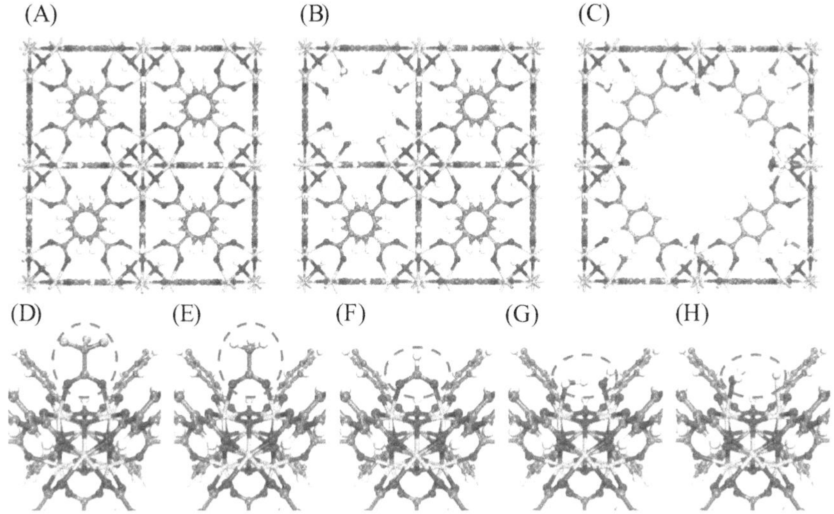

FIGURE 7.31 Illustration of (A) defect-free and (B, C) defective UiO-66 structures. The defective structures shown in (B) and (C) lose two linkers and one cluster per unit cell, respectively. (D) – (H) Schematics for compensating the unsaturated metal sites introduced by the defects, which respectively show one defect center (i.e., two unsaturated Zr atoms) coordinated with trifluoroacetate (CF_3COO^-), acetate (CH_3COO^-), formate ($HCOO^-$), hydroxide (OH^-), and chloride (Cl^-). All atomic structures are represented in the ball-stick mode with the following color codes: carbon in *gray*, hydrogen in *white*, oxygen in *red*, zirconium in *cyan*, fluorine in *magenta*, and chlorine in *green*. Source: *Reprinted with permission from a work by Lyu et al. (2019).*

7.2.2 Forward osmosis process

The FO technique is another distinguished membrane separation method. In seawater desalination via FO, water molecules are driven to flow through a semipermeable membrane from the feed solution side with a higher water chemical potential (low osmotic pressure) to the draw solution side with a lower water chemical potential (high osmotic pressure). The driving force is the difference in osmotic pressure across the membrane and is usually controlled by regulating the salt concentration in each solution. Compared with RO, the FO process has a low fouling tendency and can be scaled up conveniently.

The desalination performance of an FO system is primarily determined by the properties of the semipermeable membrane. An ideal FO membrane can facilitate both high water permeation and high salt rejection. Commercial FO membranes are typically dense polymeric membranes fabricated by attaching an active layer on a porous supporting layer. A critical issue that limits the desalination performance of these membranes is the internal concentration polarization (ICP) effect caused by the permeation of ions through them. Generally, the ICP effect occurs because of the accumulation of ions in the porous membrane, which alters the ion concentration gradient across the entire FO system, particularly at the membrane boundaries. Consequently, the ICP effect can significantly reduce the water flux of the FO system. Since the extent of the ICP effect is proportional to the membrane thickness, an effective approach for countering this effect is to reduce the membrane thickness.

MD simulations have established nanoporous monolayer graphene as a promising candidate for membranes in FO processes due to its ultrathin thickness (Gai et al., 2014). Fig. 7.32A–C shows the atomic model of an MD simulation for FO systems using a nanoporous graphene monolayer as the filtration membrane. An aqueous NaCl solution with a salt concentration in the range from 0.6 mol L^{-1} to 5 mol L^{-1} is adopted as the draw solution, while freshwater is assigned as the feed solution. The pore edges are functionalized with F (Fig. 7.32A) or N (Fig. 7.32B) atoms, and the corresponding functionalized graphene monolayers are named GF or GN. The pore diameter varies from 3.7 Å to 17.6 Å. Fig. 7.32D shows the water flow rate and salt rejection rate for both GF and GN membranes with different pore diameters. For an FO system with a membrane pore diameter of ~12 Å, complete salt rejection can be achieved with a high water flow rate of ~80 L cm^2 h for GF (or ~ 40 L cm^2 h for GN), which is over four orders of magnitude higher than that of a typical cellulose triacetate membrane. For a pore diameter over 12 Å, the water flow rate of the system can be further enhanced, but its salt rejection will be compromised, particularly when a GF membrane is adopted.

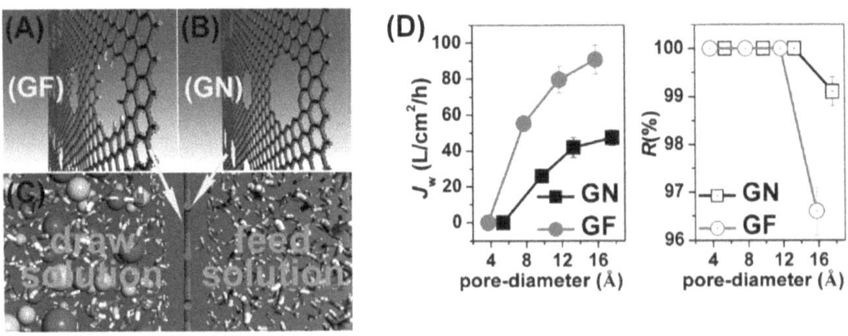

FIGURE 7.32 Atomic structure of the porous graphene monolayer with the pores functionalized by (A) F atoms and (B) N atoms. (C) MD simulation model for the FO desalination system using nanoporous graphene as the filtration membrane. (D) Dependence of the water flow and salt rejection rates on the pore diameter. Source: *Reprinted with permission from a work by Gai et al. (2014).*

Compared with the GN membrane, the GF membrane can result in a high water flow rate and a low salt rejection rate for a pore diameter exceeding 12 Å. Such differences are closely related to the adsorption energy of salt ions and water molecules at the pore edge. Fig. 7.33A shows the adsorption configuration and energy E_a of a Na^+ ion and a water molecule at a pore edge. The adsorption energy of adsorbents at the pore edges of the GF membrane is lower than that for the GN membrane. A high adsorption energy indicates that the solution interacts

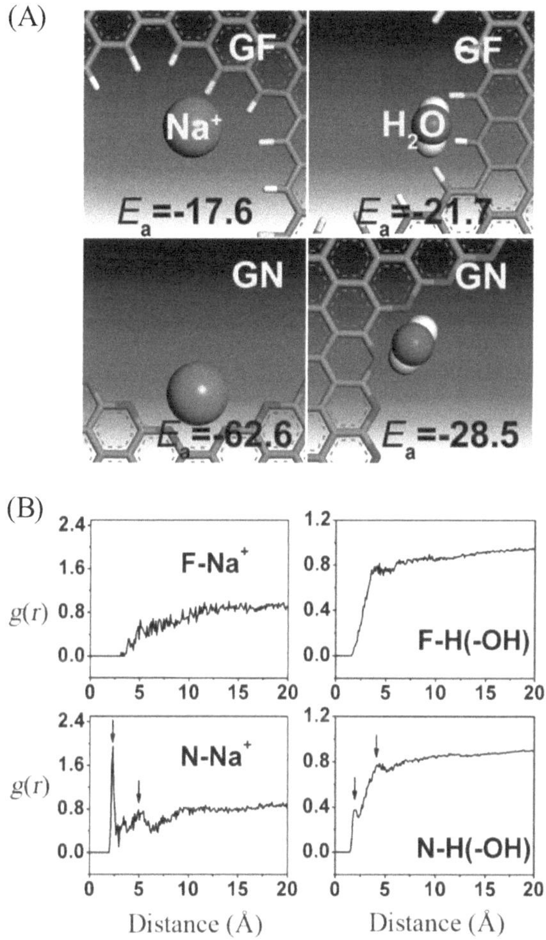

FIGURE 7.33 (A) Adsorption configurations of a Na^+ ion and a water molecule at the pore edge of the GF or GN membrane with their corresponding adsorption energy E_a. The negative sign of the adsorption energy indicates that it is an exothermic physical process. The C, F, O, H, N, and Na atoms are colored in *gray, light blue, red, white, blue,* and *purple,* respectively. (B) RDFs of the F and N atoms in the graphene pores with respect to salt ions and water molecules. Source: *Reprinted with permission from a work by Gai et al. (2014).*

strongly with the pores, and thus both the salt ions and water molecules have to overcome high-energy barriers to penetrate the membrane. Hence, when a GF filtration membrane is replaced with a GN membrane in FO, ion rejection is enhanced, while water permeation is reduced.

The stronger interactions between the target ions and the pore edges of the GN membrane can also be deduced from the radial distribution functions (RDFs) of the functional group atoms. Generally, the RDF of a practical atom pair describes the density of one atom type as a function of the distance from the reference atom type. From the RDF, the interaction strength of this atom pair can be analyzed. In Fig. 7.33B, the RDF of the F (or N) atom at the graphene pore with salt ions and water molecules is presented. Strong peaks appear on the RDF curves of the $N-Na^+$ and N–H pairs, while no distinct peaks are visible for the RDF curves of the $F-Na^+$ and F–H pairs. These observations indicate that the interaction between the GN pore and ions (or water molecules) is stronger than that for the GF pore. Hence, the GF pores should correspond to a higher water flow rate and a lower salt rejection rate.

The desalination performance of FO desalination systems utilizing multilayer graphene membranes is also highly dependent on the membrane geometry configuration (Dahanayaka et al., 2017b), which is similar to the RO desalination systems discussed in Section 7.2.1. A complete rejection of ions can be achieved for bilayer graphene membranes of which the pore diameter is less than 9 Å. As the pore diameter of each graphene layer increases, the water flow rate increases monotonically, while the ion rejection rate decreases. By increasing the number of graphene layers, a high salt rejection rate and a low water flow rate can be achieved.

7.2.3 Capacitive deionization

In a CDI system, ions with opposite charges are adsorbed on two electrodes under the effect of an external electric field, for which the electric potential typically ranges from 1.2 V to 1.5 V. This adsorption process continues until the electrodes are saturated with ions. Saturated electrodes can be cleaned by removing or reversing the external electric field to release the ions that are adsorbed on the electrodes. CDI is a membrane-free operation and is therefore a low-pressure process. Fig. 7.34A illustrates the separation mechanism of a CDI system. Energy stored in the electrodes during the deionization cycle due to the adsorption of ions with opposite charges can be recovered during the regeneration or discharge cycle. The net energy consumption is the difference between the supplied energy and recovered energy. Hence, while the CDI process is energetically favorable, efforts are needed to perform regeneration effectively.

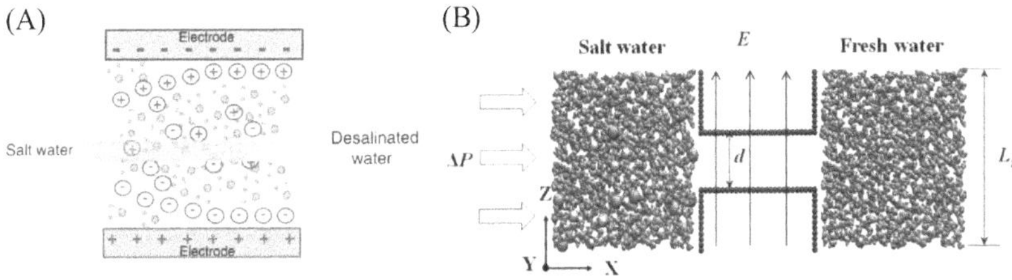

FIGURE 7.34 (A) Schematic of the CDI process. (B) MD simulation model of a CDI system using graphene monolayers as its electrodes. The *green* and *yellow* spheres represent Na^+ and Cl^- ions, respectively. Source: *Reprinted with permission from a work by Zhang et al. (2017).*

Generally, electrodes in a CDI system are fabricated using porous materials. The deionization performance of the system is determined by the properties of the electrodes, including their pore volume, pore size, pore connectivity, electronic conductivity, and electrochemical stability.

The most common precursors for the electrodes are carbon-based, such as activated carbon, carbon aerogel, ordered mesoporous carbon, and carbon black. Compared with these carbon-based materials, graphene-based electrodes have high electronic conductivity, excellent electrochemical stability, and large specific surface areas. Thus, they are promising candidates for CDI electrodes.

Fig. 7.34B shows the MD simulation model of a CDI system utilizing two graphene monolayers as the cathode and anode. This model is similar to those shown in Fig. 7.21, except that the left reservoir is filled with an aqueous NaCl solution instead of freshwater. An external pressure ΔP is applied to drive the salt solution through the nanochannel of thickness d formed by the two graphene electrodes. An external electric field E is applied along the channel thickness direction. Alternatively, the electric field can also be generated by charging the graphene electrodes with opposite charges. The simulation results show that at low electric field intensities ($0.01-0.25\ \mathrm{V\ \AA^{-1}}$) and low driving pressures (5–10 MPa), the salt ions are repelled from the nanochannel entrance and remain in the left reservoir. For high electric field intensities ($1-5\ \mathrm{V\ \AA^{-1}}$) and driving pressures (over 100 MPa), the ions are driven into the nanochannel and trapped at the surface of the electrodes. The presence of ions on the surface of the electrodes imposes a pinning effect on the water molecules passing through the nanochannel, thereby leading to a low water flow rate.

When the surface corrugation of the graphene monolayers is taken into consideration, the intensity of the electric field required to constrain the movement of the salt ions is significantly reduced due to the trapping effect resulting from the corrugation (Dahanayaka et al., 2017a), as shown in Fig. 7.35. For an ideally flat graphene electrode, the ions are distributed evenly on its surface. For a corrugated graphene electrode, the ions entering the nanochannel are concentrated near its entrance, which is beneficial to salt rejection. However, corrugation also reduces the water flow velocity in the channel because of the increased friction between the water molecules and the graphene surface, in contrast to the case of an ideal flat electrode.

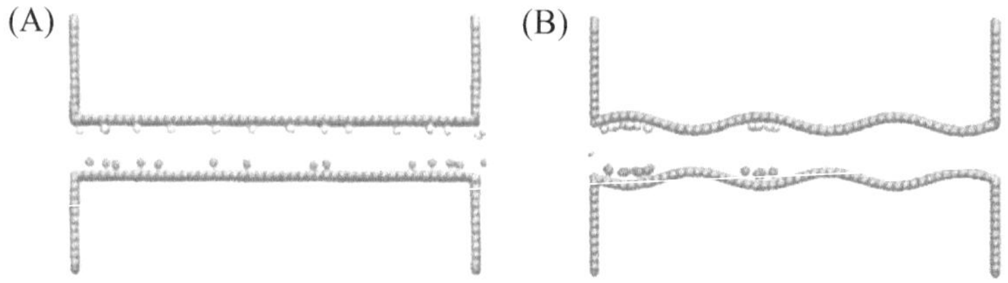

FIGURE 7.35 Distributions of ions inside a nanochannel formed by (A) ideally flat graphene electrodes and (B) corrugated graphene electrodes. Source: *Reprinted with permission from a work by Dahanayaka et al. (2017a).*

To further enhance the CDI performance of a graphene-based membrane, Ho and Striolo (2015) have designed a novel nanostructured carbon electrode using a partially porous graphene bilayer. Fig. 7.36 shows the structural schematic of a CDI system utilizing such electrodes. The space between the inner layers of the two electrodes facing each other provides the central pore for liquid transport. In addition, the space between the two graphene layers of each electrode offers a permeation passage for water molecules and ions. In the inner layers of the two electrodes, two holes with the same radius R are created to allow the side pores and central pore to exchange ions and water molecules.

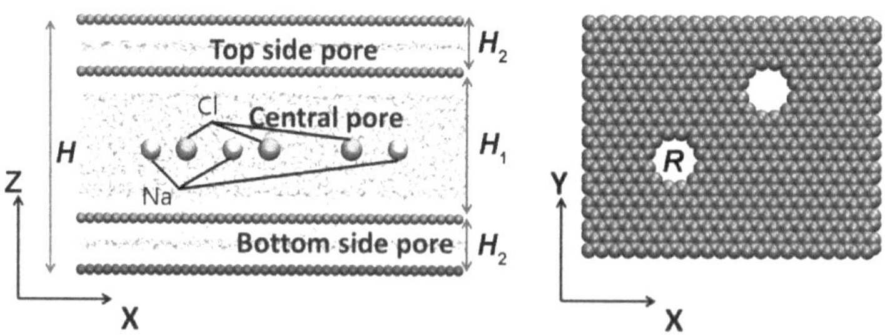

FIGURE 7.36 Schematic representation of the CDI system utilizing partially porous graphene bilayers as its electrodes. Two holes with the same radius R are created in the inner graphene layers of the electrodes to improve their CDI performance. The *yellow* and *green* spheres represent the Na^+ and Cl^- ions, respectively. Source: *Modified and reprinted with permission from a work by Ho and Striolo (2015).*

To investigate the CDI performance of the electrodes, both equilibrium and nonequilibrium MD simulations have been performed. In the equilibrium simulation, the distance between the inner layers of the two electrodes, denoted by H_1, is taken to be 20 Å, while the interlayer distance between the two graphene layers in each electrode, denoted by H_2, is taken as 7 Å. Initially, 2268 water molecules and 4 NaCl ion pairs (corresponding to a NaCl solution with a salinity of ~5.45 g L^{-1}) are placed within the central pore. A surface charge density of 3.2 μC cm^{-2} is generated by assigning a constant partial charge to each C atom. The simulation is conducted in the NVT (constant number of species, volume, and temperature) ensemble for 100 ns at a temperature of 300 K, and the system is considered to have reached an equilibrium state when the number of ions flowing in the side pores remains constant for 50 ns.

Fig. 7.37 shows the evolution of the positions of representative Na^+ and Cl^- ions along the z-direction for the first 50 ns for the cases in which $R = 5$ Å and $R = 7.5$ Å. It is observed that the Na^+ and Cl^- ions can penetrate the pores of small sizes in the inner graphene layers and remain within the side pores after ~10 ns. When the pore size is large, the Na^+ ions can diffuse and enter the side pores quickly in ~2 ns and remain inside it for the entire duration. However, the Cl^- ions require almost 20 ns to enter a side pore and will remain there for another ~20 ns before diffusing back into the central pore. These results indicate that a sufficiently small pore radius R is required to trap the salt ions within the side pores and prevent their back-diffusion due to thermal motion.

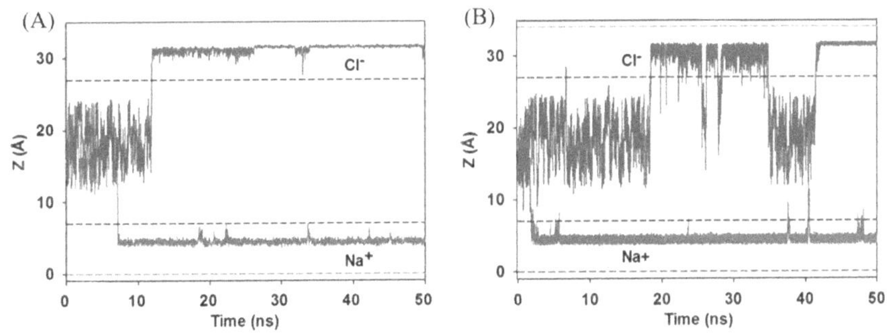

FIGURE 7.37 The z-coordinates of representative Na^+ and Cl^- ions during the first 50 ns when (A) $R = 5$ Å and (B) $R = 7.5$ Å. Source: *Modified and reprinted with permission from a work by Ho and Striolo (2015).*

In the nonequilibrium MD simulation, a constant acceleration of 2.4×10^{13} m s^{-2} (corresponding to a pressure of 100 MPa) is applied to all the water molecules and salt ions to drive them along the x-direction when the system reaches the equilibrium state after the equilibrium simulation of 100 ns. The simulation proceeds until all the water molecules and ions have almost constant velocities along the x-direction for 100 ns, as shown in Fig. 7.38A. It is apparent that the velocities of the Na^+ and Cl^- ions in the side pores are nonzero and are close to those of the water molecules, thereby indicating that the ions move with the water molecules. In addition, the Na^+ ions have higher velocities than the Cl^- ions due to the more intensive steric effect imposed on the Cl^- ions by the inner graphene layer. In the central pore, water molecules flow with a much higher velocity than the water molecules and ions in the side pores.

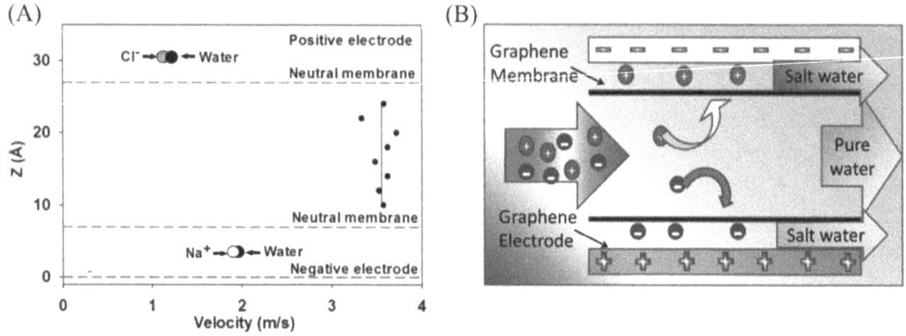

FIGURE 7.38 (A) Velocities of water molecules and Na^+ and Cl^- ions inside the CDI system using the partially porous graphene bilayers as the electrodes. (B) Schematic of the continuous operation process of the CDI system using the graphene bilayer electrode. Source: *Modified and reprinted with permission from a work by Ho and Striolo (2015).*

The slippage of salt ions with water molecules in the side pores can improve the performance of CDI devices tremendously. This improvement is attributed to desalination via conventional CDI devices or systems being achieved through ion adsorption on the surfaces of the electrodes. When the electrodes reach their adsorption saturation, electrode regeneration is required before any further desalination can occur, and thus the desalination process is discontinuous. In contrast, CDI electrodes fabricated using a graphene bilayer with a partially porous structure provide the possibility of a continuous desalination operation without the need for electrode regeneration. Such a continuous operation can be realized in the manner indicated in Fig. 7.38B. During the operation, saline water is fed continuously to the central pore; the salt ions diffuse from the central pore to the side pores under the effect of the surface charge on the electrodes. Upon entering the side pores, the salt ions flow through the system smoothly with the water molecules, allowing both purified and saline water to be simultaneously produced constantly. It is estimated that, under optimum conditions (a surface charge density of $4.2\ \mu C\ cm^{-2}$ and a salinity lower than $10.9\ g\ L^{-1}$), graphene bilayer electrodes can achieve complete salt rejection, a 70% water recovery rate, and a water permeability of 30 L cm^{-2} day^{-1} MPa^{-1}. However, the practical fabrication of partially porous graphene bilayers and their implementation in practical CDI systems still present a significant challenge.

Besides the above-mentioned graphene-based electrodes, MOF-based electrodes have been adopted to construct a CDI system for seawater desalination (Chang et al., 2015; Wang et al., 2017; Zhang et al., 2018a). During such a desalination process, salt rejection is mainly realized by salt ion adsorption on the surface of MOF-based electrodes. For salt ions to be strongly adsorbed onto MOF-based electrodes, strong interactions between the ions and the electrodes are required. In the MD study by Hong et al. (2021), three different salt solutions (NaCl, $CrCl_6$, and their mixture) are used to evaluate the salt adsorption performance of four different types of ZIFs, namely CdIF-1, ZIF-8, ZIF-67, and BeIF-1. The model shown in Fig. 7.34B is employed, in which the graphene electrodes are replaced with two ZIF electrodes. The four ZIFs share the same organic linker and rho-topology but contain different metal ions: Zn, Co, Cd, and Be for ZIF-8, ZIF-67, CdIF-1, and BeIF-1, respectively. The $CrCl_6$ solution is chosen to model the capability of the ZIF electrodes in removing heavy ions from seawater.

The simulation results show that among the four types of ZIFs, BeIF-1 exhibits the best performance in terms of salt rejection, while CdIF-1 corresponds to the highest water flow rate. A detailed analysis of the interactions of the salt ions with the ZIF electrodes reveals that the positively charged Cr^{6+} and Na^+ ions are mainly adsorbed at the locations of the N atoms in the organic linkers, while the negatively charged Cl^- ions are adsorbed at the locations of the metal ions (i.e., Zn, Co, Cd, and Be). In addition, due to their small ion radius (0.95 Å), only Na^+ ions can penetrate the ZIF electrodes under the effect of the external electric field that is applied in perpendicular to the water flow direction to drive the vertical movement of the ions. Despite having a smaller ion radius (0.44 Å) than the Na^+ ions, the Cr^{6+} ions are mainly adsorbed onto the surface of one of the ZIF electrodes because of the accumulation of Cl^- ions around them, which prevents them from entering the ZIF membrane. The accumulation effect becomes prominent when the $CrCl_6$ solution is mixed with the NaCl solution because of the abundance of Cl^- ions. The accumulation of Cl^- ions around the Cr^{6+} ions screens the effect of the electric field and lowers the ion removal efficiency of the membrane.

7.3 Liquid–vapor phase transition

The liquid–vapor phase transition is a fundamental phenomenon in nature. One of the most common liquid–vapor phase transition processes is the evaporation of a liquid droplet in a gaseous environment or a thin liquid film on a solid substrate. Evaporation is of fundamental importance in many natural and industrial processes, for example, transpiration, thermoregulation, water circulation, curing, and printing. Recently, the topic of the evaporation of a droplet has attracted increasing research interest because of its significant role in numerous novel practical applications, such as spray cooling, inkjet printing, nanofabrication, and modern biotechnology.

Many experimental and theoretical works have been devoted to understanding the underlying mechanisms of droplet evaporation and the fundamental principles of its various applications on the macroscale. However, contemporary theoretical models and experimental techniques fall short in obtaining high-resolution dynamical information on the molecular scale. In addition, for many novel applications that involve the evaporation of a liquid droplet or thin-film, the evaporation behavior occurs on the nanoscale and deviates significantly from that at the surfaces of bulk liquids because of various effects and factors, including the size effect of a nanodroplet, the texture and wettability of a substrate, as well as heat transfer at the liquid–substrate or liquid–vapor interface.

The capacity of MD simulation in providing detailed information of all the particles in a microscopic system renders this simulation technique a promising approach for investigating the evaporation of liquid droplets or thin films. The first MD modeling study of liquid droplets is attributed to Rusanov and Brodskaya (1977) who studied the effects of temperature, volume, and the number of molecules on the density and pressure profiles of a droplet system. The simulation results show that the droplet surface tension is positively proportional to the droplet size. A critical breakthrough in the MD simulation of the liquid droplet has been made by Thompson et al. (1984) who have investigated the effects of the temperature and size of the droplet on its surface tension. By focusing on mass transfer at a liquid–vapor interface, Long et al. (1996) have studied the evaporation of an Ar droplet. Based on their work, Kaltz et al. (1998) have further studied the evaporation of an O_2 droplet in a H_2 or He environment and advanced the MD study of droplet evaporation to the next level by investigating a diatomic system. Landry et al. (2007) have focused on the evaporation of an Ar droplet in a wide range of temperatures at which both subcritical and supercritical phase transitions could be observed. A subcritical transition indicates that the droplet surface temperature is lower than the critical phase-transition temperature, whereas a supercritical transition means the exact opposite.

An important parameter that determines the evaporation behavior of a liquid droplet is its Knudsen number (Kn), which is defined as $\mathrm{Kn} = \lambda/D$ with λ denoting the atomic mean free path near the droplet surface and D is the droplet diameter. When $\mathrm{Kn} << 0.1$, the droplet can be considered to exhibit continuum regime evaporation, which obeys the relationship commonly known as the D^2 law. This law states that the square of the droplet diameter possesses a linear relationship with time t, i.e., $dD^2/dt = -K$ with K denoting the evaporation constant. When the D^2 law is invoked, several assumptions are made: (1) the droplet evaporation is subcritical, (2) the liquid and vapor phases are quasi-steady, (3) the droplet remains spherical throughout its lifetime, and (4) the evaporation occurs in an environment where forced and natural convection is negligible.

When $\mathrm{Kn} \gg 1$, the droplet belongs to the free molecular regime where its evaporation behavior is significantly different from that of a droplet in the continuum regime because of the small droplet size. Under such a condition, kinetic-based models, such as the Hertz–Knudsen–Langmuir model, can be used to predict the evaporation behavior of the droplet. The Hertz–Knudsen–Langmuir model states that the droplet diameter D, instead of D^2, changes linearly with time. When this model is applied, two assumptions are made: (1) the liquid and vapor phases are quasi-steady, and (2) the velocity distributions of the vapor-phase atoms that adsorb into and evaporate from the droplet are Maxwellian in nature (Kryukov et al., 2004).

In the transitional regime, with a moderate value of Kn, a theoretical prediction of the evaporation behavior of a droplet requires solutions of the Boltzmann equation. More details of theoretical prediction models for droplet evaporation dynamics can be accessed in an early book by Frohn and Roth (2000) and another by Sirignano (1999).

The main limitation of the theoretical models developed for droplet evaporation mentioned above lies in the fact that they often require knowledge of phenomenological parameters or assumptions regarding the evaporation conditions. For instance, the evaporation and condensation coefficients are required when the Hertz–Knudsen–Langmuir model is applied, and the D^2 law is valid only when the liquid and the vapor phases are quasi-steady, and the droplet remains spherical. As the size of a droplet approaches the nanoscale, its evaporation is governed by surface heat transfer and energy balance instead of mass diffusion inside the vapor, which renders it challenging to apply existing theoretical models to describe the evaporation behavior of the droplet. MD simulation has proven to be a more fundamental approach for investigating droplet evaporation dynamics.

In the following sections, the main MD simulation techniques for studying droplet evaporation in a gaseous environment and droplet evaporation on a solid substrate are introduced with highlights on various factors, including the droplet constituents, external electric fields, substrate texture, and wettability.

7.3.1 Droplet evaporation in a gaseous environment

Droplet evaporation in an environment of its gaseous counterpart might be the simplest case of liquid evaporation for MD simulation. Fig. 7.39 shows a typical MD simulation model for the evaporation of an Ar droplet surrounded by Ar gas. PBCs are applied along the *x*-, *y*-, and *z*-directions. The first step of the simulation is to prepare a homogeneous liquid Ar at a prescribed temperature and density by placing Ar atoms at the sites of a face center cubic (FCC) lattice in a cubic simulation box. The simulation is then conducted in the NVT ensemble for a specific time. A homogeneous vapor is prepared in another system with the desired temperature and density in a similar manner. In the next step, a sphere with a desired diameter is removed from the vapor cube, and a sphere with the same diameter is selected from the liquid cube and inserted into the vapor cube to fill the region where the vapor sphere is removed, thereby forming a system in which vapor particles surround a spherical droplet. To avoid any overlap between the atoms at the liquid–vapor interface, the distance of all the atom pairs across this interface is calculated. If the distance of an atom pair is smaller than a critical value (that is usually chosen to be the zero-energy pair-separation

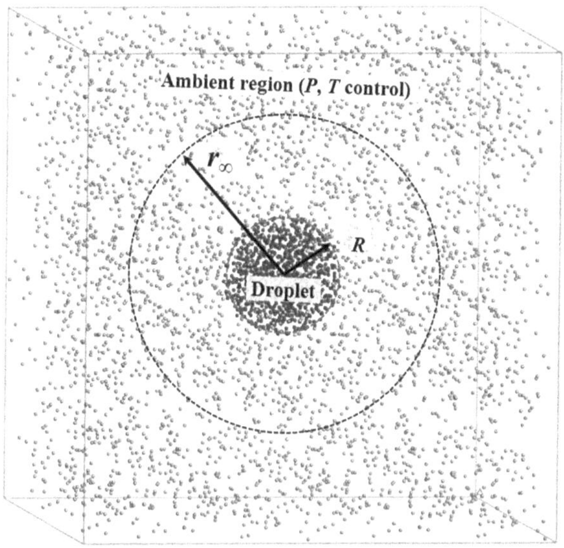

FIGURE 7.39 System configuration of the MD simulation model for the evaporation of a liquid Ar droplet. The droplet atoms are enclosed in the inner sphere region and the atoms outside this region form the vapor-phase. The variables T and P represent the ambient temperature and pressure, respectively (Landry et al., 2007).

distance σ of the LJ potential), either the atom in the liquid phase or the atom in the vapor phase will be randomly deleted.

In the simulation of the evaporation process, proper temperature and pressure control are required. To achieve this objective, the simulation cell is divided into three regions by two spheres, as indicated by the dotted lines (Fig. 7.39). The red atoms enclosed in the inner sphere constitute the liquid droplet. The region beyond a distance of r_∞ from the center of the droplet is defined as the ambient region. The pressure and temperature control of the entire simulation system can be realized through the control of this region. An appropriate value of r_∞ is required to ensure that the ambient region is adequately large to include sufficient atoms to facilitate a stable temperature and pressure control.

A temperature or pressure gradient between the liquid and vapor phases is imposed to initiate the evaporation process. Such a gradient can be created during the construction process of the simulation system. Alternatively, the temperature or pressure gradient can be imposed after the vapor and liquid phases reach an equilibrium state. The temperature gradient is imposed by abruptly increasing the temperature of the ambient region via the velocity rescaling method, while the pressure gradient is imposed by reducing the pressure of the ambient region via deleting atoms within it.

There are several methods of determining if the liquid and vapor phases have reached an equilibrium state. The most direct method would be to examine the number of atoms in the vapor phase. At the equilibrium state, the time-averaged number of atoms in the vapor phase will be constant. Another method is based on the average kinetic energy per atom. At the equilibrium state, there is no net mass transfer between the liquid phase and the vapor phase, and thus the drifting velocity of both phases is zero. Consequently, the kinetic energy per atom in both phases should equal $3k_BT/2$, which can be taken as a criterion for the equilibrium state. An equivalent criterion can be set on the mean displacement of the molecules from the center of mass of the entire system. Due to the absence of a drifting velocity at an equilibrium state, the mean displacement remains constant with time.

Once the evaporation process is initiated, the ambient temperature is kept constant through a velocity rescaling algorithm. The ambient pressure is controlled by randomly deleting or adding atoms to the ambient region so that the vapor density in this region remains a prescribed constant corresponding to the target pressure and temperature. When an atom is added to the ambient region, its location is randomly selected and verified to ensure that no atoms overlap. This pressure control method is a vital improvement to early MD simulations that model liquid droplet evaporation, in which the ambient pressure of the system is allowed to increase without any control of the vapor density. Neglecting the variation in the vapor density or ambient pressure may significantly affect the simulation results for a long simulation duration. Therefore, a pressure control strategy is crucial in achieving reliable statistical results, particularly when the ambient and vapor regions are not much larger than the liquid droplet, which is common because increasing the sizes of the ambient and vapor regions will directly result in a large system size and high computational cost. Hence, monitoring the vapor density inside the ambient region is an efficient pressure control method for the MD simulation of droplet evaporation. This control is particularly important for droplets with small diameters because their surface temperature and evaporation behavior are highly sensitive to the ambient pressure.

Note that the addition or deletion of atoms during the simulation can result in some artificial behaviors of the system at the initial stage, particularly when there is an initial large pressure gradient across the interface of the ambient and vapor regions. This pressure gradient arises from the initial pressure of the vapor region being lower than the target pressure of the ambient region and drives the atoms in the ambient region toward the droplet. This unrealistic behavior is more prominent initially when the pressure difference is large but will gradually diminish as the pressure of the ambient region eventually balances that of the vapor region.

In addition, two corrections must be made to the atomic velocities before calculating the temperatures of the regions. The first correction accounts for the drifting velocity associated with the center of mass of the system. It should be subtracted from the velocity of each atom since a uniform translational velocity shared by all the atoms in a system does not contribute to its temperature. This drifting velocity is due to the pressure control scheme in which the addition or deletion of atoms in the ambient region defies the principle of momentum conservation of the system.

The second correction accounts for the radial drift of the evaporated atoms as they leave the droplet. For this correction, a procedure proposed by Sumardiono and Fischer (2006) can be used. In this procedure, the droplet is assumed to retain its spherical symmetry during the entire evaporation process, and hence its drift velocity only has a radial component. The spherical region spanning from the center of the droplet to the edges of the simulation box is divided into n spherical shells. In each shell, the individual velocities of the atoms are decomposed into radial and tangential components. The average of the radial velocity components of all the atoms in each spherical shell corresponds to its drift velocity. When the temperature of a shell is calculated, its drift velocity is subtracted from the individual velocities of the atoms that it contains.

There are three general definitions to distinguish between the liquid and vapor phases in droplet evaporation, two of which were proposed by Thompson et al. (1984) as the "Cornell" and "Oxford" definitions. The other was proposed by ten Wolde and Frenkel (1998).

The "Cornell" definition determines whether a particle i belongs to the liquid phase by calculating the shortest distance between it and all the particles that have been identified in the liquid phase. If this distance is smaller than a predetermined critical value, the particle i is regarded as in the liquid phase. Usually, the critical distance is set at $\sim 1.5\,\sigma$ with σ being the zero-energy distance of the LJ potential that describes the interaction between an atom in the droplet and an atom in the vapor. A prerequisite for applying the "Cornell" definition is to have an initial list of particles that constitute the droplet. The procedure for implementing this definition begins with identifying the "center" of the droplet, which is taken to be the particle with the smallest distance to the center of mass of the droplet. Upon locating the "center" particle, particles within a critical distance $\sim 1.5\,\sigma$ of it are located, and they form the first shell surrounding it. Then, by scanning radially outward, particles within the same distance $\sim 1.5\,\sigma$ of the particles of the first shell are located to form the second shell. This procedure is repeated until no further particles are located within the critical distance of the outmost shell of the droplet. The list of droplet particles is then updated accordingly to prepare for the next round of liquid–vapor identification.

In the "Oxford" definition, a particle i is considered to be in the liquid phase at time t if there are at least n_c particles located within a critical distance $\sim 2.5\sigma$ of it. In the MD simulations of droplet evaporation, n_c is chosen to be much smaller than the average number of particles in a sphere with a radius of $\sim 2.5\sigma$ in a single liquid phase but is slightly larger than the corresponding number for a single vapor phase. For the Ar droplet simulated, the number n_c ranges from 4 to 7 depending on the ambient temperature.

The definition of ten Wolde and Frenkel (1998) is similar to the "Oxford" definition, and the main difference is that the critical distance in their definition is chosen as 1.5σ, which corresponds to the location of the first minimum in the RDF of the liquid. The other difference is that the number n_c in their definition has a fixed value of 4.

From a physical perspective, the "Cornell" definition is the most intuitive, but its implementation is also the most computationally intensive. This is because determining whether a particle belongs to the liquid phase or the vapor phase requires the calculation of its distances to all the particles that have been identified in the liquid phase in the previous round of identification. Hence, the identification based on the "Cornell" definition is dependent on historical data. For the other two definitions, it is possible to identify, at any given time, every particle belonging to the liquid or the vapor phase by simply calculating the number of neighbors within a certain distance of it.

The Ar droplet with an initial diameter of ~4.5 nm is taken as an example to discuss the influence of the ambient pressure on its evaporation induced by a temperature gradient. An LJ potential truncated at a distance of 2.5σ is adopted to model the interactions among the Ar atoms. The simulation time step is set at 10 fs. The liquid phase is determined based on the definition of ten Wolde and Frenkel (1998). The droplet radius R is taken to be that of a sphere surrounding the center of mass of the droplet that contains the same number of Ar atoms as that in the liquid phase. Based on the droplet radius, atoms in the liquid phase with a distance of less than $R/2$ from the center of mass of the droplet are considered to form the droplet core, while the other atoms in the liquid phase form the droplet surface. The critical temperature and density of liquid Ar are 130 K and 538 kg m^{-3}, respectively, which correspond to a critical pressure of 4.1 MPa.

Fig. 7.40 shows the temperature T_S of the droplet surface as a function of time for ambient pressures of 0.4 MPa, 1.1 MPa, 3.0 MPa, and 6.1 MPa at an ambient temperature of 500 K. The simulation time is normalized by the droplet lifetime τ_d so that all the simulated cases are comparable. Here, the droplet lifetime refers to the time required for a droplet to evaporate entirely. At a low ambient pressure of 0.4 MPa, the droplet surface temperature first decreases, resulting from the ambient pressure being lower than the initial pressure of the droplet. After $t/\tau_d \approx 0.1$, the droplet surface temperature converges to a value known as the pseudo wet-bulb temperature. The term "pseudo" is used because the droplet temperature is always nonuniform and transient, according to Givler and Abraham (1996). No droplet heating occurs at this temperature, and all the heat received by the droplet from the vapor is consumed by droplet evaporation. At the ambient pressures of 1.1 MPa and 3.0 MPa, the droplet surface temperature first increases and then converges to a pseudo wet-bulb temperature at a normalized time of 0.1 and 0.3, respectively. In both cases, the pseudo wet-bulb temperature is below the critical temperature of liquid Ar. At an ambient pressure of 6.1 MPa, which is greater than the

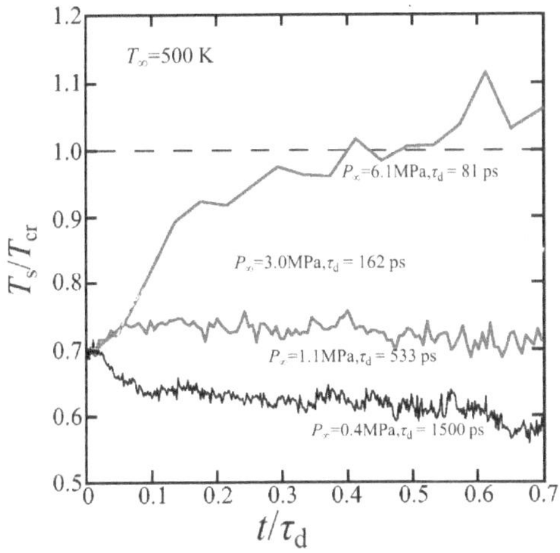

FIGURE 7.40 Surface temperature of the Ar droplet as a function of time for ambient pressures P_∞ of 0.4 MPa, 1.1 MPa, 3.0 MPa, and 6.1 MPa at an ambient temperature of 500 K. Source: *Modified and reprinted with permission from a work by Landry et al. (2007).*

critical pressure of 4.1 MPa of liquid Ar, the droplet surface temperature first increases and exceeds the critical temperature, and subsequently, it fluctuates above this temperature during the entire evaporation process.

The variations of the Ar droplet temperature during its evaporation at different ambient pressures result in distinct evaporation behaviors. For an ambient pressure of 0.4 MPa, the density of the droplet increases and its diameter decreases, which can be attributed to the initial temperature decrease. At the normalized time of 0.1 when the pseudo wet-bulb temperature is reached, the droplet diameter starts to exhibit an almost linear decrease with time. This linear trend is to be expected due to the large value of Kn. Based on the relation $\mathrm{Kn} = \lambda/D$ and the estimation of $\lambda = \left(\sqrt{2}\pi d^2 n_v\right)^{-1}$, the value of Kn for the case in which the ambient pressure is 0.4 MPa is found to be much larger than 1. Here, d denotes the effective atomic diameter, and n_v is the atomic number density of the vapor near the droplet surface. Hence, the evaporation behavior of the droplet can be described by the Hertz–Knudsen–Langmuir model. Compared with the D^2 law, the simulated evaporation rate is extremely low. This lower rate indicates that for droplet evaporation with a large value of Kn, hydrodynamic-based models, such as the D^2 law, are not suitable for describing the droplet evaporation behavior observed in the MD simulation, and kinetic-based models, including the Hertz–Knudsen–Langmuir model, should be adopted.

Under the moderate ambient pressure of 3.0 MPa, it is observed that upon reaching the pseudo wet-bulb temperature, the droplet evaporation obeys the D^2 law, that is, the rate of change of the square of the droplet diameter is a constant. As the ambient pressure increases to 6.1 MPa, which is a value larger than the critical pressure, the droplet surface temperature surpasses the critical temperature, indicating that the evaporation condition is supercritical. Under this condition, the droplet tends to lose its sphericity because of the significant decrease in its surface tension. In such circumstances, the number of droplet atoms is used to characterize the evaporation rate instead of the droplet diameter, and they decrease linearly with time for most of the evaporation process.

Note that the simulation results presented above do not provide any general rules for liquid droplet evaporation. The specific evaporation behaviors observed for different simulation systems depend not only on the ambient pressure but also on many other factors, including the ambient temperature, droplet size, and droplet species.

A more complicated case, the evaporation of a saline aqueous droplet in the presence of an electric field, can be taken as an example. The evaporation of a droplet containing dissolved salts under the effect of an intensive electric field is a critical step for the electrospinning technique, which is a direct approach for the continuous production of polymer fibers. In a typical electrospinning system, a polymer solution is pumped and ejected from a syringe needle that is placed in a high-voltage electric field. Upon ejection, the polymer solution forms a spraying stream, and the solvent begins to evaporate under the strong electric field, leaving the solute to deposit and solidify as a nonwoven fibrous structure on a ground metal plate. A rapid evaporation of the solvent is essential for the formation of ultrafine fibers with nanoscale diameters.

A typical MD simulation system for modeling the evaporation of a saline aqueous droplet under the effect of an electric field is shown in Fig. 7.41 (Wang et al., 2014). The simulation box is a cube with a side length of 12 nm. The aqueous droplet is placed in the center of the simulation box and surrounded by N_2 gas. The droplet has an initial radius of 2 nm and contains 1120 water molecules, which corresponds to a density of $1\ g\ cm^{-3}$. The surrounding nitrogen gas comprises 600 molecules, which corresponds to a density of $16.47\ kg\ m^{-3}$. Three species of salts (NaCl, LiCl, and KCl) are individually dissolved within the droplet to investigate the influence of different salt species on droplet evaporation. The number of LiCl salt molecules varies from

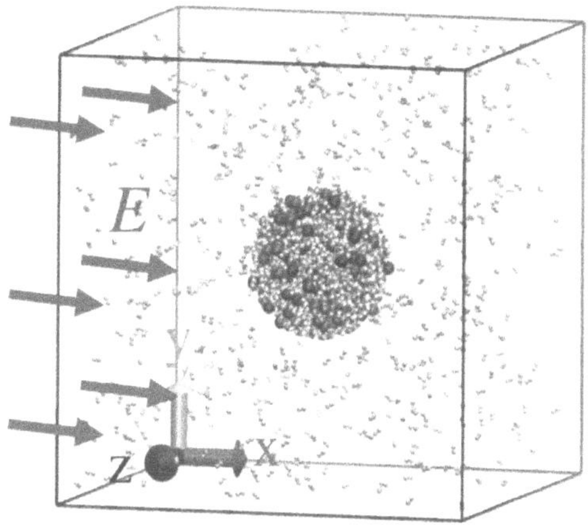

FIGURE 7.41 Simulation model for the evaporation of an aqueous droplet within a homogeneous electric field in a nitrogen gas environment. Source: *Reprinted with permission from a work by Wang et al. (2014).*

20 to 120 for studying the effect of ion concentration on droplet evaporation. A homogeneous electric field is applied along the *x*-direction, and its intensity varies in the range of 0.4–0.8 V nm^{-1}.

The SPC/E (extended simple point charge) water model is chosen to construct the water molecules. A combination of a truncated short-range LJ potential (with a truncation distance of 10 Å) and a long-range electrostatic potential is used to model the interactions among all the water molecules and salt ions. PBCs are applied along the *x*-, *y*-, and *z*-directions. Before the evaporation is initiated, the liquid droplet and vapor are first equilibrated at a temperature of 298 K in the NVT ensemble for 0.1 ns with a time step of 1 fs. The equilibrium state is considered to be reached when the number of water molecules in the droplet only fluctuates slightly around a constant value.

Upon reaching an equilibrium state, evaporation is initiated by abruptly increasing the temperature of the nitrogen gas to 600 K via the velocity rescaling method. The liquid and the vapor phase are distinguished by using the definition of ten Wolde and Frenkel (1998) with a critical length R_{cl} of 4.34 Å. Fig. 7.42 shows snapshots of the evaporation process of the water droplet containing 120 dissolved NaCl molecules at an electric field intensity of 0.8 V nm^{-1}. The distributions of the number density of the water molecules and salt ions are also shown accordingly. It is observed that in the presence of an electric field, the aqueous droplet is stretched along the direction of the field. After the water molecules

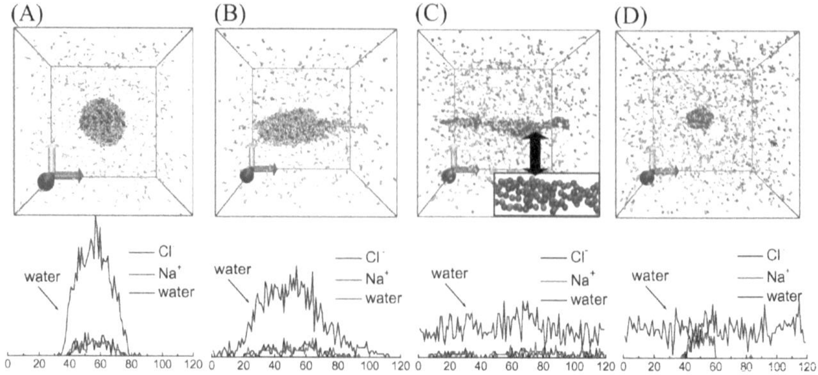

FIGURE 7.42 Simulation snapshots of an aqueous droplet containing 120 dissolved NaCl molecules and the corresponding distributions of the number density of the water molecules, Na^+ ions, and Cl^- ions at an electric field intensity of 0.8 V nm^{-1} at different time: (A) 0 ps, (B) 100 ps, (C) 200 ps, and (D) 1000 ps. Source: *Modified and reprinted with permission from a work by Wang et al. (2014).*

have fully evaporated at ~200 ps, the salt ions regroup and form an ion cluster.

Fig. 7.43A shows the number of water molecules in the pure water droplet and the aqueous NaCl droplet as a function of time at

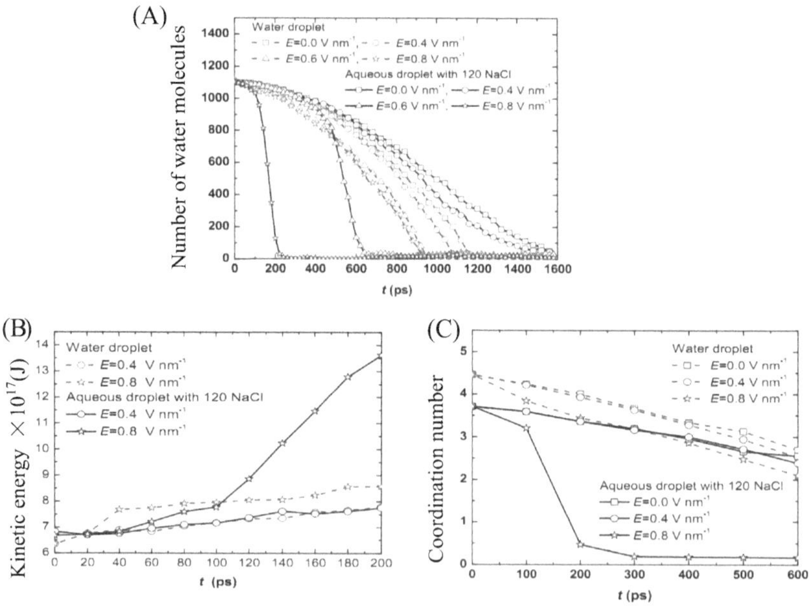

FIGURE 7.43 (A) Number of water molecules as a function of time in the pure water droplet and the aqueous droplet with 120 dissolved NaCl molecules at various electric field intensities. (B) Average kinetic energy and (C) coordination number of the water molecules for the two droplets. Source: *Reprinted with permission from a work by Wang et al. (2014).*

different electric field intensities. With the increase of the electric field intensity, the evaporation rates of both droplets increase. In the absence of the electric field or at a low field intensity, the evaporation rate of the aqueous droplet is lower than that of the pure water droplet. This result indicates that the addition of ions to a water droplet does not always enhance its evaporation rate. As the electric field intensity increases to 0.6 V nm^{-1} and beyond, the evaporation rate of the aqueous droplet increases dramatically and greatly exceeds that of the pure water droplet. In addition, most of the evaporation of the aqueous droplet rapidly occurs in a short duration. For instance, at an electric field intensity of 0.6 V nm^{-1}, the number of water molecules in the aqueous droplet decreases sharply in the time between 500 ps and 600 ps.

The lower evaporation rate of the aqueous droplet in the electric fields with field intensities lower than 0.4 V nm^{-1} is mainly because such intensities are insufficient to significantly stretch the droplet, which still retains its sphericity during the evaporation process. Due to the strong attractive forces between the salt ions and the water molecules, hydration shells form around the ions. Although the formation of hydration shells is helpful in reducing the interactions among the water molecules, the attractive forces between the ions and water molecules still impose additional energy barriers that the water molecules must overcome to escape from the droplet. Furthermore, the hydration effect becomes increasingly prominent as the evaporation process proceeds. Consequently, the evaporation rate of the aqueous droplet is lower than that of the pure water droplet.

When the electric field intensity increases to 0.6 V nm^{-1} and beyond, the droplet is stretched significantly and loses its sphericity, as shown in Fig. 7.42. This deformation increases the surface area of the droplet and facilitates heat transfer from the nitrogen gas to it, thereby boosting its evaporation rate. The strong electric field will also drive the cations and anions to move along or against it, thereby increasing the kinetic energy of the ions and water molecules, as indicated by Fig. 7.43B. With higher kinetic energy, the water molecules have higher probabilities of escaping from the droplet. Moreover, the deformation of the droplet reduces the coordination number of water molecules, as implied by Fig. 7.43C. With a smaller coordination number, each water molecule is connected to fewer neighboring water molecules, and therefore the energy barrier for water evaporation is smaller. The combination of the aforementioned effects contributes to the rapid evaporation of the aqueous droplet within a strong electric field.

Among the three salt solutions considered (KCl, NaCl, and LiCl), the KCl solution corresponds to the greatest increase in the evaporation rate of the droplet due to the weakest attractive interactions between the K^+ ions and the water molecules. Note that an optimized salt ion concentration is required to achieve the maximum evaporation rate. If the concentration exceeds the optimized value, the attractive forces between the anions and cations will suppress the driving effect of the electric field and reduce the kinetic energy of the water molecules; consequently, the evaporation rate of the droplet will decrease.

7.3.2 Liquid evaporation on a substrate

Compared with the evaporation of a liquid droplet in a gaseous environment, the evaporation of a liquid droplet or a liquid thin film on a substrate is a common phenomenon involved in many industrial applications and natural processes. For instance, there is usually a tiny amount of water existing on various biological or material surfaces, such as the surfaces of plant leaves, skin, and soil. The loss of water from these surfaces causes distinctive effects, including plant withering, skin dehydration, and soil desertification.

On the macroscale, various theories and models have been developed to describe the dynamics of droplet evaporation on substrates. For instance, three evaporation models can be adopted to describe the changes in the contact line and contact angle of a droplet on a substrate during the droplet evaporation. The first one is the constant contact radius model in which droplet evaporation does not cause the contact line to shift but reduces the contact angle instead. The second one is the constant contact angle model in which droplet evaporation leads to shrinkage of the contact line, while the contact angle remains unchanged. The third one is the mixed model in which both the contact line and contact angle change during the entire evaporation process. Birdi and Vu (1993) have found that the evaporation rate of a liquid on a substrate exhibits a linear relationship with time if the contact angle is smaller than 90°, whereas a contact angle larger than 90° results in a nonlinear relationship. For hydrophobic substrates, Golovko et al. (2009) and Xu et al. (2013) have observed a linear relationship between $m^{2/3}$ (m denotes the mass of the liquid droplet) and time. In contrast, a linear relationship between m and time has been reported by Hu and Larson (2002) for hydrophilic substrates.

When the droplet size approaches the nanoscale, the laws of evaporation for macroscopic droplets on substrates lose their validity. The mass transfer within a three-phase system is assumed to dominate the evaporation process, and heat transfer is usually neglected on the macroscale. However, on the nanoscale, heat transfer between the interfaces of different phases plays a more significant role than mass transfer. A recent simulation work by Holyst and Litniewski (2008) demonstrates that the evaporation of a nanosized droplet on a substrate is primarily regulated by heat transfer instead of mass transfer. In addition, the temperature distribution across the liquid–vapor interface can be discontinuous. This discontinuous distribution cannot be explained by the classical models developed for exploring macroscale droplet evaporation.

An MD simulation of the evaporation of a droplet on a substrate generally begins by achieving thermal equilibrium among all the phases (solid substrate, liquid droplet, and vapor) in a simulation box. The simulation box is typically a cubic box with PBCs applied along the x-, y-, and z-directions. The solid substrate is positioned at the bottom of the simulation box, and the liquid droplet is located directly above the substrate with a small distance between them. Initially, all the liquid-phase atoms are arranged to form a cubic lattice with a chosen lattice constant corresponding to the liquid density. A ceiling wall is usually added at the top of the simulation box to prevent vapor-phase atoms from passing through it and interacting unphysically with the substrate or liquid droplet because of the PBCs applied. Nevertheless, this ceiling wall may not be necessary. Without the wall, the atomic layers of the substrate (if it possesses several atomic layers) can be divided into three regions instead. The layers in the bottom region are fixed in place to eliminate any movement of the substrate. The layers in the middle region are assigned as a heat source from which thermal energy is supplied to the system. The remaining top layers are assigned as an interacting wall (without temperature control) through which heat is transmitted to the liquid droplet from the heat source. An essential requirement of the thickness of the fixed bottom region is that it should be sufficiently large to prevent the heat source from interacting with the vapor-phase atoms that have reached the bottom surface of the solid substrate by crossing the top boundary of the simulation box in the absence of a ceiling wall.

In the initial configuration, there is no gas phase. To achieve thermal equilibrium among the solid, liquid, and vapor phases, the temperature of the substrate is fixed at a constant value. After a certain duration, the droplet relaxes, and a vapor phase is formed with a temperature equal to that of the droplet. The equilibrium state of the droplet on the substrate can be verified by characterizing the number of atoms in the droplet in a manner similar to how the equilibrium state is identified for a droplet in a gaseous environment.

To initiate the evaporation process, two strategies can be adopted. The first is heating the substrate, while the second is reducing the pressure of the vapor phase by increasing the simulation box size. The first strategy is more frequently adopted in reported research works because droplet evaporation on a heated substrate is a common phenomenon in natural and industrial processes.

For the substrate heating strategy, the temperature of the substrate is rapidly increased to a high value. Once the substrate reaches the target temperature, droplet evaporation is initiated. The entire evaporation process can be divided into two stages according to the temperature of the droplet. In the first stage, the droplet temperature increases monotonically to a target value, while in the second stage, it stays invariant at this value. Note that during the evaporation process, only the temperature of the substrate is being controlled. The liquid droplet and vapor are placed in the NVE (constant number of species, volume, and energy) ensemble without temperature control. The heat transferred from the substrate to the droplet provides the energy required for the droplet evaporation.

An issue arises when light atoms are present in the droplet evaporation system, for example, the H atoms within water molecules. Atoms in the solid substrate and water droplet possess significantly different vibrational frequencies due to the huge difference in their atomic masses. To fully capture the dynamic characteristics of the light atoms, the time step of the MD simulation should be sufficiently small. However, such a tiny time step is usually unnecessary for simulating the solid substrate and is computationally costly. The high-frequency vibrations of the light atoms in such systems are commonly suppressed to lower the computational cost. If this is not feasible, the entire solid substrate is fixed in place, and the heat transfer between the solid substrate and the liquid/vapor phase is realized through a new thermal wall algorithm that was first proposed by Maroo and Chung (2010).

The fundamental notion behind this algorithm is shown in Fig. 7.44. A "critical zone" is defined as the region directly above the substrate up to a critical distance of r_{cr} away. Only the droplet particles within this critical zone can experience repulsive forces from the substrate particles. At the critical distance, a droplet particle will experience zero force from the substrate surface particle that is nearest to it. Beyond this critical distance, a droplet particle will only experience attractive forces from the substrate particles. If the distance between a droplet particle and any of the substrate particles exceeds the cutoff distance of the LJ potential that models their interaction, the droplet particle will not experience any forces from the substrate. Within the critical zone, the magnitude of the net force imposed on a droplet particle by the substrate is compared with that of the net force imposed on the droplet particle by all the other droplet particles. If the former is larger, this droplet particle will be "heated up" to the target temperature.

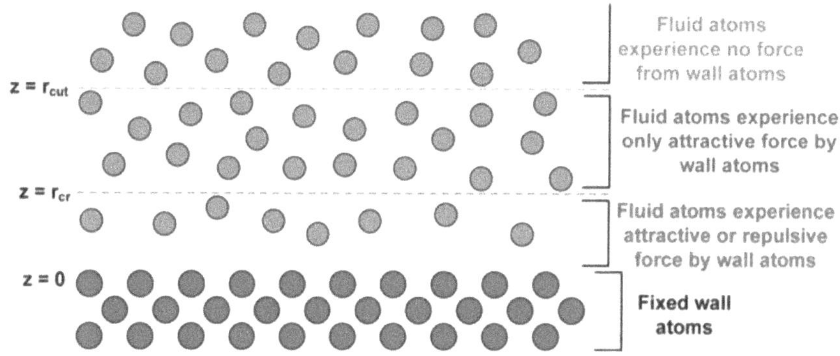

FIGURE 7.44 Illustration of the thermal wall algorithm. Source: *Reprinted with permission from a work by Maroo and Chung (2010).*

The heating process of a droplet particle under the effect of the substrate can be divided into three stages. During the first stage, the droplet particle is accelerated toward the substrate surface and enters the critical zone. In the second stage, the droplet particle gains energy from the substrate surface and is heated up. In the third stage, the droplet particle is repelled by the substrate and exits the critical zone.

Although this algorithm has proved to yield meaningful results, its implementation in practical MD simulations is tedious, particularly when droplet particles are not monoatomic. When droplet particles are monoatomic, the critical distance of the critical zone can be derived from the minimum energy of the LJ potential used to describe the interactions among the droplet particles. However, for nonmonoatomic droplet particles, it is not possible to obtain a definite value of the critical distance analytically. In such circumstances, a trial-and-error approach is usually adopted, as proposed by Sumith and Maroo (2015), who have extended the above-discussed thermal equilibrium algorithm to the study of water droplet evaporation on a platinum surface. To estimate the critical distance, a cube of water with a side length of 4 nm is relaxed and allowed to spread freely on a platinum surface. The trial values of the critical distance are selected to vary around the distance between the platinum substrate surface and the location where an individual water molecule exhibits the minimum potential energy when brought in close contact with the substrate. The equilibrated droplet temperature exhibits a linear relationship with the trial value of the critical distance. Hence, the precise value of the critical distance can be obtained by fitting the equilibrated droplet temperature to a linear function of the trial values of the critical distance and setting the droplet temperature to be the substrate temperature.

Besides the issue on the modeling of light atoms as discussed above, another two technical issues should be noted. The first pertains to the calculation of the temperature of the vapor phase. In Section 7.3.1, two corrections that must be performed during the temperature calculation of the vapor generated by a droplet in a gaseous environment have been discussed. One of these corrections addresses the drifting velocity of the vapor-phase atoms when they escape from the droplet. For droplet evaporation on a heated substrate, the vapor phase also possesses such a drifting velocity that must be subtracted from the individual velocities of the vapor particles in calculating the temperature of the vapor. Such a temperature correction is seldom performed in reported research works. One possible reason is that it is challenging to define the drifting velocity of the vapor for the evaporation of a droplet on a substrate and obtain an accurate description of it. However, for the evaporation of a thin liquid film on a substrate, the direction of the drifting velocity is usually well defined. Thus, temperature correction can be conveniently performed.

The second issue is distinguishing between the vapor and the liquid phase for droplet evaporation on a substrate. Among the three definitions of a liquid phase in the case of droplet evaporation in a gaseous environment, only the "Cornell" definition can be extended to the case of droplet evaporation on a substrate. However, implementing this definition is computationally costly. The other two definitions identify the particles in the vapor phase based on their much fewer nearest neighbors than those in the solid phase. They are inapplicable in this case because most of the particles in the liquid phase near the substrate–droplet interface also have much fewer nearest neighbors within the critical distance than those far away from the interface. However, the number density of the droplet particles exhibits a sharp decay, which is utilized to identify the vapor and the liquid phase and hence their interface.

Hereafter, some exemplifying research works regarding droplet evaporation on heated substrates are briefly discussed. Zhang et al. (2013) have studied the evaporation of a nanosized Ar droplet on a heated substrate and reported that the temperature distribution across the liquid–vapor interface is discontinuous during the initial stage of the evaporation in which the vapor temperature is always higher than that of the droplet. In addition, as the droplet–substrate interaction is artificially strengthened (by tuning the potential energy well ε of the LJ potential that is adopted to describe the droplet–substrate interaction), the evaporation rate increases because of the enhanced thermal conduction across the liquid–solid interface. The droplet evaporation dynamics always follow the mixed model, i.e., the contact angle and contact radius always decrease simultaneously regardless of the droplet–substrate interaction strength. Below a critical droplet size, the contact radius decays at a much lower rate than the contact line, whereas above this critical size, the opposite holds true. The droplet evaporation and the heat transfer to the vapor phase from the solid substrate occur preferentially in the vicinity of the three-phase contact line for cases involving strong droplet–substrate interactions. However, for cases involving weak droplet–substrate interactions, there are no preferential locations for the evaporation and heat transfer.

Zhang et al. (2015) have investigated the evaporation of saline aqueous nanodroplets on a platinum surface. As the salt concentration increases, the evaporation rate of the droplets decreases because of the hydration effect of the salt ions. After the water molecules have fully evaporated, the salt ions form two distinct types of salt crystals, clump and ringlike, depending on the solid–liquid interaction strength, the initial salt concentration, and the evaporation rate of the droplets.

In the MD simulation study conducted by Wan et al. (2015), the evaporation of water on solid surfaces with nanoscale hydrophobic–hydrophilic patterns has been investigated. Unexpectedly, water evaporation occurs much more rapidly on the patterned surfaces than on surfaces with uniform wettability. On the patterned surfaces, all the water molecules reside in the hydrophilic regions. Hence, the water layers on the patterned solid surfaces are thicker than those on solid surfaces with uniform wettability. However, the increase in the thickness of the water layers does not affect the evaporation rate significantly. Interestingly, although no water molecules reside on the surface of the hydrophobic regions, these regions contribute significantly to the overall water evaporation rate. This result can be attributed to many water molecules diffusing across the boundaries between the hydrophilic and hydrophobic regions as well as evaporating from the hydrophobic surface. Hence, by manipulating the lengths of the boundaries of such patterned surfaces, their corresponding evaporation rates can be effectively controlled.

The strategy of creating a patterned surface to enhance water evaporation on a substrate has been adopted by Kieu et al. (2019) in their MD simulation study of water evaporation on graphene oxide. In their simulation model, a periodic corrugation with a sinusoidal morphology is introduced to a graphene oxide monolayer. The simulation results show that in the presence of the corrugation, the evaporation of water is noticeably enhanced, and the evaporation rate increases linearly with the amplitude of the corrugation. The analysis of the dynamic behaviors of the water molecules indicates that the nonuniform density distribution of the water molecules caused by the corrugation is the main reason for the improved water evaporation.

By developing a coarse-grained MD (CGMD) model, Zhang et al. (2018b) have conducted simulations on water evaporation through meso-size (2–5 nm) channels formed by graphene bilayers to study membrane distillation processes, in which vertically aligned graphene layers are adopted as the membrane. In the CGMD model, a water molecule is represented as a bead with a dipole moment, and four C atoms in graphene are represented as a single bead in the coarse-grained graphene layers. The graphene membrane is in direct contact with a water droplet. After a water molecule has evaporated from the droplet, it enters the channels of the graphene bilayers and is transported within them.

The interactions between the water and graphene beads are modeled as van der Waals forces and described by the LJ potential. The relevant potential parameters are first carefully calibrated to reproduce the cohesive energy between a water molecule and a sufficiently large graphene monolayer, and these parameters are then validated by reproducing the contact angles of water droplets on a graphene monolayer and a graphite substrate. With the implementation of the developed CGMD model, the transportation of water molecules in the channel of a graphene bilayer is simulated for a duration of $\sim$100 ns, which allows for a comprehensive evaluation of the influences of channel opening, hydrostatic pressure, and temperature in a statistically reliable manner.

The CGMD simulation results show that the water permeation flux through the graphene bilayer channels is almost three orders of magnitude higher than that through the commonly used microporous polymer membranes. Moreover, the flux exhibits nonmonotonic dependence on the channel opening. It is the highest when the channel opening is 2 nm, at which the dominant transport mode of water molecules inside the channel changes from surface diffusion to Knudsen diffusion. With surface diffusion, the water molecules are adsorbed onto the surface of the graphene layers and subsequently move on the surface along the lateral direction of the graphene channels, as shown in Fig. 7.45A. The desorption of water molecules from the graphene surface occurs at a low frequency. With Knudsen diffusion, the water molecules stay away from the graphene layers most of the time and only change their movement directions once they come into contact with the graphene layers (Fig. 7.45B). Compared with surface diffusion, Knudsen diffusion corresponds to a high resistance for molecule transport within a channel. Hence, the water permeation flux for the channel opening of 5 nm is much lower than that for the opening of 2 nm.

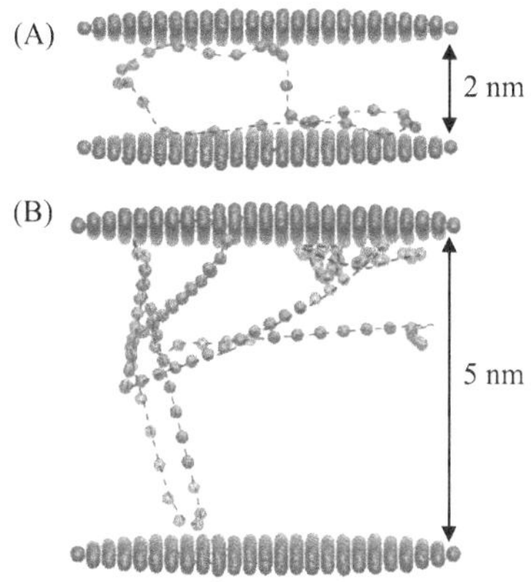

FIGURE 7.45 Movement trajectories of a water molecule being transported through a graphene channel with an opening of (A) 2 nm and (B) 5 nm. Each green dot represents a point in the trajectories of the water molecule. The *red* and *black* dashed lines indicate two independent trajectories in two separate MD simulations. The *purple* dots represent the graphene layers. Source: *Modified and reprinted with permission from a work by Zhang et al. (2018b).*

References

Babicheva, R.I., Dahanayaka, M., Liu, B., Korznikova, E.A., Dmitriev, S.V., Wu, M.S., et al., 2020. Characterization of two carbon allotropes, cyclicgraphene and graphenylene, as semi-permeable materials for membranes. Mater. Sci. Eng. B 259, 1–13.

Berezhkovskii, A., Hummer, G., 2002. Single-file transport of water molecules through a carbon nanotube. Phys. Rev. Lett. 89, 064503-1–064503-4.

Birdi, K.S., Vu, D.T., 1993. Wettability and the evaporation rates of fluids from solid surfaces. J. Adhes. Sci. Technol. 7, 485–493.

Chang, L., Li, J., Duan, X., Liu, W., 2015. Porous carbon derived from metal-organic framework (MOF) for capacitive deionization electrode. Electrochim. Acta 176, 956–964.

Cohen-Tanugi, D., Grossman, J.C., 2012. Water desalination across nanoporous graphene. Nano Lett. 12, 3602–3608.

Cohen-Tanugi, D., Lin, L.-C., Grossman, J.C., 2016. Multilayer nanoporous graphene membranes for water desalination. Nano Lett. 16, 1027–1033.

Dahanayaka, M., Liu, B., Hu, Z., Chen, Z., Law, A.W.-K., Zhou, K., 2017a. Corrugated graphene layers for sea water desalination using capacitive deionization. Phys. Chem. Chem. Phys. 19, 8552–8562.

Dahanayaka, M., Liu, B., Hu, Z., Pei, Q., Chen, Z., Law, A.W.-K., et al., 2017b. Graphene membranes with nanoslits for seawater desalination via forward osmosis. Phys. Chem. Chem. Phys. 19, 30551–30561.

De Luca, S., Todd, B.D., Hansen, J.S., Daivis, P.J., 2014. Molecular dynamics study of nanoconfined water flow driven by rotating electric fields under realistic experimental conditions. Langmuir 30, 3095–3109.

Frohn, A., Roth, N., 2000. Dynamics of Droplets. Springer, Berlin.

Gai, J.-G., Gong, X.-L., Wang, W.-W., Zhang, X., Kang, W.-L., 2014. An ultrafast water transport forward osmosis membrane: porous graphene. J. Mater. Chem. A 2, 4023–4028.

Givler, S.D., Abraham, J., 1996. Supercritical droplet vaporization and combustion studies. Prog. Energy Combust. Sci. 22, 1–28.
Golovko, D.S., Butt, H.J., Bonaccurso, E., 2009. Transition in the evaporation kinetics of water microdrops on hydrophilic surfaces. Langmuir 25, 75–78.
Gupta, K.M., Zhang, K., Jiang, J., 2015. Water desalination through zeolitic imidazolate framework membranes: significant role of functional groups. Langmuir 31, 13230–13237.
Hata, K., Futaba, D.N., Mizuno, K., Namai, T., Yumura, M., Iijima, S., 2004. Water-assisted highly efficient synthesis of impurity-free single-walled carbon nanotubes. Science 306, 1362–1364.
Heffelfinger, G.S., van Swol, F., 1994. Diffusion in Lennard-Jones fluids using dual control volume grand canonical molecular dynamics simulation (DCV-GCMD). J. Chem. Phys. 100, 7548–7552.
Hinds, B.J., Chopra, N., Rantell, T., Andrews, R., Gavalas, V., Bachas, L.G., 2004. Aligned multiwalled carbon nanotube membranes. Science 303, 62–65.
Ho, T.A., Striolo, A., 2015. Promising performance indicators for water desalination and aqueous capacitors obtained by engineering the electric double layer in nano-structured carbon electrodes. J. Phys. Chem. C. 119, 3331–3337.
Holt, J.K., Park, H.G., Wang, Y., Stadermann, M., Artyukhin, A.B., Grigoropoulos, C.P., et al., 2006. Fast mass transport through sub-2-nanometer carbon nanotubes. Science 312, 1034–1037.
Holyst, R., Litniewski, M., 2008. Heat transfer at the nanoscale: evaporation of nanodroplets. Phys. Rev. Lett. 100, 055701-1–055701-4.
Hong, T.Z.X., Dahanayaka, M., Liu, B., Law, A.W.-K., Zhou, K., 2021. Zeolitic imidazolate frameworks as capacitive deionization electrodes for water desalination and Cr(VI) adsorption: a molecular simulation study. Appl. Surf. Sci. 546 (149080), 1–12.
Hu, H., Larson, R.G., 2002. Evaporation of a sessile droplet on a substrate. J. Phys. Chem. B 106, 1334–1344.
Hu, Z., Chen, Y., Jiang, J., 2011. Zeolitic imidazolate framework-8 as a reverse osmosis membrane for water desalination: Insight from molecular simulation. J. Chem. Phys. 134, 134705-1–134705-6.
Huang, H., Song, Z., Wei, N., Shi, L., Mao, Y., Ying, Y., et al., 2013. Ultrafast viscous water flow through nanostrand-channelled graphene oxide membranes. Nat. Commun. 4, 1–9.
Hummer, G., Rasaiah, J.C., Noworyta, J.P., 2001. Water conduction through the hydrophobic channel of a carbon nanotube. Nature 414, 188–190.
Insepov, Z., Wolf, D., Hassanein, A., 2006. Nanopumping using carbon nanotubes. Nano Lett. 6, 1893–1895.
Joseph, S., Aluru, N.R., 2008. Why are carbon nanotubes fast transporters of water? Nano Lett. 8, 452–458.
Kalra, A., Garde, S., Hummer, G., 2003. Osmotic water transport through carbon nanotube membranes. Proc. Natl. Acad. Sci. U S Am. 100, 10175–10180.
Kaltz, T.L., Long, L.N., Micci, M.M., Little, J.K., 1998. Supercritical vaporization of liquid oxygen droplets using molecular dynamics. Combust. Sci. Technol. 136, 279–301.
Kieu, H.T., Zhou, K., Law, A.W.-K., 2019. Surface morphology effect on the evaporation of water on graphene oxide: a molecular dynamics study. Appl. Surf. Sci. 488, 335–342.
Konatham, D., Yu, J., Ho, T.A., Striolo, A., 2013. Simulation insights for graphene-based water desalination membranes. Langmuir 29, 11884–11897.
Kryukov, A.P., Levashov, V.Y., Sazhin, S.S., 2004. Evaporation of diesel fuel droplets: kinetic vs hydrodynamic models. Int. J. Heat. Mass. Transf. 47, 2541–2549.

Landry, E.S., Mikkilineni, S., Paharia, M., Mcgaughey, A.J.H., 2007. Droplet evaporation: a molecular dynamics investigation. J. Appl. Phys. 102, 124301-1–124301-7.

Liang, Z., Tsai, H.L., 2012. A method to generate pressure gradients for molecular simulation of pressure-driven flows in nanochannels. Microfluid. Nanofluid. 13, 289–298.

Liu, C., Li, Z., 2010. Molecular dynamics simulation of composite nanochannels as nanopumps driven by symmetric temperature gradients. Phys. Rev. Lett. 105, 174501-1–174501-4.

Liu, B., Zhou, K., 2019. Recent progress on graphene-analogous 2D nanomaterials: properties, modeling and applications. Prog. Mater. Sci. 100, 99–169.

Liu, C., Lv, Y., Li, Z., 2012. Fluid transport in nanochannels induced by temperature gradients. J. Chem. Phys. 136, 114506-1–114506-5.

Liu, Y.-C., Shen, J.-W., Gubbins, K.E., Moore, J.D., Wu, T., Wang, Q., 2008. Diffusion dynamics of water controlled by topology of potential energy surface inside carbon nanotubes. Phys. Rev. B 77, 125438-1–125438-7.

Liu, B., Wu, R., Baimova, J.A., Wu, H., Law, A.W.-K., Dmitriev, S.V., et al., 2016a. Molecular dynamics study of pressure-driven water transport through graphene bilayers. Phys. Chem. Chem. Phys. 18, 1886–1896.

Liu, B., Wu, R., Law, A.W.-K., Feng, X.-Q., Bai, L., Zhou, K., 2016b. Channel morphology effect on water transport through graphene bilayers. Sci. Rep. 6, 1–14.

Liu, Y., Xie, D., Song, M., Jiang, L., Fu, G., Liu, L., et al., 2018. Water desalination across multilayer graphitic carbon nitride membrane: insights from non-equilibrium molecular dynamics simulations. Carbon 140, 131–138.

Long, L.N., Micci, M.M., Wong, B.C., 1996. Molecular dynamics simulations of droplet evaporation. Comp. Phys. Commun. 96, 167–172.

Lyu, Q., Deng, X., Hu, S., Lin, L.-C., Ho, W.S.W., 2019. Exploring the potential of defective UiO-66 as reverse osmosis membranes for desalination. J. Phys. Chem. C. 123, 16118–16126.

Majumder, M., Chopra, N., Andrews, R., Hinds, B.J., 2005. Nanoscale hydrodynamics: enhanced flow in carbon nanotubes. Nature 438, 44.

Maroo, S.C., Chung, J.N., 2010. A novel fluid-wall heat transfer model for molecular dynamics simulations. J. Nanopart. Res. 12, 1913–1924.

Miller, S.A., Young, V.Y., Martin, C.R., 2001. Electroosmotic flow in template-prepared carbon nanotube membranes. J. Am. Chem. Soc. 123, 12335–12342.

Nair, R.R., Wu, H., Jayaram, P.N., Grigorieva, I.V., Geim, A.K., 2012. Unimpeded permeation of water through helium-leak-tight graphene-based membranes. Science 335, 442–444.

Perreault, F., De Faria, A.F., Elimelech, M., 2015. Environmental applications of graphene-based nanomaterials. Chem Soc Rev 44, 5861–5896.

Rinne, K.F., Gekle, S., Bonthuis, D.J., Netz, R.R., 2012. Nanoscale pumping of water by AC electric fields. Nano Lett. 12, 1780–1783.

Rusanov, A.I., Brodskaya, E.N., 1977. The molecular dynamics simulation of a small drop. J. Colloid Interface Sci. 62, 542–555.

Sirignano, W., 1999. Fluid Dynamics and Transport of Droplet and Sprays. Cambridge University Press, Cambridge.

Sokhan, V.P., Nicholson, D., Quirke, N., 2002. Fluid flow in nanopores: accurate boundary conditions for carbon nanotubes. J. Chem. Phys. 117, 8531–8539.

Striolo, A., 2006. The mechanism of water diffusion in narrow carbon nanotubes. Nano Lett. 6, 633–639.

Suk, M.E., Aluru, N.R., 2010. Water transport through ultrathin graphene. J. Phys. Chem. Lett. 1, 1590–1594.

Sumardiono, S., Fischer, J., 2006. Molecular simulations of droplet evaporation processes: adiabatic pressure jump evaporation. Int. J. Heat. Mass. Transf. 49, 1148–1161.

Sumith, Y.D., Maroo, S.C., 2015. Surface-heating algorithm for water at nanoscale. J. Phys. Chem. Lett. 6, 3765–3769.

Takaba, H., Onumata, Y., Nakao, S.-I., 2007. Molecular simulation of pressure-driven fluid flow in nanoporous membranes. J. Chem. Phys. 127, 054703-1–054703-8.
Talapatra, S., Kar, S., Pal, S.K., Vajtai, R., Ci, L., Victor, P., et al., 2006. Direct growth of aligned carbon nanotubes on bulk metals. Nat. Nanotechnol. 1, 112–116.
ten Wolde, P.R., Frenkel, D., 1998. Computer simulation study of gas-liquid nucleation in a Lennard-Jones system. J. Chem. Phys. 109, 9901–9918.
Thomas, J.A., McGaughey, A.J.H., 2009. Water flow in carbon nanotubes: transition to subcontinuum transport. Phys. Rev. Lett. 102, 184502-1–184502-4.
Thompson, S.M., Gubbins, K.E., Walton, J.P.R.B., Chantry, R.A.R., Rowlinson, J.S., 1984. A molecular dynamics study of liquid drops. J. Chem. Phys. 81, 530–542.
Wang, B.-B., Wang, X.-D., Duan, Y.-Y., Chen, M., 2014. Molecular dynamics simulation on evaporation of water and aqueous droplets in the presence of electric field. Int. J. Heat. Mass. Transf. 73, 533–541.
Wan, R., Wang, C., Lei, X., Zhou, G., Fang, H., 2015. Enhancement of water evaporation on solid surfaces with nanoscale hydrophobic-hydrophilic patterns. Phys. Rev. Lett. 115, 195901-1–195901-5.
Wang, Z., Yan, T., Shi, L., Zhang, D., 2017. In situ expanding pores of dodecahedron-like carbon frameworks derived from MOFs for enhanced capacitive deionization. ACS Appl. Mater. Interfaces 9, 15068–15078.
Wang, Y., Zhao, Y., Huang, J., 2011. Giant pumping of single-file water molecules in a carbon nanotube. J. Phys. Chem. B 115, 13275–13279.
Xie, J.-F., Cao, B.-Y., 2017. Fast nanofluidics by travelling surface waves. Microfluid. Nanofluid. 21, 1–14.
Xu, W., Leeladhar, R., Kang, Y.T., Choi, C.H., 2013. Evaporation kinetics of sessile water droplets on micropillared superhydrophobic surfaces. Langmuir 29, 6032–6041.
Yamada, T., Namai, T., Hata, K., Futaba, D.N., Mizuno, K., Fan, J., et al., 2006. Size-selective growth of double-walled carbon nanotube forests from engineered iron catalysts. Nat. Nanotechnol. 1, 131–136.
Zhang, J., Borg, M.K., Sefiane, K., Reese, J.M., 2015. Wetting and evaporation of salt-water nanodroplets: a molecular dynamics investigation. Phys. Rev. E 92, 052403-1–052403-11.
Zhang, J., Fang, J.H., Han, J., Yan, T., Shi, L., Zhang, D., 2018a. N, P, S co-doped hollow carbon polyhedra derived from MOF-based core-shell nanocomposites for capacitive deionization. J. Mater. Chem. A 6, 15245–15252.
Zhang, J., Leroy, F., Müller-Plathe, F., 2013. Evaporation of nanodroplets on heated substrates: a molecular dynamics simulation study. Langmuir 29, 9770–9782.
Zhang, H., Liu, B., Kieu, H.T., Wu, M.S., Zhou, K., Law, A.W.-K., 2018b. Coarse-grained molecular dynamics study of membrane distillation through meso-size graphene channels. J. Membr. Sci. 558, 34–44.
Zhang, H., Liu, B., Wu, M.S., Zhou, K., Law, A.W.-K., 2017. Transport of salty water through graphene bilayer in an electric field: a molecular dynamics study. Computat. Mater. Sci. 131, 100–107.

CHAPTER

8

Application of molecular dynamics simulation in other problems

With the development of new molecular dynamics (MD) simulation techniques, the range of its applications has been extended to many aspects of materials science and mechanical engineering. This chapter extends the discussion of MD simulation on the mechanical, thermal, and mass transport problems in the previous chapters to other problems, including simple chemical reactions, irradiation processes, and material crystallization. The chapter begins with an introduction to the reactive force field (ReaxFF), followed by two detailed examples of its application in modeling chemical reactions. Afterward, basic simulation techniques for both electron and ion irradiation processes are discussed. Finally, the application of MD simulation in material crystallization is highlighted by briefly discussing the main simulation procedures.

8.1 Reactive molecular dynamics simulations

Chemical reactions generally involve the breaking and forming of chemical bonds caused by the transfer of electrons. Methods based on the principles of quantum mechanics, such as the ab initio method, can offer valuable theoretical guidance on the electronic level, but they are often computationally intense for simulations exploring the full dynamic evolution of a system. In contrast, MD simulations based on classical Newtonian mechanics require significantly less computational resources, and thus they can better describe dynamic processes over larger temporal and spatial scales. However, in MD simulations, empirical potential energy functions are generally adopted, which typically

DOI: https://doi.org/10.1016/B978-0-12-816419-8.00013-1

require predefined treatments of the chemical bonding between the atoms. Therefore the investigation of reactive events involving bond breaking and forming is precluded.

8.1.1 ReaxFF force field

The aforementioned dilemma motivates the development of reactive force fields that introduce connection-dependent terms to describe bond breaking and forming events. One of the most popular force fields for reactive MD simulations is ReaxFF. In principle, ReaxFF describes reactive events in a molecular system through a bond order formalism, in which the bond order is calculated empirically from interatomic distances. Electronic interactions driving chemical bond breaking and forming are treated implicitly, allowing an MD model to simulate chemical reactions without having to address explicit quantum mechanical calculations. In other words, chemical reactions are treated classically with the implementation of ReaxFF, which opens the door to studies of numerous chemical phenomena that could not be modeled previously by MD simulation.

In particular, ReaxFF enables simulations involving reactive events that occur at interfaces between solid, liquid, and gas phases because the information of each atomic element in ReaxFF is transferable across its different phases. For example, O atoms existing in a gas phase of O_2, a liquid phase of H_2O, or a solid oxide are all treated by the same mathematical formalism in ReaxFF. Such transferability arises from the fact that ReaxFF does not employ the strategies of atom classification, i.e., ReaxFF does not classify atoms of the same element into different atom types when these atoms are placed in different chemical environments, which contrasts popular nonreactive force fields such as AMBER (assisted model building with energy refinement) and CHARMM (Chemistry at Harvard Macromolecular Mechanics). This transferability, coupled with a lower computational cost that allows for longer simulation timescales, enables ReaxFF to model phenomena that are dependent not only on the reactivities of the involved species but also on their dynamic factors (e.g., diffusivity and solubility) that affect how they migrate within the system (Senftle et al., 2016).

Since the development of the first version of ReaxFF in 2001 for simulating hydrocarbon systems, successive versions have been developed over the years to explore various material systems such as clay, coal, and protein. Presently, ReaxFF contains parameter sets for over half of the existing elements (Senftle et al., 2016; Psofogiannakis and Van

Duin, 2016; Nayir et al., 2018), as shown in Fig. 8.1. The elements for which a ReaxFF parameter set has been developed are highlighted in green in the periodic table. Note that although a ReaxFF description exists for these elements, one cannot simply combine these parameter sets in a random manner and expect to obtain satisfactory transferability. There are currently two major combinations of parameter sets that are intratransferable with one another: (i) the combustion branch and (ii) the aqueous branch (i.e., the ReaxFF branches). For more details of these two branches, please refer to the work by Senftle et al. (2016).

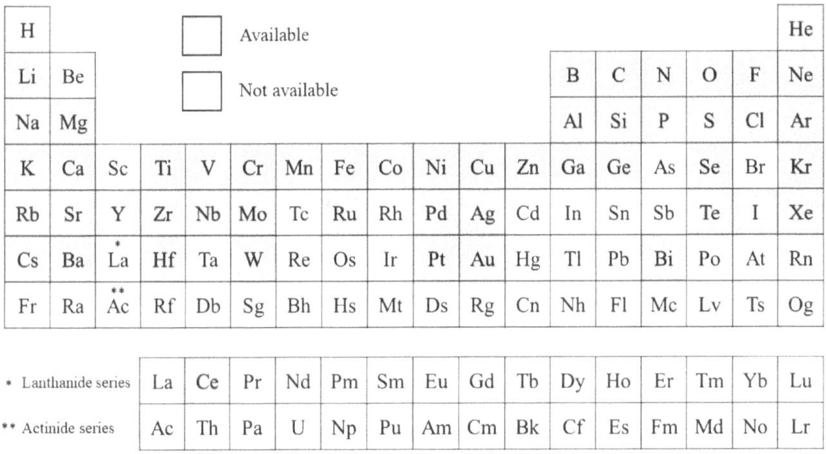

FIGURE 8.1 Overview of elements with available parameter sets in ReaxFF.

ReaxFF enables the simulation of chemical reactions by implementing a bond order formalism in conjunction with polarizable charge descriptions to model both reactive and nonreactive interactions between atoms. In ReaxFF, the total energy of a material system is expressed as the sum of various components as

$$E_{system} = E_{bond} + E_{angle} + E_{tors} + E_{over} + E_{vdW} + E_{Coul} + E_{specific}. \quad (8.1)$$

Here, E_{bond} denotes the potential energy associated with the bonding between pairs of atoms and is described by a continuous function of interatomic distances; E_{angle} and E_{tors} represent the potential energy related to three-body (bending of a bond angle) and four-body (twist of a dihedral angle) interactions, respectively; E_{over} denotes an energy penalty term that prevents the overcoordination of atoms and is calculated based on atomic valence rules; E_{vdW} and E_{Coul} describe the energy corresponding to nonbonding van der Waals and electrostatic interactions, respectively; E_{specific} represents system-specific energy terms that are generally excluded, except in specific cases in which they are required to capture features unique to the system of interest, such as lone-pair, conjugation, and hydrogen bonding. For the complete functional forms of ReaxFF, one should refer to a work by Chenoweth et al. (2008).

The first four terms of Eq. (8.1) are all dependent on the bond order, which is assumed to be determined solely by the interatomic distances and can be directly calculated by the following empirical formula:

$$\begin{aligned} \text{BO}_{ij} &= \text{BO}_{ij}^{\sigma} + \text{BO}_{ij}^{\pi} + \text{BO}_{ij}^{\pi\pi} \\ &= \exp\left[p_{\text{bo1}}\left(\frac{r_{ij}}{r_o^{\sigma}}\right)^{p_{\text{bo2}}}\right] + \exp\left[p_{\text{bo3}}\left(\frac{r_{ij}}{r_o^{\pi}}\right)^{p_{\text{bo4}}}\right] + \exp\left[p_{\text{bo5}}\left(\frac{r_{ij}}{r_o^{\pi\pi}}\right)^{p_{\text{bo6}}}\right], \end{aligned} \tag{8.2}$$

where BO_{ij} denotes the bond order between the ith and the jth atom; r_{ij} is the interatomic distance between the ith and the jth atom; r_o^{σ}, r_o^{π}, and $r_o^{\pi\pi}$ are the equilibrium bond lengths for the σ, π, and $\pi\pi$ bonds, respectively; the six p_{bo} terms are empirical parameters that can be obtained through calculations based on quantum mechanical methods. The continuous characteristic of the functional expression for the bond order in Eq. (8.2) enables a differentiable potential energy surface to be obtained, which is required to calculate interatomic forces. Since no cut-off distance has been defined for calculating the bond order, Eq. (8.2) also accommodates long-range covalent interactions in transition states, allowing ReaxFF to efficiently predict reaction energy barriers and reaction pathways. In MD simulations that employ the ReaxFF, the bond order is typically calculated within a distance of 0–5 Å, which is sufficiently large to capture even the weakest covalent interactions between most elements. However, such a distance can be increased for certain cases, for example, a molecular system that contains elements exhibiting large covalent radii. However, in these cases, the inclusion of this long-range covalent interaction necessitates a bond order correction to remove spurious bonding characteristics of the interactions among nonbonded atoms, such as neighboring H atoms in a methane molecule. In MD simulations employing the ReaxFF force field, the first four energy terms in Eq. (8.1) that depend on the bond order are first calculated. Subsequently, a charge equilibrium scheme is implemented at each time step to determine partial atomic charges based on the Coulombic interactions calculated.

8.1.2 Application of ReaxFF in reactive MD simulations

In this section, two examples of MD simulations employing ReaxFF are presented. The first example pertains to the chemisorption and decomposition of a hydrocarbon on the surface of a Ni nanoparticle, and it has been selected to highlight the capacity of ReaxFF in modeling reactive chemistry.

Nickel (Ni) is a common catalyst utilized in many vital industrial processes, such as the synthesis of ammonia and higher hydrocarbons. It has been reported that Ni nanoparticles can also catalyze the growth of carbon nanotubes (CNTs) from hydrocarbons (Paillet et al., 2004). To understand the fundamental chemical processes of hydrocarbon molecules reacting on the surfaces of a Ni crystal (particularly the low-index and the defect-rich surfaces) and the role of these processes in CNT growth, Mueller et al. (2010) have conducted a reactive MD simulation study on the chemisorption and decomposition of six types of simple hydrocarbons on the surface of a Ni nanoparticle using the ReaxFF force field.

The six types of hydrocarbons selected are methane (CH_4), ethyne (C_2H_2), ethene (C_2H_4), propene (C_3H_6), benzene (C_6H_6), and cyclohexane (C_6H_{12}). Among them, only methane and cyclohexane are saturated hydrocarbons, whereas the others are unsaturated. In the study, ethene and ethyne are chosen to compare the reactivities of species with different numbers of π bonds (one for ethene and two for ethyne). Benzene is selected to consider the effects of aromaticity and ring structures, while propene is included to investigate the effects of weak allylic C–H bonds. The saturated hydrocarbons, methane, and cyclohexane are chosen to examine the initial cleavage of the C–H bond within hydrocarbon molecules that do not exhibit a strong chemisorbing tendency toward Ni.

In the simulation, a Ni nanosphere with a diameter of 21 Å (corresponding to 468 Ni atoms) is positioned in the center of a cubic simulation box with a side length of 80 Å. The hydrocarbon molecules are then randomly distributed throughout the simulation box under the condition that a distance of at least 3 Å is maintained between the hydrocarbon molecules and the surface of the nanosphere. The temperature of the system is set at 500 K initially, and it is gradually increased at a constant rate of 20 K ps^{-1} until it reaches 2500 K. Such a high temperature is adopted to accelerate the rate of the chemisorption and decomposition processes so that sufficient dynamic information of these reactions can be collected within a practicable simulation time. By monitoring the population of each chemical species over time, the evolution of the catalytic processes can be described, and a reaction network indicating the transformations of various intermediates can be established throughout the simulation.

Take benzene C_6H_6 as an example. The reactive MD simulation results show that the chemisorption of C_6H_6 molecules on the surface of the Ni nanosphere is realized by binding the C–C π orbitals to the surface Ni atoms, thereby leading to the benzene rings remaining parallel to the nanoparticle surface. At an initial temperature of 500 K, the decomposition of benzene is not observed. As the system temperature is increased to 900 K (in 19 ps), an adsorbed C_6H_6 molecule is observed to lose a H atom to form a C_6H_5 molecule. As the temperature is further increased, additional dehydrogenation occurs, during which di-σ

bonded C_6H_4 is formed at 45 ps and 1400 K, tri-σ bonded C_6H_3 at 49 ps and 1500 K, and eventually 1,2,3,4-C_6H_2 at 63 ps and 1750 K, which exhibits four bonds with the nanosphere surface. Subsequently, the C−C bonds begin to break as C_6H_3 is converted to C_5H_3, which further decomposes as either C−C or C−H bonds are broken. The breaking of a C−C bond is always accompanied by the separation of a single C atom or a C pair from a longer hydrocarbon chain. Fig. 8.2 shows the

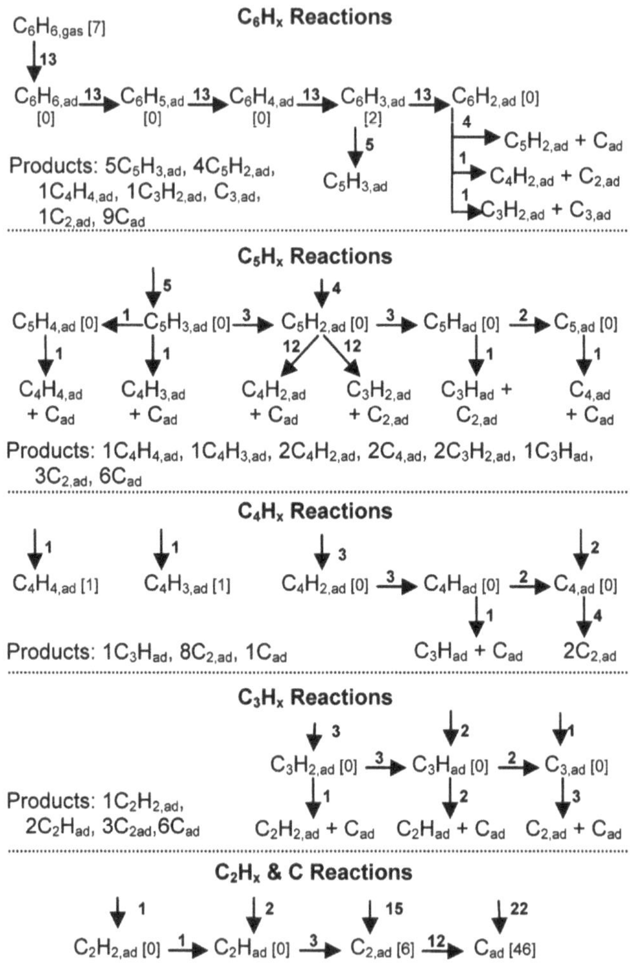

FIGURE 8.2 Reaction network for the chemisorption and decomposition of benzene molecules on the surface of a Ni nanosphere. The number in each bracket of the reaction network is the final population of each species, and the number beside each reaction arrow is the number of times this reaction occurs in the entire reaction network. Source: *Reprinted with permission from a work by Mueller et al. (2010).*

full reaction network of the initial 20 benzene molecules drawn from the reactive MD simulation.

A further examination of the atomic trajectories provides the reason that the reaction proceeds along the pathway presented in Fig. 8.2. For instance, the dehydrogenation of C_6H_6 terminates at C_6H_2, and no C_6H or C_6 rings are observed during the simulation, which can be explained by the specific spatial configuration of the C_6H_6 molecules after their chemisorption on the Ni nanosphere surface. When the C_6H_6 molecules are initially adsorbed onto the nanosphere, their rings are parallel to the nanosphere surface (Fig. 8.3A). Hence, the breaking of one C−H bond frees up a C atom in the benzene ring to form a σ bond with the Ni nanosphere, but at the same time, the planar shape of the ring is distorted. The formation of a second σ bond with the nanosphere surface, which corresponds to the loss of another H atom of the benzenide (C_6H_5) ring, results in a 1,2-benzyne (C_6H_4) ring that is perpendicular to the nanosphere surface to which it is bonded. With the loss of one more H atom, C_6H_4 is transformed into C_6H_3 with the perpendicular orientation retained (Fig. 8.3B). With such an orientation, only two of the four remaining H atoms of the 1,2-benzyne ring will be close enough to the nanosphere surface to react with it. Consequently, neither C_6H nor C_6 rings are observed.

Several general conclusions can be drawn by comparing the reaction networks and atomic trajectories of the six types of hydrocarbons. First, unsaturated hydrocarbons (molecules with π bonds) are adsorbed and decomposed much more readily than saturated hydrocarbons

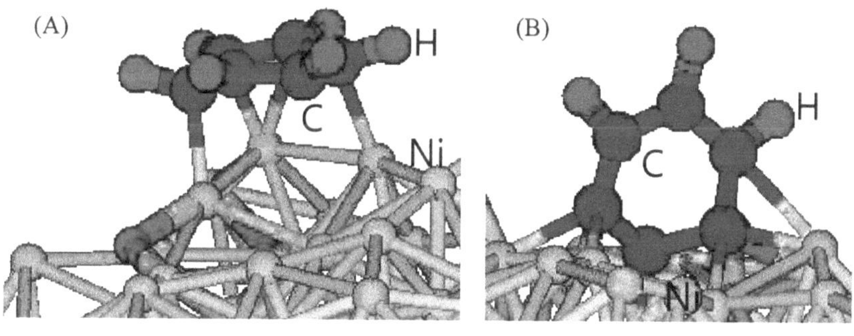

FIGURE 8.3 Snapshots of atomistic configurations of a benzene (C_6H_6) molecule during the simulation of its decomposition on the surface of a Ni nanoparticle. (A) The C_6H_6 molecule is chemisorbed on the nanoparticle surface and stands in parallel with the particle surface. (B) After losing three H atoms, the C_6H_6 molecule is transformed into a C_6H_3 molecule that stands in perpendicular to the nanoparticle surface. Color code: C, green; H, red; Ni: ice blue. Source: *Modified and reprinted with permission from a work by Mueller et al. (2010).*

(molecules with only σ bonds), which might be a significant factor in selecting feedstock species for low-temperature CNT growth. Second, once the hydrocarbon molecules are chemisorbed to the Ni nanosphere, their dehydrogenation always occurs before their decomposition into lower-order hydrocarbons. In other words, the cleavage of the C–C bonds only occurs when one of the C atoms forming such a bond is denuded of H atoms. Third, the breaking of C–H bonds occurs more readily when they are close to the surface of the nanosphere. Finally, the cleavage of the C–C bonds requires a bare C atom at the end of a hydrocarbon chain. The cleavage is realized by inserting a Ni atom into the C–C bond, and the C atom at the end of the chain is drawn into the nanosphere subsurface, where it is stabilized by forming four bonds with neighboring Ni atoms.

The second example is the combustion process of a hydrocarbon, which has been chosen to illustrate the capability of reactive MD simulation in investigating chemical reactions under high-temperature and/or high-pressure conditions.

Fossil fuels play an essential role in our daily lives. Detailed research of the pyrolysis behavior of fossil fuels is essential to achieving an in-depth understanding of their combustion phenomena, improving the efficiency of existing current combustors, and reducing the production of pollutants. However, pyrolysis is a highly complex, nonequilibrium process that usually involves hundreds of species and thousands of reactions. Characterizing such nonequilibrium processes, particularly those with complex mechanical conditions and a plethora of short-lived intermediates, is virtually impossible with most contemporary experimental and theoretical methods.

Practical fuels generally contain a large number of hydrocarbon compounds. Taking *n*-heptane (C_7H_{16}, one of the main components of fuels) as an exemplar of practical fuels, Ding et al. (2013) have conducted a reactive MD simulation study on the pyrolysis of *n*-dodecane using the ReaxFF. The intermediate mechanism of the reaction, its intermediate/product distributions, and corresponding kinetic behaviors were systematically studied.

In the simulation, 16 C_7H_{16} molecules are placed in a periodic cubic simulation box with a side length of 25 Å, which corresponds to a density of $0.17\ \mathrm{g\ cm^{-3}}$. The system temperature is initially set at 1000 K, and an annealing process is subsequently simulated by gradually increasing the system temperature to 3000 K at a heating rate of $5\ \mathrm{K\ ps^{-1}}$. Afterward, the system is equilibrated at 3000 K for 100 ps. The simulation is repeated 12 times to obtain an adequate collection of the intermediate processes and products, which is useful for a detailed examination of the pyrolysis of *n*-dodecane.

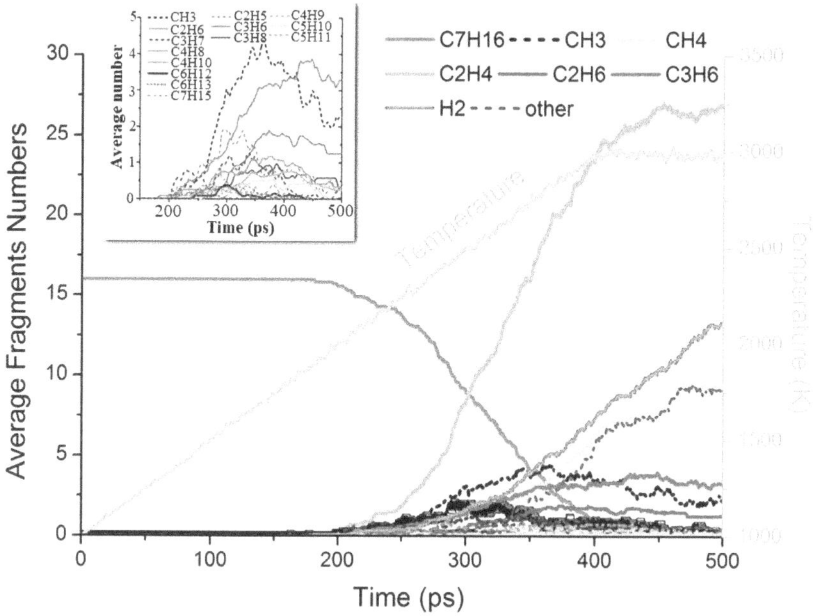

FIGURE 8.4 Evolution of the number of the major species during the MD simulation as a function of time. All the minor intermediates and products are considered one entity and denoted by the label "other." The dashed and solid lines represent the free radicals and stable products, respectively. The pink line represents the system temperature as a function of time. *MD*, Molecular dynamics. Source: *Reprinted with permission from a work by Ding et al. (2013).*

Fig. 8.4 presents the evolution of the number of major species that appear in the simulation as a function of time. It is observed that the decomposition of *n*-heptane only occurs after 180 ps at a temperature of ~1750 K. Various species, including saturated, unsaturated, short-lived, and long-lived compounds, are produced. Ethylene is the most abundant product whose amount increases rapidly with time and reaches a plateau by the end of the simulation. Many unstable radical intermediates are produced during the pyrolysis process, for example, CH_3, C_2H_5, C_3H_7, and C_4H_9, as indicated by the inset of Fig. 8.4. In addition, the number of these unstable intermediates reaches a maximum in the middle of the simulation and gradually decreases afterward.

According to the decomposition rate of *n*-heptane, the entire pyrolysis process can be divided into three stages. The first stage begins with the decomposition of *n*-heptane and lasts for 243 ps. In this stage, the system temperature is kept at 2000 K, and the decomposition rate is relatively low; only 12.5% of the *n*-heptane molecules decompose. In the second stage that lasts from 243 ps to 365 ps, with the system temperature ranging from

2000 K to 2500 K, the time-averaged decomposition rate is much higher than that in the first stage, and most of the *n*-heptane molecules (68.8%) decompose. The final stage is from 365 ps to 420 ps, during which the remaining 18.7% of the *n*-heptane molecules decompose. After 420 ps, the pyrolysis process reaches the end. The intermediates produced during the three stages are gradually consumed by the continuous reactions to generate small fragments.

A detailed analysis has indicated that 370 individual reactions occur in total during the annealing process, which can be divided into two groups: intramolecular dissociation reactions numbering 200 and intermolecular reactions numbering 170. During these reactions, a total of 151 main species, including stable products and unstable intermediates, are observed, although most of these intermediates have a very short lifetime. These 151 species can be divided into 11 categories based on the number of carbon atoms in their molecular skeleton, i.e., C_0, C_1, C_2, C_3, etc. The majority of these species have a low carbon content (C_0, C_1, C_2, C_3), with the C_2 species exhibiting the largest carbon content.

Alternatively, according to their reaction behaviors, these species can also be classified into four types: long-chain alkyl free radicals, small free radicals, stable products, and highly unsaturated products. The long-chain alkyl free radicals play an essential role in the pyrolysis of hydrocarbon, and its representative examples include $C_2H_5^\bullet$, $C_3H_7^\bullet$, $C_4H_9^\bullet$, and $C_5H_{11}^\bullet$. These free radicals are generated once the decomposition of *n*-heptane molecules begins, and most of them are produced via intramolecular reactions. For example, the main intramolecular reactions that produce $C_2H_5^\bullet$ include the β-scission reactions of alkyl free radicals (e.g., $C_4H_9^\bullet$), the H-addition reactions of C_2H_4, and the $C-C$ bond fission reactions of stable alkanes (e.g., *n*-heptane). Since most of the free radicals are unstable, the long-chain alkyl free radicals further dissociate rapidly and form smaller radicals or stable species. At the end of the pyrolysis process, only a small number of the long-chain alkyl free radicals remain.

The small free radicals, such as H and methyl free radicals ($CH_3^\bullet$), are abundant and are observed in all stages of the *n*-heptane pyrolysis process due to the intramolecular reactions involved and the direct decomposition of *n*-heptane. These small radicals are highly reactive, and most of the intermolecular reactions related to the intermediates are caused by these radicals. It is observed that the H free radicals are more readily produced than consumed. The ethyl radical β-scission reactions are identified as the main H producing reactions, while the reverse reactions constitute the main channel for H consumption.

Various stable products are generated during the pyrolysis process, including H_2, alkanes, and alkenes. Among them, ethylene is the most abundant and reaches the maximum quantity after *n*-heptane has completely decomposed. The main reactions that produce ethylene are the alkyl free radical β-scission reactions. The product evolution is highly dependent on the number of free radicals generated from the intramolecular reactions.

The highly unsaturated intermediates include methylene free radicals, vinyl-free radicals, and acetylene, and they appear mainly in the final stages of the pyrolysis process. For instance, C_2H_2 is the most frequently observed during the pyrolysis process and is generated mainly from the secondary dissociation of the α-olefins through reaction pathways such as $CH_2=CH_2 \rightarrow CH_2=CH^{\bullet} + H$ and $CH_2=CH^{\bullet} \rightarrow CH \equiv CH + H$. Note that the yields of the final products depend highly on the reaction conditions. More unsaturated radicals can be generated near the end of the pyrolysis process if the system temperature is increased.

By investigating the time evolution of the quantities of these four types of species, it is observed that the reduction of n-heptane molecules during the initiation stage is mainly due to intramolecular reactions through the fission of C–C and C–H bonds. In addition, the fission of C–C bonds occurs more frequently than that of C–H bonds and accounts for 56% of the decomposed n-heptane molecules. Moreover, the fission of C–C bonds occurs preferentially at the center of the n-heptane chains instead of its ends. During the second stage of the pyrolysis process, the intermediate reactions are observed to be dominated by those involving free radicals, and most of these reactions are intramolecular. A large number of free radicals are produced during this stage but are consumed immediately by the following intermolecular and intramolecular reactions. The hydrogen abstraction reaction (e.g., $CH_4 + H \rightarrow CH_3^{\bullet} + H_2$) is the most dominant intermolecular reaction during the second stage.

The large variety of species and interactions identified in the reactive MD simulation of the pyrolysis of n-heptane indicate that the process is complicated. However, such complexity lends credibility to the capacity of reactive MD simulation in exploring the detailed mechanisms behind chemical reactions. Aside from the two examples listed above, the reactive MD simulation approach has been adopted to investigate many other chemical reaction processes.

For instance, ReaxFF has been used to model the formation of defects during the reduction of graphene oxide (GO) (Bagri et al., 2010). Based on the simulation results, reorganization patterns of the C atoms in a graphene sheet during its reduction can be illustrated, allowing the identification of more efficient pathways for minimizing defect formation during the reduction of GO. Similarly, ReaxFF has been adopted to study the hydrogenation mechanism of graphene (Flores et al., 2009). By studying the configurations of the H atoms involved in detail, the hydrogenation process of graphene has been meticulously explored.

As another example, the dissociation of water on an aluminum nanocluster surface has been investigated by Russo et al. (2011) who identified a unique two-step mechanism for the dissociation process that involves both the solvation of water molecules within the nanocluster and the adsorption of water molecules on its surface. In addition, the effects of the initial concentration of water molecules on their dissociation have been investigated to show that an optimal concentration exists, at which the highest dissociation efficiency can be obtained. These MD simulation results are potentially useful for designing and applying active metal–water systems for usage as light, portable, and on-demand energy sources via rapid hydrogen gas production.

Bedrov et al. (2012) have conducted reactive MD simulations to study reactions involving singly reduced ethylene carbonate in lithium battery electrolytes. The possible influence of local radicals and ions on the reaction mechanisms has been provided based on the simulation results.

Despite the wide range of applications for reactive MD simulation based on ReaxFF, extensive efforts are still required to refine the technique further and obtain more accurate descriptions of charges, explicit descriptions of electrons, and optimized parameter sets of interatomic interactions.

8.2 Irradiation processes

Irradiation of materials plays a vital role in determining their atomic structures and properties. Some of the most profound effects of irradiation on materials are observed in the cores of nuclear power reactors. Atoms comprising the structural components of such reactors are displaced consistently due to their collision with high-energy radiation particles, including electrons, protons, neutrons, and ions. A similar situation is encountered by the components of a spacecraft in outer space exposed to a plethora of high-energy particles. Exposure to radiation by high-energy particles may introduce significant microstructural damage to a variety of materials. Such damage is closely related to the formation, distribution, and interaction of point defects (vacancies and interstitials), point defect clusters (e.g., Frenkel pairs, vacancy clusters, and interstitial loops), dislocation segments and networks, inert gas bubbles, and voids.

Radiation is not always destructive and can serve as an efficient approach for various types of material processing. For instance, the electron irradiation generated by high-resolution transmission electron microscopes (HRTEM) has been utilized for the creation of single vacancies and other atomic-scale defects (Rodriguez-Manzo and Banhart,

2009; Kotakoski et al., 2011; Susi et al., 2012), the formation of new nano-objects (Chuvilin et al., 2010), chemical reactions (Zoberbier et al., 2012), and irradiation-activated molecular motion (Khlobystov et al., 2004). Ion irradiation can be adopted to fabricate nanostructures by welding low-dimensional nanomaterials, such as CNTs and graphene nanosheets (Jung et al., 2004; Liu et al., 2015; Taguchi et al., 2019). In addition, ion irradiation can be employed to modify material properties through defect engineering (Tolvanen et al., 2009; Krasheninnikov and Nordlund, 2010) and manufacture nanowires via selectively sputtering using the masking technique (Krasheninnikov et al., 2002). Hence, understanding radiation processes is vital for designing materials with a high radiation resistance for advanced nuclear energy and aerospace applications and may also lead to the advent of novel material processing and manufacturing techniques.

However, the interaction between radiation particles and the atoms in a material is a complex process that occurs on temporal and spatial scales that are generally inaccessible to contemporary experimental approaches. From an atomic point of view, the effects of irradiation on materials are rooted in the initial collision event in which an energetic projectile strikes a target atom. With an energy that exceeds the threshold displacement energy of the atom, such an initial collision displaces the target atom in a crystalline material from its equilibrium lattice site, leaving a vacancy site behind. The displaced atom eventually comes to rest at a location between lattice sites and becomes an interstitial atom. The vacancy–interstitial pairs (also known as the Frenkel pairs) and their interactions play a crucial role in determining the physical and mechanical properties of the irradiated crystalline material. Besides the atomic displacement, an energetic charged particle propagating within the material provides energy to electrons via the electron stopping power. To capture the detailed dynamics of the displaced atoms and the associated energy transfer, MD simulation has been increasingly adopted to investigate phenomena pertaining to irradiation processes. In the following sections, a few examples of MD simulations of electron and ion irradiation are presented.

The first example is the MD simulation study of CNT welding under electron beam irradiation conducted by Jang et al. (2004). Due to their unique electronic properties and nanoscale dimensions, CNTs are promising candidates as nanowires, and appropriately jointed single-walled CNTs can serve as the building block for various electronic devices. A high-quality junction between two CNTs is crucial for preserving their excellent electronic properties. Compared with other techniques of creating junctions between CNTs, electron beam irradiation boasts the advantages of reduced contamination and a highly focused beam spot of a subnanometer size.

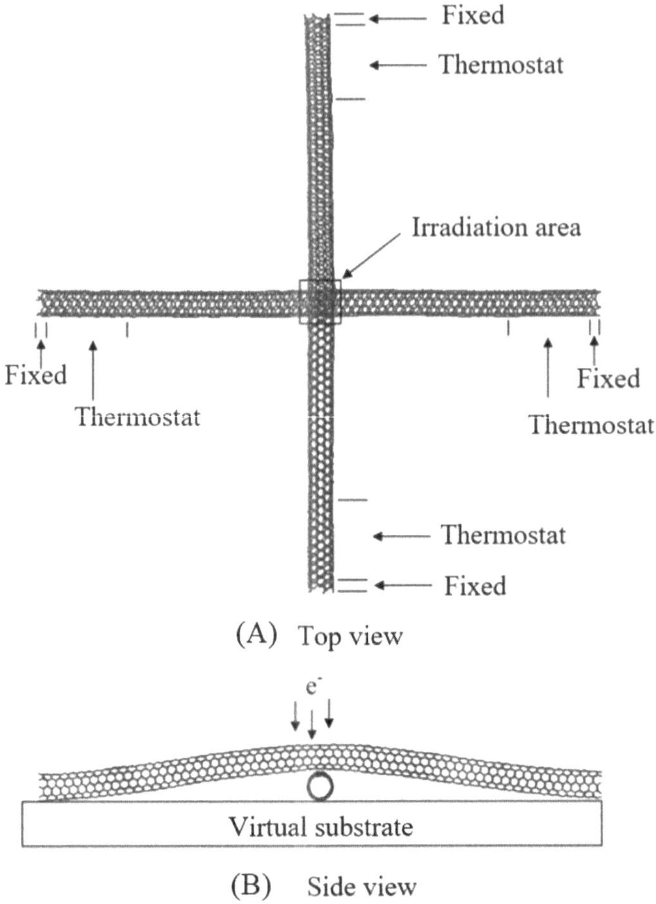

FIGURE 8.5 Molecular dynamics simulation model of the welding of two carbon nanotubes by electron beam irradiation: (A) top view, (B) side view. Source: *Modified and reprinted with permission from a work by Jang et al. (2004).*

To provide a detailed structural characterization and thermal stability analysis of CNT junctions obtained through electron beam irradiation, an MD simulation model consisting of two CNTs perpendicular to each other was constructed by Jang et al. (2004), as presented in Fig. 8.5. In the model, the atoms at the ends of the two CNTs are fixed. A region covering a length of 20 Å next to the fixed ends of each CNT is defined as the thermobath region and is coupled with an appropriate thermostat at a fixed temperature of 300 K to allow for the dissipation of energy associated with the irradiation events. The atoms in the other regions are treated as Newtonian atoms and allowed to move freely. For computational efficiency, instead of its actual atomistic counterpart, an imaginary substrate is placed directly below the CNTs to restrict their motion in the direction perpendicular to the substrate surface. The REBO (reactive empirical bond order) potential, which allows for the modeling of bond breaking and forming, is adopted to describe the C–C interactions involved.

Since MD simulation cannot model electrons explicitly, the interactions between the electron beam and the CNTs are modeled by periodically assigning a velocity to a randomly selected primary knockout atom (PKA) in the irradiation area. Here, the term "PKA" refers to the first atom that a projectile particle encounters in the target material during an irradiation process. Due to the characteristically small timescale of MD simulations, only "head-on" collisions with large energy transitions in the direction of the electron beam are considered, whereas "side collisions" with less significant energy transitions are not. The kinetic energy assigned to a PKA is set at 10 eV, which corresponds to head-on collisions between the atom and an electron beam with an energy of ~50 KeV. The velocity assignment of a randomly selected PKA is performed every 10 timesteps for a total of 2 ps, followed by a cooling process lasting for 1 ps to prevent the system from melting or undergoing thermal dissociation as a result of the increase in temperature at the junction area caused by the irradiation. This irradiation–cooling process is repeated 20 times, followed by 20 ps of annealing at 600 K and a final cooling process to 300 K.

The simulation results indicate that after the electron beam irradiation, a significant degree of disordered covalent joining forms at the junction area instead of complete merging. Compared with the theoretical "ideal" junction with a minimal number of bonding defects, the junctions obtained by electron beam irradiation have low strength but high ductility.

Note that in this example, the irradiated electrons are modeled implicitly, and their effects on the C atoms are described by periodically assigning a velocity to a randomly selected PKA in the irradiation area. The advantage of this approach lies in its simple implementation in MD simulations. However, it is always necessary to exercise caution in determining the magnitude of the velocity assigned to the PKA to avoid both under- and overestimations of interactions between the irradiated electrons and atoms in the irradiation area.

The second example is modeling the damage to a polyimide (PI) system caused by electron beam irradiation. This example is selected as it models the electron beam explicitly and presents a novel approach for modeling the interaction between an electron beam and a target material. PI is a polymer that is extensively used in the spacecraft industry.

Unfortunately, the properties of PI change significantly when it is exposed to high-energy electrons and protons, and heavy ions in outer space, which inevitably induces structural deformations such as chain scission and cross-linking.

To investigate the effects of electron irradiation on the breaking of intermolecular bonds and the creation of radicals within PI, Rahnamoun et al. (2019) have constructed an MD simulation model using a polarizable ReaxFF that permits the study of the chemical dynamics introduced to PI by an electron beam. In the simulation, 20 PI chains are constructed, each consisting of 7 repeating units ($-C_{22}H_{10}O_5N_2-$). The PI model is first equilibrated at a temperature of 300 K before being exposed to an electron beam. After equilibration, the electron beam is introduced, which is modeled as a straight row of immobile massless pseudoparticles with zero initial charges distributed uniformly with a fixed interval of 1 Å. The length of the row is set to be that of the simulation box. A total of 25 rows are created in the system before starting the simulation (Fig. 8.6) instead of being inserted into the system sequentially during the simulation.

Generally, the damage in a material induced by electron beam irradiation results from Coulombic interactions between high-energy electrons in the irradiation beam and electrons in the target material. Since

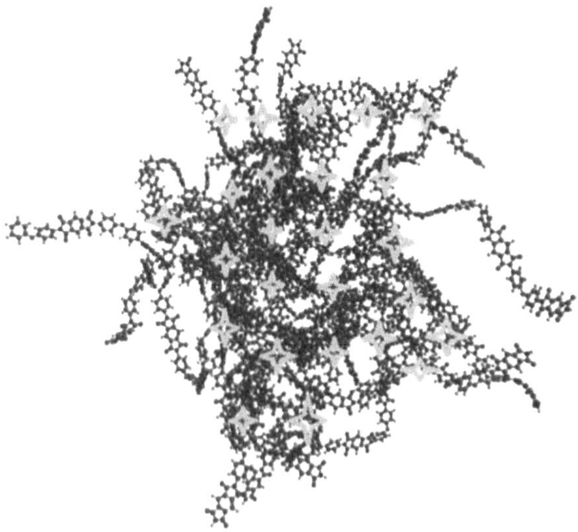

FIGURE 8.6 Schematic of the MD simulation model for the PI molecular system under electron beam irradiation. Each yellow star represents the location of an electron beam, which is modeled by a string of massless pseudoparticles. *MD*, Molecular dynamics. Source: *Reprinted with permission from a work by Rahnamoun et al. (2019).*

ReaxFF includes a polarized charge function capable of describing local charge distributions and their changes during the simulation, the Coulombic interactions between the electron beam and the PI molecules can be effectively simulated. In the MD simulation, the effect of an electron beam is modeled by changing the charge of each of its constituent pseudoparticles from 0 to −1. Each modeled electron beam is activated for 2 fs sequentially. Before activating each electron beam, the PI molecule is allowed to relax for 10 fs in the microcanonical ensemble. After the final electron beam is deactivated, the simulation continues in the microcanonical (NVE) ensemble for another 0.5 ns for the structure of the PI system to evolve.

The simulation results indicate that the atomic structure of the PI molecules is modified during and after the electron beam irradiation. During the irradiation process, H atoms are separated from different parts of the PI molecule through the cleavage of C—H bonds. After the electron beam is removed, a structural rearrangement of the PI molecules occurs, which mainly includes imide-ring opening, chain scission at the ether bridges, the formation of carbonyl-like moieties, and cross-linking between the chains.

The third example pertains to creating defects in supported graphene nanosheets by exposing them to energetic heavy ion irradiation. Graphene has a broad range of applications due to its unique mechanical and physical properties. Engineering graphene by ion beam irradiation is considered rather promising because the direct doping of foreign atoms can be conveniently realized, and defects can be created in a controllable manner for further functionalization. The substrate that supports the graphene nanosheets has a significant impact on their defect formation process. To understand the underlying defect formation mechanism and the effects of the substrate, Zhao et al. (2012) have performed MD simulations on the ion irradiation process of graphene nanosheets supported by a SiO_2 substrate.

In the simulation system, a graphene nanosheet with dimensions of 68.17×39.36 Å^2 is deposited on top of a SiO_2 substrate with a thickness of ~30 Å, as shown in Fig. 8.7. Periodic boundary conditions (PBCs) are applied in the *x*-, *y*-, and *z*-directions. The initial distance between the graphene nanosheet and the substrate is set at 2.90 Å. To prevent the system from being displaced by ion bombardment, the atoms in the bottom layer of the substrate are fixed during the simulation. Both Ar and Si ions are assigned as the irradiation particles.

The AIREBO (adaptive intermolecular reactive empirical bond order) potential is adopted to describe the C—C interactions in the graphene nanosheet. In contrast, the Tersoff potential is adopted for the atomic interactions within SiO_2. The interactions between the graphene nanosheet and the SiO_2 substrate are considered to be van der Waals

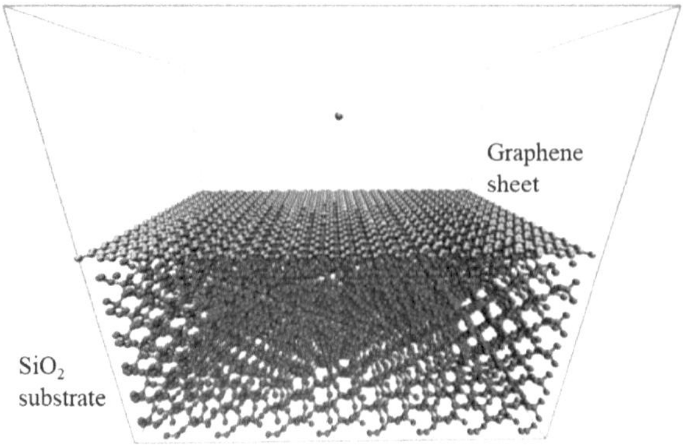

FIGURE 8.7 Illustration of the simulation setup for the irradiation of a graphene nanosheet supported by a SiO_2 substrate. The indigo, red, and yellow spheres represent the C, O, and Si atoms, respectively. The incident ion is denoted by the pink sphere located above the graphene nanosheet. Source: *Modified and reprinted with permission from a work by Zhao et al. (2012).*

forces and described by the Lennard–Jones (LJ) potential. To realistically model energetic collisions, the Tersoff potential for Si and O is smoothly integrated with the Ziegler–Biersack–Littmark (ZBL) universal repulsive potential using a Fermi function. Both the interactions between the radiating Ar ions and the C, Si, and O atoms and the interactions between the radiating Si ions and the C atoms are modeled by the ZBL repulsive potential only.

The entire system is first fully relaxed at a temperature of 300 K prior to the irradiation process. An incident ion is initially positioned at a distance of 50 Å above the graphene nanosheet to prevent spurious interactions. Meanwhile, the initial lateral position of the ion is randomly selected but restricted to within a rectangular impact area in the center of the system with dimensions of 20×10 Å^2. Upon the relaxation of the system, an amount of kinetic energy in the range from 0.1 keV to 1000 keV is randomly assigned to the ion, and the incident direction is set to be normal to the graphene nanosheet. For every value of the kinetic energy considered, 100 independent simulation runs are performed for statistical averaging.

The borders of the graphene nanosheet are coupled with a Berendsen thermostat to prevent waves induced by the ion from returning to the impact area over the periodic boundaries. During the collision between the incident ion and the graphene nanosheet, the NVE ensemble is used. The Newtonian equations of motion are integrated with a variable timestep that depends on the kinetic energy of the incident ion. The simulation of the collision event proceeds until the maximum kinetic energy of the atoms in the system is less than 0.8 eV. Afterward, the entire system is relaxed at a temperature of 300 K for 50 ps with a time step of 1 fs.

Fig. 8.8 shows the damage probabilities of the supported graphene nanosheets exposed to Ar and Si ion irradiation as functions of the incident kinetic energy. For comparison, the damage probabilities of suspended graphene nanosheets under Ar and Si ion irradiation are also plotted. The damage probability is defined as the number of defects of any type created for all the 100 independent collision events. It is observed that the damage probabilities for the supported graphene nanosheets are lower than those of the suspended graphene nanosheets if the incident ion possesses a low kinetic energy. As the kinetic energy of the incident ion increases, the damage probabilities of the suspended graphene nanosheets begin to decrease and eventually become lower than those of the supported graphene nanosheets.

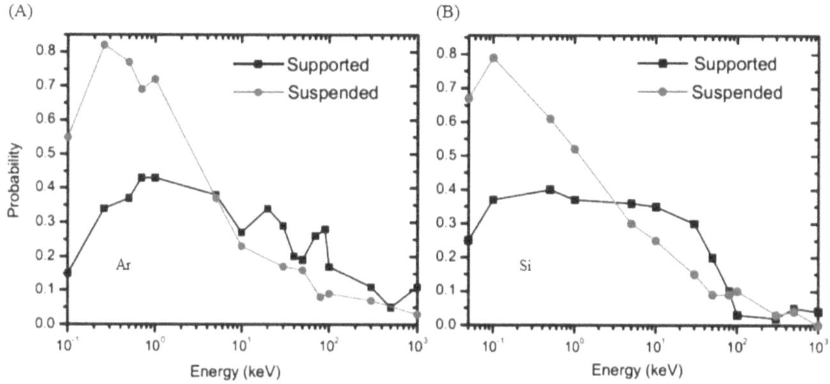

FIGURE 8.8 Damage probabilities of the supported and suspended graphene nanosheets under (A) Ar ion and (B) Si ion irradiation as functions of the kinetic energy of the incident ion. Source: *Reprinted with permission from a work by Zhao et al. (2012).*

The low damage probability for a supported graphene nanosheet at low incident energy is attributed to the high threshold displacement energy T_d of its constituent C atoms. The displaced C atoms in the graphene nanosheet first collide with the substrate, and most of them are reflected, which renders it more difficult for them to be displaced from the nanosheet. Meanwhile, since the C atoms in the suspended graphene nanosheet that are displaced by the sputtering ions will be sputtered away, each of these C atoms should correspond to the formation of a defect.

The high damage probability in the supported graphene nanosheet at a high incident energy is closely related to the backscattered ions, O atoms, and Si atoms sputtered from the substrate. If the kinetic energy of the incident ion is high, it can easily penetrate the graphene nanosheet. Afterward, the ion is either reflected by the substrate or embedded into it. Consequently, some O or Si atoms may be sputtered out before further colliding with the graphene nanosheet. These sputtered atoms usually possess a sufficiently high kinetic energy to create defects in the graphene nanosheet. The most abundant type of defect created in the supported graphene nanosheet is the single-vacancy defect, whereas that in the suspended graphene nanosheet is the vacancy cluster. Other types of defects, such as the Stone–Wales defect, structure distortions, and nonhexagonal rings, are also identified in the supported graphene nanosheets. In addition, the sputtered atoms from the substrate are observed to be incorporated into the graphene nanosheet by interacting with the defects created within the nanosheet, which provides a new route to the controllable doping of graphene.

Based on the above discussion, it is clear that MD simulation can be adopted to model both the irradiation processes induced by electrons and heavy ions. With a proper description of the interactions between electrons (or ions) and atoms in a material that is irradiated, various irradiation processes that have practical applications, such as joining and defect engineering, can be well simulated.

8.3 Material crystallization

Material crystallization is an essential physical process that frequently occurs in many industrial processes and our daily lives. It occurs during the formation of solid particles from a vapor, the solidification of a liquid melt, and the formation of dispersed solids in a liquid solution. The crystallization process consists of two major events. The first is nucleation, which refers to the emergence of crystalline particles from a supercooled liquid or a supersaturated solution. The second is crystal growth, during which the size of the particles increases, and a crystal state is formed.

The properties of crystal products rely highly on their crystallization conditions. Experimental approaches for investigating material crystallization processes usually encounter difficulties in reproducing extreme crystallization conditions, limitations in in situ observations of crystal nucleation and growth processes, as well as high costs. On the other hand, MD simulation permits versatile control of the system conditions (e.g., temperature and pressure) in exploring crystallization processes and provides detailed information on the phase evolutions involved. Hence, the crystallization processes of various systems have been studied via MD simulation.

For instance, Zhang and Larson (2018) have performed an all-atom MD simulation to explore the effects of short-chain branching in a polyethylene (PE) oligomer on its crystal nucleation. In the simulation study, the OPLS–AA (optimized potentials for liquid simulation–all atoms) force field is used to model PE molecules of various sizes and degrees of alkyl branching. The intermolecular interactions are modeled by a combination of van der Waals forces and Coulombic interactions. The initial atomic configuration of the simulation model is prepared by loosely packing 400 PE chains (each consisting of 50 C atoms in the backbone) in a cubic box with a side length of 10 nm. The system is then subjected to a pressure of 1 bar at 600 K to form an isotropic polymer melt. The process is performed in the isothermal–isobaric (NPT) ensemble coupled with an isotropic Berendsen barostat. Here, the term "isotropic" indicates that the pressures in the x-, y-, and z-directions are tuned uniformly. The polymer melt is quenched to a temperature of 350 K, which is below the melting temperature of PE. Upon reaching 350 K, the system is relaxed at this temperature in the NPT ensemble coupled with an anisotropic Parrinello–Rahman barostat for 20 ns. Here, the term "anisotropic" indicates that the dimensions of the simulation box can be adjusted independently in the x-, y-, and z-directions.

Based on the analysis of the incubation time for nucleation, the degree of ordering of the backbone atoms, and the evolution of the overall crystallinity, it is found that short alkyl branches, such as methyl and ethyl, can only retard the nucleation kinetics of the PE system mildly. However, bulkier side chains such as propyl, butyl, and hexyl branches can significantly suppress the nucleation of the PE crystals by disrupting the arrangement of the backbone atoms of the PE chains and preventing them from joining a growing crystal. In addition, the degree of branching and the distributions of the alkyl branches significantly impact the crystallization kinetics. Compared with branches that are closely grouped along the backbone of the PE chains, well-spaced branches are discovered to exert a much more significant effect on the crystallization of the PE system.

In most MD simulations of polymeric systems, including the simulation discussed above, the lengths of polymer chains are usually much shorter than those in practical polymer systems. Furthermore, physical entanglement among the chains is generally not considered, which is mainly due to the limited system size in MD simulations. To increase the chain length of a modeled polymer system, the coarse-grained MD (CGMD) technique is usually adopted (refer to Chapter 4, Advanced Ab Initio Molecular Dynamics and Coarse-Grained Molecular Dynamics, for more discussions on CGMD). For example, Hall et al. (2019) adopted a Shinoda–Devane–Klein (SDK) coarse-grained model to investigate the effects of chain entanglement on the crystallization of PE polymer. In the SDK model, the PE chains are represented by two types of beads that correspond to $-(CH_2)_3-$ and $-CH_2CH_2CH_3$, respectively. The stretching of the bond between two connected beads and the bending of the bond angle formed by three consecutive beads along the backbone of a PE chain are modeled by two different harmonic potential functions. Interactions among beads that belong to different PE chains and interactions among beads belonging to the same chain that are separated by more than two bonds are modeled by the LJ potential.

The simulation procedure is similar to that in the study conducted by Zhang and Larson (2018), in which a loose and low-density PE system is first constructed and then melted by gradually increasing the system temperature above its melting point. Afterward, the PE melt is subjected to isobaric compression via the NPT ensemble. The crystallization of the PE melt is then simulated by quenching the system to a temperature below its melting point and then keeping it at such a temperature for PE crystals to form. Due to the reduced degrees of freedom of the coarse-grained model, the maximum number of atoms included in this simulation exceeds 10^6, and a period of $\sim 10\,\mu s$ can be simulated in the CGMD model. Structural (e.g., the mean square radius of gyration of polymer chains) and dynamical (e.g., the mean square displacement of polymer monomers) quantities obtained from the simulation indicate that the MD simulation based on the SDK model accurately captures the thermodynamical properties of the PE melt and its crystalline phases, molecular-level structure, and dynamical characteristics.

Besides polymeric systems, MD simulation is frequently employed to study the crystallization of metallic systems. The simulation procedures for the crystallization of metallic systems are nearly identical to those for polymeric systems. However, the analysis of microstructure evolution during the crystallization processes is different because polymers and metals have vastly different crystal structures. For metallic systems, quantities such as the system energy, pair distribution function, atomic diffusion coefficient, and structure entropy are usually selected to characterize their microstructure evolution.

References

Bagri, A., Mattevi, C., Acik, M., Chabal, Y.J., Chhowalla, M., Shenoy, V.B., 2010. Structural evolution during the reduction of chemically derived graphene oxide. Nat. Chem. 2, 581–587.

Bedrov, D., Smith, G.D., van Duin, A.C.T., 2012. Reactions of singly-reduced ethylene carbonate in lithium battery electrolytes: a molecular dynamics simulation study using the ReaxFF. J. Phys. Chem. A 116, 2978–2985.

Chenoweth, K., van Duin, A.C.T., Goddard, W.A., 2008. ReaxFF reactive force field for molecular dynamics simulations of hydrocarbon oxidation. J. Phys. Chem. A 112, 1040–1053.

Chuvilin, A., Kaiser, U., Bichoutskaia, E., Besley, N.A., Khlobystov, A.N., 2010. Direct transformation of graphene to fullerene. Nat. Chem. 2, 450–453.

Ding, J., Zhang, L., Zhang, Y., Han, K.-L., 2013. A reactive molecular dynamics study of n-heptane pyrolysis at high temperature. J. Phys. Chem. A 117, 3266–3278.

Flores, M.Z.S., Autreto, P.A.S., Legoas, S.B., Galvao, D.S., 2009. Graphene to graphane: a theoretical study. Nanotechnology 20, 1–16.

Hall, K.W., Sirk, T.W., Klein, M.L., Shinoda, W., 2019. A coarse-grain model for entangled polyethylene melts and polyethylene crystallization. J. Chem. Phys. 150, 244901-1–244901-14.

Jang, I., Sinnott, S.B., Danailov, D., Keblinski, P., 2004. Molecular dynamics simulation study of carbon nanotube welding under electron beam irradiation. Nano Lett. 4, 109–114.

Jung, Y.J., Homma, Y., Vajtai, R., Kobayashi, Y., Ogino, T., Ajayan, P.M., 2004. Straightening suspended single walled carbon nanotubes by ion irradiation. Nano Lett. 4, 1109–1113.

Khlobystov, A.N., Porfyrakis, K., Kanai, M., Britz, D.A., Ardavan, A., Shinohara, H., et al., 2004. Molecular motion of endohedral fullerenes in single-walled carbon nanotubes. Angew. Chemie-Int (Ed.) 43, 1386–1389.

Kotakoski, J., Krasheninnikov, A.V., Kaiser, U., Meyer, J.C., 2011. From point defects in graphene to two-dimensional amorphous carbon. Phys. Rev. Lett. 106, 105505-1–105505-4.

Krasheninnikov, A.V., Nordlund, K., 2010. Ion and electron irradiation-induced effects in nanostructured materials. J. Appl. Phys. 107, 071301-1–071301-70.

Krasheninnikov, A.V., Nordlund, K., Keinonen, J., 2002. Carbon nanotubes as masks against ion irradiation: an insight from atomistic simulations. Appl. Phys. Lett. 81, 1101–1103.

Liu, K., Hsin, C.-L., Fu, D., Suh, J., Tongay, S., Chen, M., et al., 2015. Self-passivation of defects: effects of high-energy particle irradiation on the elastic modulus of multilayer graphene. Adv. Mater. 27, 6841–6847.

Mueller, J.E., van Duin, A.C.T., Goddard, W.A., 2010. Application of the ReaxFF reactive force field to reactive dynamics of hydrocarbon chemisorption and decomposition. J. Phys. Chem. C. 114, 5675–5685.

Nayir, N., van Duin, A.C., Erkoc, S., 2018. Development of a ReaxFF reactive force field for interstitial oxygen in germanium and its application to GeO_2/Ge interfaces. J. Phys. Chem. C. 123, 1208–1218.

Paillet, M., Jourdain, V., Poncharal, P., Sauvajol, J.L., Zahab, A., Meyer, J.C., et al., 2004. Versatile synthesis of individual single-walled carbon nanotubes from nickel nanoparticles for the study of their physical properties. J. Phys. Chem. B 108, 17112–17118.

Psofogiannakis, G., Van Duin, A.C., 2016. Development of a ReaxFF reactive force field for Si/Ge/H systems and application to atomic hydrogen bombardment of Si, Ge, and SiGe (100) surfaces. Surf. Sci. 646, 253–260.

Rahnamoun, A., Engelhart, D.P., Humagain, S., Koerner, H., Plis, E., Kennedy, W.J., et al., 2019. Chemical dynamics characteristics of Kapton polyimide damaged by electron beam irradiation. Polymer 176, 135–145.

Rodriguez-Manzo, J.A., Banhart, F., 2009. Creation of individual vacancies in carbon nanotubes by using an electron beam of 1 angstrom diameter. Nano Lett. 9, 2285–2289.

Russo, M.F., Li, R., Mench, M., van Duin, A.C.T., 2011. Molecular dynamic simulation of aluminum-water reactions using the ReaxFF reactive force field. Int. J. Hydrogen Energy 36, 5828–5835.

Senftle, T.P., Hong, S., Islam, M.M., Kylasa, S.B., Zheng, Y., Shin, Y.K., et al., 2016. The ReaxFF reactive force-field: development, applications and future directions. NPJ Computat. Mater. 2, 1–14.

Susi, T., Kotakoski, J., Arenal, R., Kurasch, S., Jiang, H., Skakalova, V., et al., 2012. Atomistic description of electron beam damage in nitrogen-doped graphene and single-walled carbon nanotubes. ACS Nano 6, 8837–8846.

Taguchi, T., Yamamoto, S., Ohba, H., 2019. Synthesis of novel hybrid carbon nanomaterials inside silicon carbide nanotubes by ion irradiation. Acta Materialia 173, 153–162.

Tolvanen, A., Buchs, G., Ruffieux, P., Groning, P., Groning, O., Krasheninnikov, A.V., 2009. Modifying the electronic structure of semiconducting single-walled carbon nanotubes by Ar + ion irradiation. Phys. Rev. B 79, 125430-1–125430-13.
Zhang, W., Larson, R.G., 2018. Direct all-atom molecular dynamics simulations of the effects of short chain branching on polyethylene oligomer crystal nucleation. Macromolecules 51, 4762–4769.
Zhao, S., Xue, J., Wang, Y., Yan, S., 2012. Effect of SiO_2 substrate on the irradiation-assisted manipulation of supported graphene: a molecular dynamics study. Nanotechnology 23, 1–8.
Zoberbier, T., Chamberlain, T.W., Biskupek, J., Kuganathan, N., Eyhusen, S., Bichoutskaia, E., et al., 2012. Interactions and reactions of transition metal clusters with the interior of single-walled carbon nanotubes imaged at the atomic scale. J. Am. Chem. Soc. 134, 3073–3079.

Index

Note: Page numbers followed by "*f*" refer to figures.

N

FSC
www.fsc.org
MIX
Papier | Fördert
gute Waldnutzung
FSC® C083411

Kolibri 360 GmbH
Ferdinand-Jühlke-Straße 7
99095 Erfurt, Deutschland
produktsicherheit@kolibri360.de